Principles of Statistics

Principles of Statistics

Edited by **Calanthia Wright**

Willford Press
New York

Published by Willford Press,
118-35 Queens Blvd., Suite 400,
Forest Hills, NY 11375, USA
www.willfordpress.com

Principles of Statistics
Edited by Calanthia Wright

International Standard Book Number: 978-1-68285-191-3 (Hardback)

Printed in China.

Contents

Preface

In my initial years as a student, I used to run to the library at every possible instance to grab a book and learn something new. Books were my primary source of knowledge and I would not have come such a long way without all that I learnt from them. Thus, when I was approached to edit this book; I became understandably nostalgic. It was an absolute honor to be considered worthy of guiding the current generation as well as those to come. I put all my knowledge and hard work into making this book most beneficial for its readers.

Collection, interpretation and organization of data fall under the domain of statistical studies. The applications of statistics cover a wide range of scientific, corporate and societal problems and databases. It helps in systematic presentation of data and determination of errors. The book provides an insight into various methodologies of representing data through descriptive and inferential statistics. It includes development of statistical distributions, statistical modelling and data analysis related to statistical distribution. This book deals with principles of statistics and is an excellent reference for students and researchers engaged in this field.

I wish to thank my publisher for supporting me at every step. I would also like to thank all the authors who have contributed their researches in this book. I hope this book will be a valuable contribution to the progress of the field.

Editor

Modelling changes in Arctic Sea Ice Cover: an application of generalized and inflated beta and gamma densities

Peter Congdon

Correspondence:
p.congdon@qmul.ac.uk
School of Geography, Queen Mary University of London, London E1 4NS, UK

Abstract
A modelling framework for changing Arctic sea ice extent is developed reflecting different trends and seasonal extremes in nine Arctic sub-regions. Core sub-regions retain partial ice cover throughout the year, and in winter show complete ice cover, while in peripheral sub-regions, winter coverage is not complete, and there is no ice cover at all in the summer. A generalized beta representation is developed for monthly ice extents in core sub-regions, with inflation to model maximum winter extents. For peripheral sub-regions, a gamma time series with excess zeroes (representing summer sea ice absence) is developed. Different trend representations (deterministic vs. stochastic) are compared for non-extreme observations. Other potential applications of the generalized beta density allowing zero or maximum inflation are discussed.

Keywords: Sea ice extent; Seasonal extremes; Inflated generalized beta; Inflated gamma; Bayesian

1 Introduction

In recent years, Arctic sea ice has been declining with wider climatic implications. The latter are multi-faceted and subject to discussion and uncertainty; see, for example, (Screen et al. 2013). However, one implication follows from the role of sea ice in regulating the global temperature via its ability (compared to the ocean surface) to reflect the sun's radiation. The albedo of snow-covered sea ice is 0.90, meaning it reflects 90% of the sun's radiation, whereas the ocean surface reflects only 10%. Less sea ice and more ocean surface will lead to a warmer Arctic, and contribute to global warming. There is also evidence that decline in Arctic sea ice is influencing atmospheric circulation (including the jet stream) within and beyond the Arctic, with impacts including winter cold surges in Europe and North America (Liu et al. 2012).

Satellite records of Arctic sea ice extent (available with full annual coverage since 1979) show a downward trend, albeit with fluctuations such as unusually low points in 1990 (Serreze et al. 1995) and in 2007 (Comiso et al. 2008). For the Arctic as a whole, the rate of decline in summer ice extent has been greater (Stroeve et al. 2012), with an 8.2% linear rate of decline (per decade) in September ice extent between 1979 and 2011, as against a 3.9% linear rate of decline for March ice extent (see Table 1).

Table 1 Arctic sea ice extent

	(millions of km²)	
	March	September
1979	16.4	7.2
1980	16.1	7.8
1985	16.1	6.9
1990	15.9	6.2
1995	15.3	6.1
2000	15.3	6.3
2005	14.8	5.6
2006	14.5	5.9
2007	14.7	4.3
2008	15.2	4.7
2009	15.2	5.4
2010	15.1	4.9
2011	14.3	4.6
2012	15.2	3.6

However, the Arctic extent totals in Table 1 aggregate over sub-regions with dissimilar trends and seasonal extremes; for example, some sub-regions show substantial falls in extent in both winter and summer months. The present study accordingly reviews issues in modelling sea ice trends for Arctic sub-regions, and includes short term projections three years beyond the range of the observed data. While a number of forecasts of sea ice cover have been developed (e.g. Stroeve et al. 2007), these generally consider the entire Arctic, and do not consider sub-regional differences. The observed data are monthly NASA satellite records from 1979 to 2011 (396 observations), disaggregated to nine Arctic sub-regions. These are the central Arctic Ocean, the Canadian Archipelago, Hudson Bay, Sea of Okhotsk, Bering Sea, Baffin Bay, Greenland Sea, Kara-Barents Sea, and Gulf of St. Lawrence (see Figure 1). The observed data were extracted in June 2013 at http://neptune.gsfc.nasa.gov/csb/index.php?section=59.

There are important regional differences in trends and seasonal patterns according to location (e.g. Arctic core vs. periphery) that are relevant to statistical analysis and forecasts. These include modelling seasonal extremes. For central regions there is still some summer cover but a reversion to complete cover (maximum possible extent) in winter months. To represent this pattern, a generalized beta representation is applied, including "maximum inflation" to account for winter total coverage.

By contrast, in some non-core regions such as the Sea of Okhotsk and Bering Sea, winter extents do not cover the entire sea region, and there is no ice cover in the summer. For these regions, a gamma density with excess zeroes (representing summer sea ice absence) is developed. In both situations, regression models including trend and seasonal components are applied to represent the change in mean sea ice extent, while a separate logit regression is applied to the probability of inflation. Contrasting stochastic and deterministic trend representations are evaluated. Estimation is based on monthly observations in the first 31 years of satellite observations (1979-2009) with cross-validation using the remaining two years of observed data as a test sample. Bayesian estimation and inference are implemented using the Winbugs package via Markov chain Monte Carlo (MCMC) methods (Lunn et al. 2009).

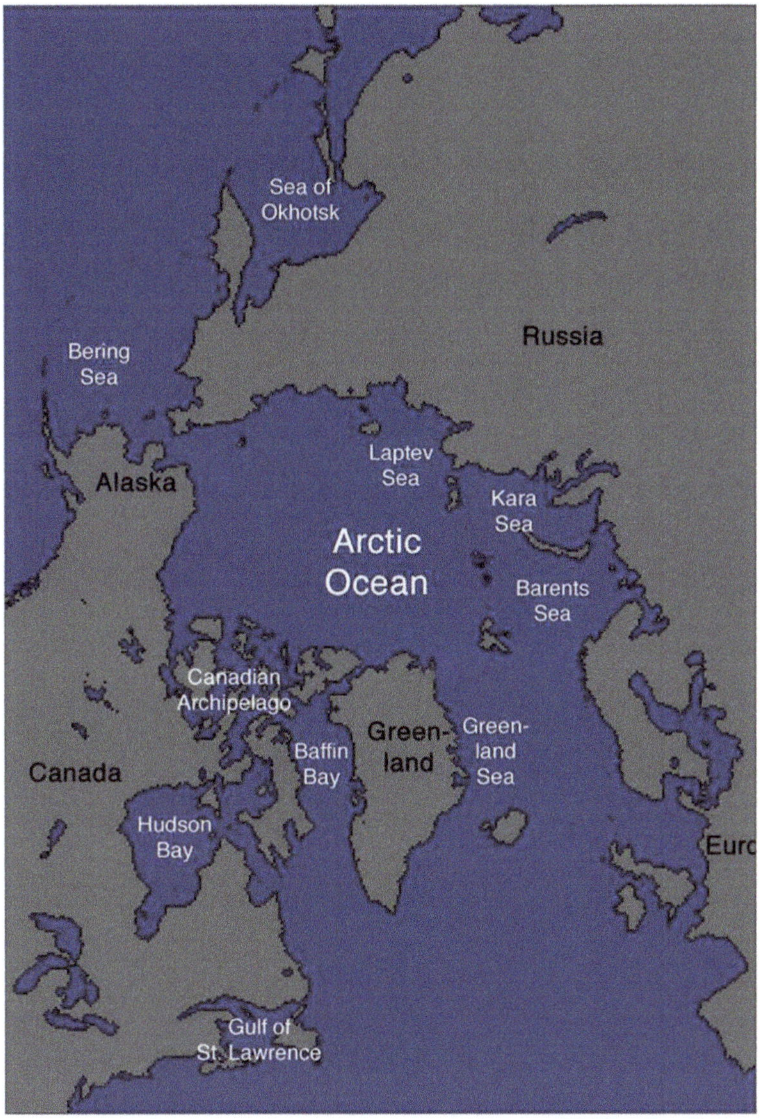

Figure 1 Map of case study sub-regions (image courtesy of National Snow and Ice Data Center (2012), University of Colorado, Boulder).

2 Generalized beta time series regression

2.1 Application of the generalized beta to sea ice extents

As mentioned above, while there is decline in Arctic sea ice overall, there are considerable regional differences in trends and seasonal ice extent. What are here denoted "core" Arctic sea regions (namely the central Arctic Ocean, the Canadian Archipelago, and Hudson Bay) retain partial ice cover throughout the year, and in winter months show complete ice cover. For example, in 2011, the central Arctic had readings of 7.158mn km^2 throughout January to March while the Canadian Archipelago had readings of 0.751 mn km^2 for January through to April. Figure 2 shows monthly extent totals for the central Arctic ocean during 2007-11 and illustrates maximum inflation in winter. The trend in these three regions is similar to that in the Arctic ocean considered as an aggregate, namely stronger declines in summer extent.

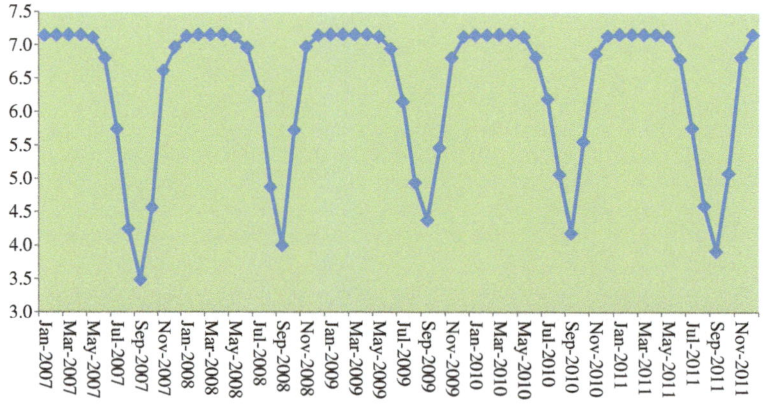

Figure 2 Observed monthly extents, millions km^2, 2007-11, central Arctic ocean.

A time series representation needs to express the annual reversion to complete cover (maximum recurrence) in winter months, together with the irregular trend for declining extent in non-winter months. The application of a generalized form of the beta density is motivated by the fact that the observed ice extents y can be seen as ratios $r = y/d$ to a maximum possible extent d, though substantive interest is in trends in extents y. It is important that the bounded nature of the response is included in any model. Another possibility might be some form of truncated sampling mechanism (e.g. a log-normal for extent readings y with ceiling d) but this precludes any analysis of the factors producing seasonal extremes.

Consider the beta distribution on (0, 1), with density function given by

$$f(z|a,b) = \frac{1}{B(a,b)} z^{a-1}(1-z)^{b-1},$$

with $a > 0, b > 0$. An alternative representation (Ospina and Ferrari 2010) involves mean and precision parameters (μ, ϕ), where $a = \mu\phi$, $b = (1-\mu)\phi$, namely

$$f(z|\mu,\phi) = \frac{1}{B(\mu\phi, \phi - \phi\mu)} z^{\mu\phi-1}(1-z)^{(1-\mu)\phi-1},$$

with $\phi > 0$, and $0 < \mu < 1$. This form facilitates separate modelling of mean and variance trends (Huang and Oosterlee 2008). For values (a, b) apart from $a = b = 1$, the $Beta(a, b)$ density has zero mass at the extreme values 0 and 1, and zero-inflated or one-inflated versions of the beta need to be applied (Ospina and Ferrari 2010). Let $g = 0$ or 1, then inflation at either boundary is achieved by the mechanism

$$f(z|\alpha_g, \mu, \phi) = \begin{cases} \alpha_g, & z = g \\ (1-\alpha_g) f(z|\mu,\phi), & z \in (0,1), \end{cases}$$

where α_g is an inflation probability.

The generalized beta is obtained by extending the support interval to an arbitrary bounded interval (c, d) (with $d > 0$) via a linear transformation $y = c + z(d-c)$ (Pham-Gia and Duong 1989), so that

$$f(y|a,b,c,d) = \frac{1}{B(a,b)(d-c)^{a+b-1}} (y-c)^{a-1}(d-y)^{b-1},$$

with equivalent representation

$$f(y|\mu,\phi,c,d) = \frac{1}{B(\mu\phi,\phi-\phi\mu)(d-c)^{\phi-1}}(y-c)^{\mu\phi-1}(d-y)^{(1-\mu)\phi-1}, \quad (1)$$

with mean $c+(d-c)\mu$, and variance $(d-c)^2\mu(1-\mu)/(\phi+1)$.

For the generalized beta in (1), inflation will need to be applied for values occurring at the boundary points, when $y=c$ or $y=d$ (this may be termed minimum and maximum inflation). The maximum inflated version of the generalized beta is particularly relevant to the sea ice application and has the form

$$f(y|\alpha_d,\mu,\phi,c,d) = \begin{cases} \alpha_d, & y=d \\ (1-\alpha_d)f(y|\mu,\phi,c,d), & y\in(c,d). \end{cases}$$

In the generalized beta applied to core Arctic regions, $c=0$ while d is the maximum winter extent (namely d_1=7.158 mn km^2 for the central Arctic ocean, d_2=0.751mn km^2 for the Canadian Archipelago, and d_3=1.233 mn km^2 for Hudson Bay).

While summer minimum extents in the central Arctic and Canadian Archipelago remain well in excess of zero, those in Hudson Bay are becoming relatively small, e.g. $y=0.025$ mn km^2in September 2010. This raises the possibility of needing to represent both maximum and minimum inflation in the generalized beta. This can be handled by the mechanism

$$f(y|\alpha_c,\alpha_d,\mu,\phi,c,d) = \begin{cases} \alpha_c, & y=c \\ \alpha_d, & y=d \\ (1-\alpha_c-\alpha_d)f(y|\mu,\phi,c,d), & y\in(c,d), \end{cases}$$

where the vector of probabilities $(\alpha_c,\alpha_c,1-\alpha_c-\alpha_d)$ should be modelled using a multinomial logistic.

2.2 Generalized beta time series regression with maximum inflation for sea ice extents

Let $\{\mu_t,\phi_t,\alpha_{dt}\}$ denote the series of parameters underlying the y_t series, $m=m_t$ represent the month that observation t corresponds to, and $s=s_t$ represent the year corresponding to observation t (e.g. $s=2$ in 1980 for observations $t=13,..,24$ and $s=31$ for observations $t=361,..,372$). A parsimonious time series model is sought (Ledolter and Abraham 1981), combining close fit with low predictive variability, especially for cross-validatory and out-of-sample predictions. As discussed in Section 4, these aspects of fit are assessed using a posterior predictive fit criterion (Laud and Ibrahim 1995). A parsimonious model for the level of the series is expressed by a logit regression in μ_t,

$logit(\mu_t)=\Delta_{ms}+\vartheta_m$, where Δ_{ms} represents trend for each combination of month m and year s, and ϑ_m represents seasonal effects.

Two options for the trend are considered. One option is a stochastic trend, with random variation around a central linear trend. To allow for steeper declines in some months, a discrete mixture is implemented via

$$\begin{aligned} \Delta_{ms} &= \rho_m\varphi_{1s}+(1-\rho_m)\varphi_{2s}, \\ \rho_m &\sim Bern(\pi_m), \\ \varphi_{0s} &= \delta_{00}+\delta_{01}s+u_s, \\ \varphi_{1s} &= \delta_{10}+\delta_{11}s+u_s, \\ u_s &\sim N\left(u_{s-1},\sigma_u^2\right), \end{aligned} \quad (2)$$

with $\rho_m \sim Bern(\pi_m)$ being binary indicators. With the constraint $\delta_{11} < \delta_{01}$, φ_{1s} represents the stronger downward trend.

The other form of trend assumption (deterministic) involves a linear trend in years s_t combined with a short term AR1 lag effect in extents y_t. The latter represents carry over between successive months; for example, if September extent is relatively low in a particular year, then October extent may also be relatively low. The linear trend may vary between months and by broad sub-period. For example, (Comiso et al. 2008) report a stronger decline during 1996-2007 than 1979-96. Here we consider three sub-periods $p = 1, ..., 3$ of 12 years, including out-of-sample years (2012-14), namely 1979-1990, 1991-2002, and 2003-2014. The trend parameter for a particular time point is chosen by a monthly specific discrete mixture between guide linear trend parameters, specific to broad period, Γ_{0p} and Γ_{1p}, with $\Gamma_{1p} < \Gamma_{0p}$. The AR1 lag effect is also taken to vary by month, with normal priors for each monthly lag parameter. Thus for month $m = m_t$, and year s_t,

$$logit(\mu_t) = \Delta_t + \vartheta_m,$$
$$\Delta_t = \gamma_0 + \gamma_{1m} s_t + \gamma_{2m} y_{t-1},$$

where the linear trend γ_{1m} for month $m = m_t$ in period p is chosen using a discrete mixture

$$\gamma_{1m} = \rho_m \Gamma_{0p} + (1 - \rho_m)\Gamma_{1p}; \qquad \rho_m \sim Bern(\pi_m).$$

Remaining aspects of the model are applicable across different representations of trend. Seasonal (monthly) effects are represented by a Fourier series (Höhle and Paul 2008),

$$\vartheta_m = \sum_{j=1}^{J_1} (\beta_{2j-1} \cos(\omega j m) + \beta_{2j} \sin(\omega j m)), \tag{3}$$

where $\omega = 2\pi/M$, with $M = 12$, and J_1 is the number of harmonics. To allow for changing precision it is assumed that

$$log(\phi_t) = \eta_1 + \eta_{2m} s, \tag{4}$$

namely a linear trend (varying by month) in year units. For example, (Stroeve et al. 2012) find evidence of increased variability in overall Arctic sea ice extent, especially in summer months.

To represent extreme data (complete winter coverage), a logit regression is used to model the probabilities α_{dt} of maximum inflation, with form

$$logit(\alpha_{dt}) = \xi_0 + \sum_{j=1}^{J_2} (\xi_{2j-1} \cos(\omega j m_t) + \xi_{2j} \sin(\omega j m_t)).$$

A trend element in the inflation probability is not included as it would be confounded with the trend model in the mean.

2.3 Other generalized beta applications

While the application here focuses on sea ice extent and a time series application, the generalized inflated beta with mechanisms or regressions for both extreme and non-extreme observations has potential applications in other settings where the observations can be regarded as ratios $r_i = y_i/d_i$ of actual extents to a maximum extent d_i, but substantive interest is in the extents y_i. The extents may be, inter alia, expressed in spatial units (e.g.

areas in millions of square kilometres) or time units (e.g. durations in hours). As an example with time extents, one might analyse hours with cloud cover y_i in relation to daylight hours d_i, while a spatial application might consider desertified extents y_i in relation to total area extents d_i.

Data of this form can be considered as a form of compositional data, and widely used methods (Aitchison and Egozcue 2005, Butler and Glasbey 2008) focus on the ratios r_i, or more specifically the log ratios. Spatial applications involving compositional data and zero inflation have been described in (Leininger et al. 2013), but also focus on the ratios.

However, for policy purposes, the interest may be in trends or patterns in the extents themselves (i.e. the y-data), rather than in the ratios, as is the case with the sea ice extents. Alternatively stated, "compositions provide information only about the relative magnitudes of the compositional components and so interpretations involving absolute values ... cannot be justified" (Aitchison and Egozcue 2005) p. 839. Thus in the case of desertification (e.g. Zhao et al. 2010), the substantive focus may be on spreading desertification, implying analysis of desertified extents y_i. Some areas may be totally desertified with $y_i = d_i$ (maximum inflation). Regression modelling of desertification extents would then need to include a mechanism or regression describing maximum inflation, as would spatial forecasting (or interpolation) of desertified extents in situations where comprehensive assessment of desertification status is only available for some area units.

The generalized beta density with maximum or zero inflation might be potentially extended to Dirichlet density applications, and to generalized inflated Dirichlet densities parallel to equation (1), where there are more than two categories and where extreme observations can occur. For example, with three categories the observations would be (y_{1i}, y_{2i}, y_{3i}), with $\sum_k y_{ki} = d_i$, and maximum inflation occurring when any $y_{ki} = d_i$. The inflation probabilities can be modelled using a multiple logistic. For example, in the sea ice application, one may distinguish by sea-ice type (e.g. Fissel et al. 2011) between perennial multi-year sea ice and first-year ice, so that sub-region observations become (y_1, y_2, y_3) for area covered by multi-year ice, area covered by first-year ice, and area without ice cover respectively.

3 Inflated gamma time series regression

In what are here denoted "peripheral" Arctic regions (Sea of Okhotsk, Bering Sea, Gulf of St. Lawrence), the winter ice extents do not provide complete ice cover of the sea region, while in at least one peak summer month (July, August, September) there is no ice cover at all. For example, Figure 3 shows the extent series for the Bering Sea for 2007-11. Let y_t denote the observed extents (mn km^2) for these regions following a gamma density,

$$f(y|\mu,\phi) = \frac{(\phi/\mu)^{\phi}}{\Gamma(\phi)} y^{\phi-1} e(-\phi y/\mu),$$

with mean μ and variance μ^2/ϕ. A zero inflated version of the gamma is motivated by the need to represent absence of ice cover in summer months, as in the mechanism

$$f(y|\alpha_0,\mu,\phi) = \begin{cases} \alpha_0, & y = 0 \\ (1-\alpha_0) f(y|\mu,\phi), & y \in (0,\infty). \end{cases}$$

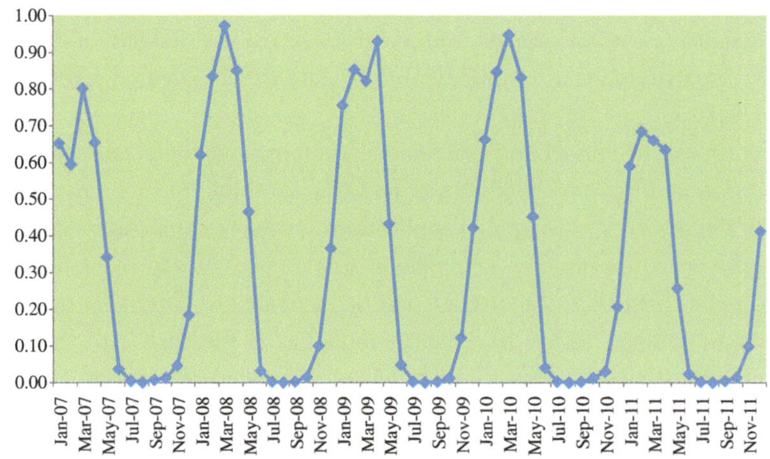

Figure 3 Observed monthly extents, millions km^2, 2007-11, Bering sea.

Let $\{\mu_t, \phi_t, \alpha_{0t}\}$ denote the series of parameters underlying the observations y_t. The link for μ_t depends on which of two options for trend is applied.

One option is a stochastic trend, with log-link

$$log(\mu_t) = \Delta_{ms} + \vartheta_m, \tag{5}$$

where Δ_{ms} is as in (2) above. The other option, a deterministic trend, combines a linear trend in years s with a short term lag effect in the response. The linear trend may vary between months, and also between sub-period p. As above, the trend parameter for a particular time point is chosen according to a monthly specific discrete mixture between guide linear trends, specific to broad period, Γ_{0p} and Γ_{1p}, with $\Gamma_{1p} < \Gamma_{0p}$. Lag effects are also taken to vary by month. To avoid an explosive impact of the lag effect (Blundell et al. 2002, Jung et al. 2006), the following link scheme provides the deterministic trend option,

$$\mu_t = \exp(\zeta_0 + \zeta_{1m} s_t + \vartheta_m) + \zeta_{2m} y_{t-1}, \tag{6}$$

with the constraint $\zeta_{2m} \geq 0$. For example, one may take ζ_{2m} to have exponential or gamma priors.

Remaining aspects of the model are applicable across different representations of trend. Seasonal effects ϑ_m are represented by a Fourier series, as in (3), and precision parameters modelled using the regression form (4). To represent summer extremes (no ice cover at all), a logit regression for the probability of zero inflation is used, with

$$logit(\alpha_{0t}) = \xi_0 + \sum_{j=1}^{J_2} (\xi_{2j-1} \cos(\omega j m_t) + \xi_{2j} \sin(\omega j m_t)).$$

For the remaining three Arctic sub-regions (Baffin Sea, Greenland Sea, Kara-Barents), the data series demonstrate neither maximum recurrence in winter, nor complete summer ice disapperance as yet. The maximum values observed for these three series (in millions km^2) are 1.766 (Baffin Sea, March 1993), 1.115 (Greenland Sea, January 1982) and 2.168 (Kara-Barents, April 1979), whereas recent winter maxima are below these levels. Recent summer extents might be taken to indicate incipient ice disappearance in summer, such as a reading of 0.04 mn km^2 in the Baffin Sea in August 2011. However, for these regions a gamma regression may be adopted, without any inflation mechanism. The time series regression therefore involves $\{\mu_t, \phi_t\}$, and stochastic and deterministic trend options, as in (5) and (6) respectively, may be compared.

4 Model fit and comparison

4.1 Analysis framework

To illustrate a full comparative analysis, three sub-regions are considered as representative of the three sets of regions described above: the central Arctic, the Sea of Okhotsk and the Greenland Sea. Different models are applied to monthly observations from January 1979 to December 2009 (namely for $t = 1, .., T$ with $T = 372$), with in-sample cross-validation for the two year period 2010-2011, namely for $t = T + 1, ..., T_1$ with $T_1 = 396$. Out of sample predictions are made for a further 36 months, January 2012 to December 2014. Posterior inferences are based on the second halves of two chain runs of 50,000 iterations, with convergence assessed using Brooks-Gelman-Rubin statistics (Brooks and Gelman 1998).

4.2 Generalized beta regression with maximum inflation

The central Arctic is one of three sea regions with recurrent winter maxima and partial summer ice cover, and Figure 4 shows the trend in yearly averages for the central Arctic, with the anomalous low cover years (1990, 2007) apparent. In applying the generalized beta regression, normal priors with mean zero and variance 1000 are adopted on fixed effect parameters $\{\beta, \xi, \eta, \Gamma, \delta, u_1, \gamma_0, \gamma_1\}$, while the lag parameters γ_{2m} are assigned N(0,1) priors. A gamma prior with shape 1 and scale 0.001 is adopted for the precision $1/\sigma_u^2$, while $Beta(1, 1)$ priors are adopted for the π_m. The Fourier seasonal representations for the means μ_t and inflation probabilities α_{dt} have $J_1 = J_2 = 3$ harmonics, as insignificant regression effects $\{\beta, \xi\}$ were obtained at higher numbers of harmonics.

Predictions are obtained by generating replicate inflation indicators $D_{new,t} \sim Bern(\alpha_{dt})$, and replicate scaled beta values $dz_{new,t}$, where $z_{new,t} \sim Beta(\mu_t\phi_t, \phi_t - \phi_t\mu_t)$. Predictions are then

$$y_{new,t} = D_{new,t}d + (1 - D_{new,t})dz_{new,t}.$$

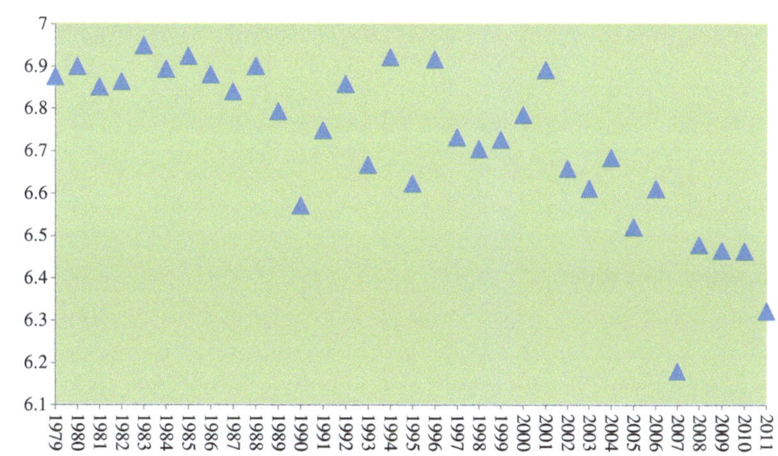

Figure 4 Yearly averages (mn km²), observed extents 1979-2011, central Arctic.

Let $M_{new,t}$ and $V_{new,t}$ be the posterior means and variances of $y_{new,t}$. In-sample predictive fit (within the training dataset from January 1979 through to the end of 2009) is based on the L-criterion (Laud and Ibrahim 1995), namely the square root of

$$L^2 = \sum_{t=1}^{T} V_{new,t} + \sum_{t=1}^{T} (M_{new,t} - y_t)^2.$$

Cross-validatory predictions beyond $T = 372$ (for the 24 months in 2010 and 2011) are made by treating $\{y_{T+1}, ..., y_{T_1}\}$ as missing data, and generating predictions $\{y_{new,T+1}, ..., y_{new,T_1}\}$. These can be compared with observed test data $y_{obs,t}$ for 2010 and 2011. Cross validatory fit using the L criterion, denoted CV-L, is based on

$$\text{CV-L}^2 = \sum_{t=T+1}^{T+24} V_{new,t} + \sum_{t=T+1}^{T+24} \left(M_{new,t} - y_{obs,t}\right)^2 = P_1 + P_2,$$

which will reflect the precision of future predictions (P_1) as well as the fit (P_2). To further assess predictive performance, a check is made whether observed extents are within 95% intervals of y_{new}. In a model that effectively reproduces the data, predictive coverage is at or above 95% (Gelfand 1996, p. 158).

Table 2 shows posterior means and variances during the cross-validation period, while Table 3 summarises the L-criteria for both cross-validation periods (test data) and the in-sample estimation (training data). It is apparent that the deterministic trend has a better cross-validation performance but a worse in-sample performance. In particular, the deterministic trend is less adaptable to anomalous observations such as occurred during the summers of 1990 and 2007. Both models are effective in reproducing the original data satisfactorily in terms of coverage obtained by $y_{new,t}$.

The better out-of-sample cross validation under the deterministic trend is due to lower predictive variability ($P_1 = 2.06$ under the deterministic trend, whereas $P_1 = 6.13$ under the stochastic trend); pure fit is broadly similar between the two options ($P_2 = 0.41$ under deterministic trend, as against $P_2 = 0.34$ under stochastic trend). Figure 5 shows observations and predictions for 2010-11, as well as out-of-sample predictions through to December 2014, under the deterministic trend option. Posterior mean predictions for September extent in 2012, 2013 and 2014 are respectively 3.88, 3.78 and 3.68mn km^2. Table 4 summarizes selected parameters from the deterministic trend model. The strongest downward trends in mean extent, as represented by the guide parameters Γ_{0p} and Γ_{1p}, are for 1979-90 and for the partially observed third period (2003-14), with estimates based on observations for 2003-09. Parameters η_{2m} for trends in precision are nonsignificant in summer months (i.e the 95% credible intervals straddle zero), though the 95% credible intervals for η_{2m} in July and August are biased to negative values, namely (-0.052,0.009) and (-0.058,0.009).

4.3 Gamma regression time series

The Sea of Okhotsk is one of three regions with recurrent summer disappearance of ice cover. Table 5 shows that unlike the trend for the whole Arctic as an aggregate (e.g. see Table 1), the most pronounced declines in average extent in the Sea of Okhotsk have been in the first quarter months (January, February, March). In applying gamma time series regression with zero inflation, priors are as in Section 4.2, except that exponential $E(1)$ priors are adopted for the lag parameters ζ_{2m}, in order to ensure that μ_t is positive - see

Table 2 In-sample cross validation (for 2010 and 2011)

		Stochastic trend			Deterministic trend		
Month	Actual (Test data)	Posterior mean	Posterior median	Posterior variance	Posterior mean	Posterior median	Posterior variance
Jan-2010	7.145	7.148	7.158	0.0004	7.148	7.158	0.0004
Feb-2010	7.154	7.154	7.158	0.0001	7.151	7.158	0.0006
Mar-2010	7.158	7.157	7.158	0.0000	7.157	7.158	0.0000
Apr-2010	7.157	7.155	7.158	0.0001	7.155	7.158	0.0000
May-2010	7.130	7.112	7.119	0.0013	7.120	7.122	0.0003
Jun-2010	6.814	6.823	6.865	0.0419	6.881	6.901	0.0173
Jul-2010	6.192	6.077	6.147	0.2322	6.206	6.234	0.1054
Aug-2010	5.061	4.886	4.944	0.5817	4.923	4.917	0.2896
Sep-2010	4.181	4.174	4.200	0.7522	4.072	4.074	0.2886
Oct-2010	5.555	5.586	5.667	0.4638	5.248	5.278	0.2480
Nov-2010	6.864	6.808	6.863	0.0693	6.828	6.863	0.0406
Dec-2010	7.137	7.108	7.138	0.0061	7.127	7.144	0.0022
Jan-2011	7.158	7.145	7.158	0.0007	7.148	7.158	0.0004
Feb-2011	7.158	7.153	7.158	0.0001	7.151	7.158	0.0003
Mar-2011	7.158	7.157	7.158	0.0000	7.157	7.158	0.0000
Apr-2011	7.157	7.154	7.158	0.0001	7.155	7.158	0.0000
May-2011	7.126	7.105	7.118	0.0029	7.119	7.121	0.0003
Jun-2011	6.786	6.777	6.856	0.0962	6.872	6.889	0.0173
Jul-2011	5.758	5.983	6.140	0.4925	6.185	6.208	0.1069
Aug-2011	4.587	4.792	4.907	1.0733	4.872	4.867	0.3136
Sep-2011	3.911	4.091	4.130	1.3271	3.984	3.981	0.3035
Oct-2011	5.082	5.479	5.628	0.8492	5.150	5.181	0.2711
Nov-2011	6.819	6.756	6.848	0.1338	6.808	6.845	0.0478
Dec-2011	7.155	7.100	7.139	0.0097	7.124	7.142	0.0024

equation (6). The Fourier seasonal effects series included in representations for μ_t and the inflation probabilities α_{0t} have $J_1 = J_2 = 3$ harmonics. Predictions are obtained by generating replicate inflation indicators $D_{new,t} \sim Bern(\alpha_{0t})$, and replicate gamma values $w_{new,t} \sim Gamma(\phi_t, \phi_t/\mu_t)$. Predictions are then obtained as

$$y_{new,t} = 0 \times D_{new,t} + (1 - D_{new,t})w_{new,t}.$$

Table 6 summarises the L-criteria for the cross-validation period (when predictions are compared to test data) and for the in-sample estimation period (when predictions compared to training data). The deterministic trend has better performance (see Figure 6 for

Table 3 Predictive fit (L-criterion) and predictive checks, sea ice extent models, central Arctic ocean

	Trend model	
	Stochastic	Deterministic
In-sample fit (1979-2009)	2.77	3.96
	Stochastic	Deterministic
Cross validation fit (2010-11)	2.54	1.57
	% of observations within 95% credible intervals of posterior predictions	
	Stochastic	Deterministic
In-sample period	97.8	97.6
Cross-validation period	100.0	100.0

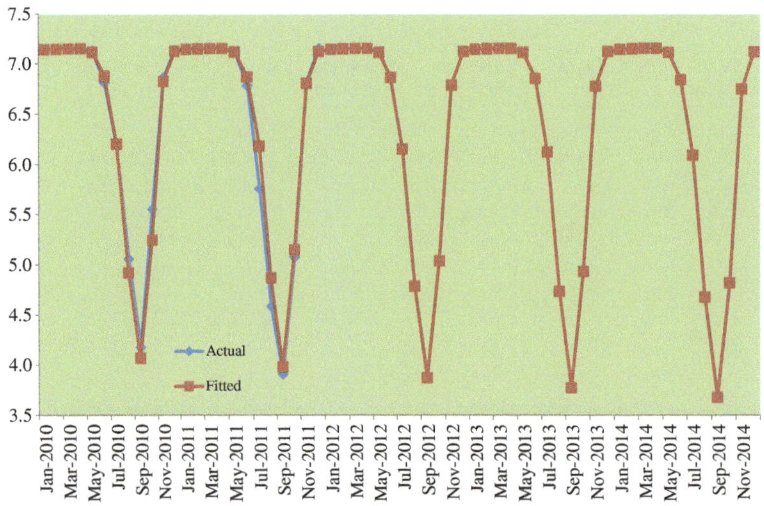

Figure 5 Central Arctic (mn km^2), cross-validation period (2010-11) actual vs fitted, and out-of-sample predictions to end 2014.

actual and fitted series). March extents (in mn km^2) are predicted to fall to 0.895 in 2012, 0.889 in 2013 and 0.879 in 2014 as compared to 0.964 in 2011. The strongest downward trend in mean extent as shown by the deterministic trend guide parameters $\{\Gamma_{0p}, \Gamma_{1p}\}$ is actually in the first sub-period (1979-1990). It may be noted that the actual data for the Sea of Okhotsk show a fall in annual average extent (mn km^2) from 0.58 in 1979 to 0.37 in 1990. The posterior means (sd's) for Γ_{1p} are respectively -0.016 (0.010), -0.0085 (0.0044), and -0.014 (0.005).

For the three remaining sea regions (e.g. the Greenland Sea) with partial summer cover and no winter maximum recurrence, a gamma regression may be applied. In fact, like the Sea of Okhotsk, the Greenland Sea has declines in extent across both winter and summer months. For example, the average March extent (in mn km^2) fell from 0.98 in 1979-89 to 0.82 in 1990-2000 and 0.78 in 2001-11, while the average September extent fell from 0.37 (1979-89) to 0.34 (1990-2000) and 0.25 (2001-11). For this sub-region, a deterministic trend model has better in-sample fit and cross-validatory fit (see Figure 7). March extents are predicted as 0.767 in 2012, 0.766 in 2013 and 0.762 in 2014 as compared to 0.752 in 2011, and 0.764 in 2010. September extents are predicted as 0.294 in 2012, 0.290 in 2013 and 0.290 in 2014 as compared to 0.332 in 2011, and 0.264 in 2010.

4.4 Implications for prediction over entire Arctic

It is relevant to assess how far aggregated predictions (totalling over models for each of the nine regions) compare with predictions from a single model applied to the extent time series for the entire Arctic sea (i.e. to the data summarised in Table 1). The aggregated predictions are based on training data for 1979-2009 and combine

a) generalized beta regressions (with maximum inflation) applied to extent data for the Central Arctic, Canadian Archipelago, and Hudson Bay;

b) zero inflated gamma regressions applied to extent data for the Sea of Okhotsk, Bering Sea, and Gulf of St.Lawrence;

Table 4 Deterministic trend parameters

	Posterior summary		
Short term lag parameters	Mean	2.5%	97.5%
γ_{21}	-0.475	-1.971	0.833
γ_{22}	-0.455	-2.074	0.960
γ_{23}	0.344	-0.310	0.947
γ_{24}	0.166	-0.227	0.668
γ_{25}	0.823	0.305	1.303
γ_{26}	0.516	0.191	0.939
γ_{27}	0.907	0.348	1.821
γ_{28}	2.004	1.453	2.431
γ_{29}	3.222	2.580	3.926
$\gamma_{2,10}$	2.999	1.552	4.394
$\gamma_{2,11}$	1.110	-0.186	2.390
$\gamma_{2,12}$	0.545	-1.110	1.875
Linear trend guide parameters by period	Mean	2.5%	97.5%
Γ_{11}	-0.028	-0.048	-0.007
Γ_{12}	-0.019	-0.029	-0.008
Γ_{13}	-0.028	-0.036	-0.019
Γ_{21}	-0.037	-0.061	-0.017
Γ_{22}	-0.025	-0.036	-0.016
Γ_{23}	-0.034	-0.043	-0.026
Precision trends η_{2m}	Mean	2.5%	97.5%
η_{21}	0.028	-0.024	0.072
η_{22}	0.092	0.033	0.134
η_{23}	0.233	0.122	0.308
η_{24}	0.068	0.011	0.113
η_{25}	0.064	0.030	0.097
η_{26}	0.000	-0.034	0.031
η_{27}	-0.021	-0.052	0.009
η_{28}	-0.023	-0.058	0.009
η_{29}	0.005	-0.035	0.041
$\eta_{2,10}$	-0.017	-0.059	0.016
$\eta_{2,11}$	-0.026	-0.060	0.005
$\eta_{2,12}$	0.001	-0.033	0.031

Table 5 Trends in extent by month, sea of Okhotsk

	Period		
Month	1979-89	1990-2000	2001-2011
January	0.91	0.73	0.73
February	1.22	1.03	1.05
March	1.24	1.10	1.06
April	0.91	0.73	0.71
May	0.34	0.27	0.29
June	0.11	0.11	0.10
July	0	0	0
August	0	0	0
September	0	0	0
October	0.05	0.05	0.04
November	0.08	0.08	0.06
December	0.35	0.36	0.30
All Year	0.44	0.37	0.36

Table 6 Posterior predictive loss, sea ice extent model, sea of Okhotsk

	Trend model	
	Stochastic	Deterministic
In-sample fit (1979-2009)	2.95	2.43
	Stochastic	Deterministic
Cross validation fit (2010-11)	1.11	0.68
	% of observations within 95% credible intervals of posterior predictions	
	Stochastic	Deterministic
In-sample period	97.6	97.6
Cross-validation period	100.0	95.8

c) gamma regressions without inflation applied to extent data for the Baffin Sea, Greenland Sea, and Kara-Barents.

The deterministic trend model is adopted, and MCMC iterations 25,000-26,000 are cumulated over the nine models, providing posterior summary statistics (mean, variance, 2.5% percentile, 97.5% percentile) for cross-validatory predicted extents in 2010-11 across the entire Arctic region. By contrast, the single region approach (applied to the extent series for 1979-2009 encompassing the entire Arctic region) involves a gamma regression with deterministic trend - since the stochastic trend option provided much less precise predictions to 2010-11.

The comparative L-criteria for cross-validatory fit (over the 24 months in 2010-11) are 2.43 (aggregated predictions over sub-regions) and 3.70 (single region approach). Figure 8 shows the cross-validatory predictions (2010-11) and out-of-sample predictions (2012-14) for the combined forecast aggregating over sub-region models. March extents (in mn km^2) are predicted as 14.42 in 2012, 14.35 in 2013 and 14.31 in 2014, as compared to actual extents of 14.34 in 2011, and 14.88 in 2010. September extents (in mn km^2) are predicted as 4.97 in 2012, 4.84 in 2013 and 4.71 in 2014, as compared to actual totals of 4.60 in 2011, and 4.90 in 2010.

Although sub-region data are not (at the time of writing) available beyond 2011, totals for the entire Arctic are available from ftp://sidads.colorado.edu/DATASETS/NOAA/G02135/. (Note that the latter differ slightly before 2012 from entire Arctic

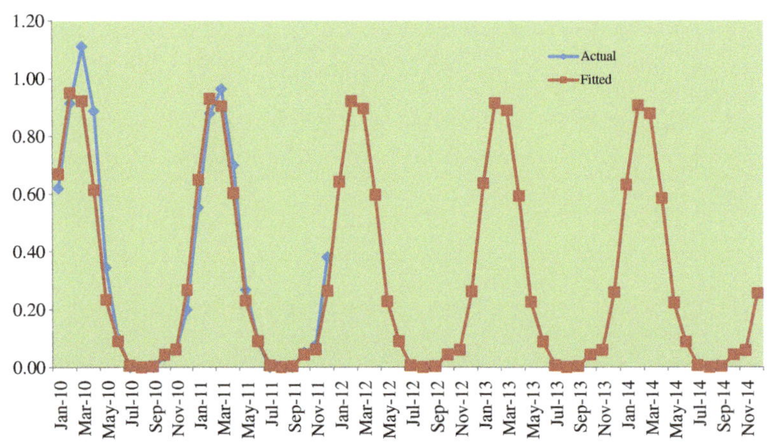

Figure 6 Sea of Okhotsk (mn km^2), cross-validation period (2010-11) actual vs fitted, and out-of-sample predictions to end 2014.

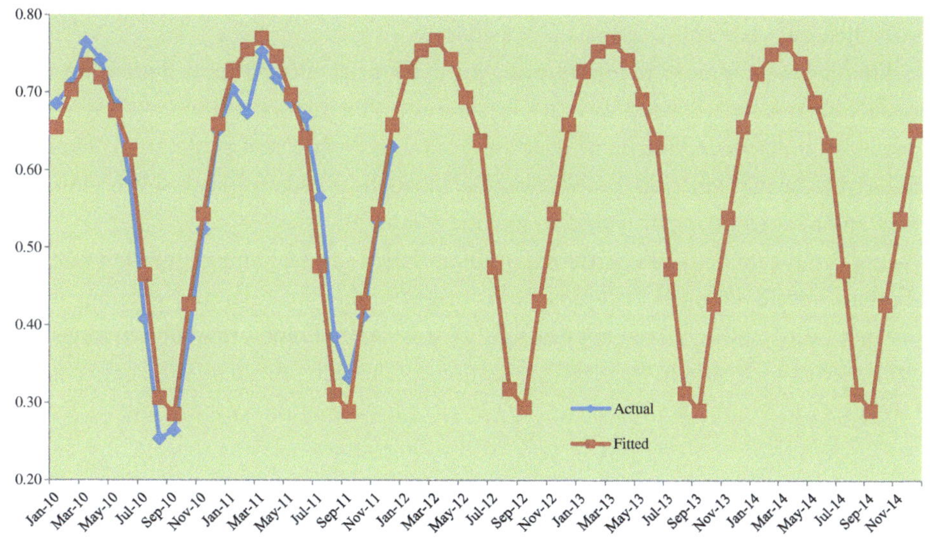

Figure 7 Greenland Sea (mn km^2), cross-validation period (2010-11) actual vs fitted, and out-of-sample predictions to end 2014.

extents based on totalling the sub-regional series at http://neptune.gsfc.nasa.gov/csb/index.php?section=59). In fact, the March 2012 figure of 15.21 mn km^2 (NSIDC, 2012) was the highest March average ice extent since 2008. By contrast, the September 2012 figure was anomalously low at 3.61 mn km^2.

5 Summary and conclusions

This paper has considered sub-regional aspects of the observed downward trend in Arctic ice sea cover. Arctic sub-regions differ in observed trends and also in seasonal extremes. Thus some regions still have total winter cover but partial summer cover, while other regions have partial winter cover and no ice cover in summer. While downward trends

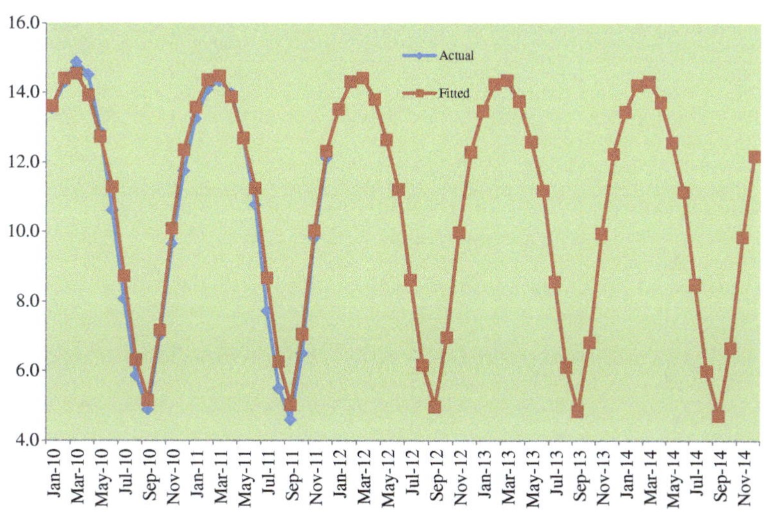

Figure 8 Entire Arctic combined forest (mn km^2), cross-validation period (2010-11) actual vs fitted, and out-of-sample predictions to end 2014.

in sea ice extent across the Arctic as a whole show a stronger summer decline, this is not necessarily the case when sub-regions are considered.

These differences are important for choosing the appropriate distribution, and inflation mechanism. Sub-regional differences in ice loss may also be important for assessing relatively localized impacts on climate or economic activities (e.g. Fissel et al. 2011). There may also be benefits in prediction (see Section 4.4) through considering such sub-regional differences and in aggregating over region-specific models.

Here generalized beta densities (with maximum inflation) and gamma regression (with zero inflation) have been used to represent recurrent winter maxima and summer ice disappearance respectively. Other applications of this methodology may be envisaged, outside time series applications. Possible applications of the generalized inflated beta density are discussed in Section 2.3. It may be noted that transformations of ice extent such as $w_t = log(1 + y_t)$ might be envisaged as ways of avoiding beta or gamma regressions, and instead leading possibly to modelling using normal or skewnormal likelihoods. However, the problem of seasonal extreme inflation at particular values remains, and the methodology proposed in the paper provides a suitable representation for such extremes or for explaining them.

Another approach, as in compositional data analysis, is to focus on the ratios $r_t = y_t/d_t$ to the maximum, in the case when the data has two categories (e.g. area covered by sea ice y_t, remaining area $d_t - y_t$). Generally compositional data analysis involves Gaussian likelihoods applied to log-transformed ratios. This method can adjust for zero inflation (e.g. Butler and Glasbey, 2008), but does not generate direct inferences or out-of-sample predictions for the extent data themselves.

Competing interests

The authors declare that they have no competing interests.

References

Aitchison, J, Egozcue, J: Compositional data analysis: Where are we and where should we be heading? Math. Geol. **37**, 829-850 (2005)

Blundell, R, Griffith, R, Windmeijer, F: Individual effects and dynamics in count data models. Journal of Econometrics. **108**, 113–131 (2002)

Brooks, S, Gelman, A: General methods for monitoring convergence of iterative simulations. J. Comput. Graph.l Stat. **7**, 434–455 (1998)

Butler, A, Glasbey, C: A latent Gaussian model for compositional data with zeros. J. R. Stat. Soc. Series C. **57**, 505–520 (2008)

Comiso, J, Parkinson, C, Gersten, R, Stock, L: Accelerated decline in the Arctic sea ice cover. Geophys. Res. Lett. **35**, L01703 (2008). doi:10.1029/2007GL031972

Fissel, D, de Saavedra Álvarez, M, Kulan, N, Mudge, T, Marko, J: Long-term trends for Sea Ice in the Western Arctic Ocean: implications for shipping and offshore oil and gas activities. In: *Proceedings of the Twenty-first 2011 International Offshore and, Polar Engineering Conference*. Maui, Hawaii, USA, (2011). International Society of Offshore and Polar Engineers (ISOPE). http://www.isope.org/publications/ISOPEproceedinglist.htm

Gelfand, A: Model determination using sampling based methods. In: W. Gilks, S. Richardson, D. Spiegelhalter (eds.) Markov Chain Monte Carlo in Practice, Chapman and Hall/CRC, Boca Raton, pp. 145–157 (1996)

Höhle, M, Paul, M: Count data regression charts for the monitoring of surveillance time series. Comput. Stat. & Data Anal. **52**, 4357–4368 (2008)

Huang, X, Oosterlee, C: Generalized beta regression models for random loss-given-default. Department of Applied Mathematical Analysis, Delft University of Technology, Report 08–10 (2008)

Jung, R, Kukuk, M, Liesenfeld, R: Time series of count data: modeling, estimation and diagnostics. Comput. Stat. Data Anal. **51**, 2350–2364 (2006)

Laud, P, Ibrahim, J: Predictive model selection. J R Stat Soc. **57B**, 247–262 (1995)

Ledolter, J, Abraham, B: Parsimony and its importance in time series forecasting. Technometrics. **23**, 411–414 (1981)

Leininger, T, Gelfand, A, Allen, J, Silander, J: Spatial regression modeling for compositional data with many zeros. J. Agricultural Biol. Environ. Stat. **18**(3), 314–334 (2013)

Liu, J, Curry, J, Wang, H, Song, M, Horton, R: Impact of declining Arctic sea ice on winter snowfall. Proc. Nat. Acad. Sci. **109**, 4074–4079 (2012)

Lunn, D, Spiegelhalter, D, Thomas, A, Best, N: The BUGS project: Evolution, critique and future directions. Stat. Med. **28**, 3049–3067 (2009)

National Snow and Ice Data Center (2012). http://nsidc.org/arcticseaicenews/2012/04/arctic-sea-ice-enters-the-spring-melt-season/

Ospina, R, Ferrari, S: Inflated beta distributions. Stat. Papers. **51**, 111–126 (2010)

Pham-Gia, T, Duong, Q: The generalized beta- and F-distributions in statistical modelling. Math. Comput. Modell. **12**, 1613–1625 (1989)

Screen, J, Deser, C, Simmonds, I, Tomas, R: Atmospheric impacts of Arctic sea-ice loss 1979–2009: separating forced change from atmospheric internal variability. Climate Dynamics (2013). doi:10.1007/s00382-013-1830-9

Serreze, M, Maslanik, J, Key, J, Kokaly, R, Robinson, D: Diagnosis of the record minimum in arctic sea ice area during 1990 and associated snow cover extremes. Geophys. Res. Lett. **22**, 2183–2186 (1995)

Stroeve, J, Holland, M, Meier, W, Scambos, T, Serreze, M: Arctic sea ice decline: faster than forecast. Geophys. Res. Lett. **34**, L09501 (2007)

Stroeve, J, Serreze, M, Holland, M, Kay, J, Malanik, J, Barrett, A: The Arctic's rapidly shrinking sea ice cover: a research synthesis. Climatic Change. **110**, 1005–1027 (2012)

Zhao, X, Luo, Y, Wang, S, Huang, W, Lian, J: Is desertification reversion sustainable in northern China: a case study in Naiman county, part of a typical agro-pastoral transitional zone in Inner-Mongolia, China. Global Environ. Res. **14**, 63–70 (2010)

2

Recent developments on the construction of bivariate distributions with fixed marginals

Gwo Dong Lin[1*], Xiaoling Dou[2], Satoshi Kuriki[3] and Jin-Sheng Huang[4]

*Correspondence:
gdlin@stat.sinica.edu.tw
[1] Institute of Statistical Science, Academia Sinica, Taiwan, Republic of China
Full list of author information is available at the end of the article

Abstract

Constructing a bivariate distribution with specific marginals and correlation has been a challenging problem since 1930s. In this survey we shall focus on the recent developments on the FGM-related distributions, including Sarmanov and Lee's distributions, Baker's distributions and Bayramoglu's distributions. This complements the most recent works of (i) the review by Sarabia and Gómez-Déniz (2008, SORT) and (ii) the monograph by Balakrishnan and Lai (2009, Springer). Some new results are provided.

Mathematics Subject Classification (2000): 62H20; 62H86; 62G30; 60E05; 62E10

Keywords: FGM distribution; Sarmanov and Lee's distribution; Baker's distribution; Bayramoglu's distribution; Bivariate order statistics; Fréchet–Hoeffding bounds; Hoeffding's identity for covariance; Weak convergence; Pearson's correlation; Spearman's rho; Kendall's tau; Chebyshev's inequality for integrals; Euler–Maclaurin summation formula

1 Introduction and a brief history

In both statistical theory and practice, we often need to construct a joint distribution with specific marginals and correlation. Ever since Wigner (1932) and Eyraud (1936), it has been an active and challenging topic. Its field of applications is wide ranging. It encompasses physics, economics, engineering, risk analysis, medicine etc. There is no dearth of interesting real life examples. Only recently Takeuchi (2010) constructed a bivariate distribution having specific marginals and certain degrees of correlation to model the joint behavior of far-infrared and far-ultraviolet galaxy luminosity. Danaher and Smith (2011) studied the interaction between a firm's total amount of purchase (log-normal distribution) and the time since last purchase (exponential distribution). For applications in reliability theory, see for example Li et al. (2013) and the references therein.

Since 1990 there have been seven conferences devoted to this topic: 1990 (Rome, Italy), 1993 (Seattle, USA), 1996 (Prague, Czech Republic), 2000 (Barcelona, Spain), 2004 (Québec, Canada), 2007 (Tartu, Estonia), and 2010 (São Paulo, Brazil), averaging one in every three to four years. The papers presented in the conferences during the period 1990-2000 have all appeared in monographs, edited by Dall'Aglio et al. (1991), Rüschendorf et al. (1996), Beněs and Štěpan (1997), and Cuadras et al. (2002), respectively. As for the 2004 and 2007 ones, they were published as special issues of *Insurance: Mathematics and Economics* (August 2005), *The Canadian Journal of Statistics* (September 2005) and

Journal of Statistical Planning and Inference (November 2009). We have yet to find any documentation for the 2010 conference.

There were three survey papers: Lai (2004; 2006) and Sarabia and Gómez-Déniz (2008). The latest monograph by Balakrishnan and Lai (2009) provided an excellent overview. As for the continuous multivariate version, there were Kotz et al. (2000) and Kotz and Nadarajah (2004). For the discrete version, see Johnson et al. (1997), Gómez-Déniz et al. (2012) and Sarabia and Gómez-Déniz (2011), among others. Then there were Cuadras (1992) and Dolati and Úbeda-Flores (2005) for constructing distributions with given multivariate marginals and given dependence structure.

Section 2 is a review of some basic properties of the bivariate distributions with fixed marginals. Then, in Sections 3 to 6, we will focus on the recent developments on the FGM-related distributions, including Sarmanov and Lee's distributions, Baker's distributions and Bayramoglu's distributions. Some new results are provided. This complements the most recent works mentioned above. Finally, in Section 7 we briefly discuss some other related distributions.

2 The natural bounds of the correlations

We first recall an important fundamental result due to Hoeffding (1940) and Fréchet (1951) independently. Let $H(x,y) = \Pr(X \le x,\ Y \le y)$ be the joint distribution of any pair (X, Y) of random variables whose marginals are $F(x) = \Pr(X \le x)$ and $G(y) = \Pr(Y \le y)$. We write $(X, Y) \sim H,\ X \sim F,\ Y \sim G$. Then the bivariate distribution H satisfies the following inequality:

$$\begin{aligned} &\max\{0,\ F(x) + G(y) - 1\} \equiv H_-(x,y) \\ &\le H(x,y) \le H_+(x,y) \equiv \min\{F(x),\ G(y)\},\quad x,y \in R \equiv (-\infty,\infty), \end{aligned} \tag{1}$$

where the extremal distributions H_+ and H_- are known as the Fréchet–Hoeffding upper and lower bounds, respectively.

To generate a pair of random variables obeying the extremal distributions, we start with $Z \sim U(0,1)$, the uniform distribution on $(0,1)$, and note that $(F^{-1}(Z),\ G^{-1}(Z)) \sim H_+$ and $(F^{-1}(Z),\ G^{-1}(1-Z)) \sim H_-$, where $F^{-1}(t) = \inf\{x : F(x) \ge t\}$, $t \in (0,1)$, is the quantile function of F. Take for example $F = G = U(0,1)$. Then $(Z,Z) \sim H_+$ with $H_+(x,y) = \min\{x,y\}$ and $(Z,\ 1-Z) \sim H_-$ with $H_-(x,y) = \max\{0, x+y-1\}$. We see that for this special case, both H_+ and H_- are *singular* bivariate distributions, with the entire mass concentrated on the lines $y = x$ or $y = -x$, respectively (see Lin and Huang 2010, and the references therein).

The inequality (1) for H implies that $\mathrm{Cov}(H_-) \le \mathrm{Cov}(H) \le \mathrm{Cov}(H_+)$ by the Hoeffding representation for covariance:

$$\mathrm{Cov}(H) \equiv \mathrm{Cov}(X,Y) = \int_{-\infty}^{\infty}\int_{-\infty}^{\infty} [H(x,y) - F(x)G(y)]\,dxdy, \tag{2}$$

whenever the double integral exists and is finite. It then follows that

$$\rho(H_-) \le \rho(H) \le \rho(H_+), \tag{3}$$

where $\rho(H)$ denotes the correlation of any pair of random variables $(X,Y) \sim H$ (or, simply, the correlation of H). Writing μ and σ for the mean and the standard deviation of the random variable appearing as a subscript, we see that the maximum correlation can be expressed explicitly in terms of the quantile functions of the marginals:

$$\rho(H_+) = \left(\int_0^1 F^{-1}(t)G^{-1}(t)dt - \mu_X\mu_Y\right)/(\sigma_X\sigma_Y) \leq 1.$$

Moreover, $\rho(H_+) = 1$ iff the marginals F and G are of the same type, namely, $F(x) = G(ax + b)$ on R for some $a > 0$ and $b \in R$. Similarly, we have

$$\rho(H_-) = \left(\int_0^1 F^{-1}(t)G^{-1}(1-t)dt - E(X)E(Y)\right)/(\sigma_X\sigma_Y) \geq -1,$$

and $\rho(H_-) = -1$ iff the distributions of X and $-Y$ are of the same type.

3 FGM distributions

To construct a bivariate distribution with fixed marginals, a well known method is the FGM distribution, dating back to Eyraud (1936), Farlie (1960), Gumbel (1960) and Morgenstern (1956). Let $(X, Y) \sim H$, $X \sim F$ and $Y \sim G$. The bivariate distribution

$$H(x,y) = F(x)G(y)\{1 + \alpha\overline{F}(x)\overline{G}(y)\}, \quad x, y \in R, \tag{4}$$

is called the FGM distribution, where $\overline{F} = 1 - F$, $\overline{G} = 1 - G$ and α is a parameter such that H is a *bona fide* bivariate distribution.

The advantage of the FGM distribution is its simplicity. Its drawback is the lack of flexibility. As was pointed out by Schucany et al. (1978), for *absolutely continuous* marginals the correlation in FGM is restricted to the interval $[-1/3, 1/3]$, which is a far cry from the full range $[-1, 1]$ enjoyed by the bivariate normal. Schucany et al.'s condition of *absolute continuity* was later relaxed by Lin (1987) to just *continuity*.

To alleviate the deficiency, Johnson and Kotz (1975) proposed the 'iterated' FGM distribution:

$$H_k(x,y) = F(x)G(y) + \sum_{j=1}^{k} \alpha_j (F(x)G(y))^{\lfloor j/2\rfloor+1}(\overline{F}(x)\overline{G}(y))^{\lfloor (j+1)/2\rfloor}, \quad x, y \in R,$$

where $\lfloor z\rfloor$ denotes the greatest integer less than or equal to z, and $k = 1, 2, 3, \ldots$.

For $k = 1$, it reverts back to the plain FGM, where $|\alpha_1| \leq 1$ if both F and G are continuous. For $k = 2$ the iterated FGM takes the form (Huang and Kotz 1984, 1999)

$$H(x,y) = F(x)G(y) + \alpha_1 F(x)G(y)\overline{F}(x)\overline{G}(y) + \alpha_2[F(x)G(y)]^2\,\overline{F}(x)\overline{G}(y), \quad x, y \in R.$$

To ensure that H be a *bona fide* distribution function, the range of the parameters, however, is more complicated, namely, for continuous marginals,

$$|\alpha_1| \leq 1, \quad \alpha_1 + \alpha_2 \geq -1, \quad \alpha_2 \leq 2^{-1}\left\{3 - \alpha_1 + (9 - 6\alpha_1 - 3\alpha_1^2)^{1/2}\right\}.$$

(See also Lin 1987.) Take $F = G = U(0, 1)$ for the sake of comparison, the iteration succeeded in extending the range to $-\frac{1}{3} \leq \rho \leq 0.434$ from the previous $-\frac{1}{3} \leq \rho \leq \frac{1}{3}$ of the plain FGM.

The iteration, however, was little studied beyond the $k = 2$ case due to the difficulty in determining the admissible range of the parameters α_j.

There were several other extensions of the FGM distribution, but the improvement in correlation was still limited (see, e.g., Balakrishnan and Lai 2009). To accommodate higher flexibility we turn to two other classes: (i) Sarmanov and Lee's distribution (Sarmanov

1966, Lee 1996) and its generalization and (ii) Baker's distribution (Baker 2008) and its generalization.

4 Sarmanov and Lee's distributions

The joint density of a Sarmanov–Lee distribution is given by

$$h(x,y) = f(x)g(y)\{1 + \alpha\theta_1(x)\theta_2(y)\}, \quad x, y \in R, \tag{5}$$

where f and g are the densities of the marginals F and G, while θ_1 and θ_2 are measurable functions satisfying the condition

$$1 + \alpha\theta_1(x)\theta_2(y) \geq 0 \text{ and } E(\theta_1(X)) = E(\theta_2(Y)) = 0,$$

which serves to ensure that h is a *bona fide* joint density.

When the marginals F and G are absolutely continuous, setting $\theta_1 = 1 - 2F$ and $\theta_2 = 1 - 2G$ shows that FGM is a special case of the Sarmanov–Lee.

Shubina and Lee (2004) showed that the maximum positive correlation of (5) is effected by concentrating all its mass on the (NE-SW) quadrants: $\{(x,y) : (x - x_0)(y - y_0) \geq 0\}$, and for the negative, on the (NW-SE) quadrants: $\{(x,y) : (x - x_0)(y - y_0) \leq 0\}$, for some real numbers x_0 and y_0.

The improvement in correlation is substantial. The maximum correlation, for the uniform marginals case, is 3/4 as opposed to 1/3 of the plain FGM and 0.434 of the ($k = 2$) iterated FGM.

To further extend the Sarmanov–Lee family (5), a 'generalized' Sarmanov–Lee was proposed by Bairamov et al. (2001),

$$h(x,y) = f(x)g(y)\left\{1 + \alpha T(F(x), G(y))\right\}, \quad x, y \in R, \tag{6}$$

where the product function $\theta_1(x)\theta_2(y)$ in (5) is replaced by $T(F(x), G(y))$ in which T is an integrable bivariate function on $[0,1]^2$, satisfying

$$\int_0^1 T(u,v)du = \int_0^1 T(u,v)dv = 0 \text{ and } 1 + \alpha T \geq 0 \text{ on } [0,1]^2, \tag{7}$$

and the parameter α satisfies

$$-\left[\max_{(u,v)\in D^+} T(u,v)\right]^{-1} \leq \alpha \leq \left[\max_{(u,v)\in D^-}(-T(u,v))\right]^{-1}, \tag{8}$$

where $D^+ = \{(u,v) : T(u,v) > 0\}$ and $D^- = \{(u,v) : T(u,v) < 0\}$. Again, the constraints (7) and (8) together guarantee that the function h in (6) is a *bona fide* bivariate density with marginal densities f and g.

The new class (6) is so rich it contains members arbitrarily close to H_+ and H_-, the two extremal distributions (1) of Fréchet–Hoeffding. It is thus flexible enough to accommodate nearly the maximum positive (negative) correlation. Some simple examples help demonstrate the idea (more can be found in Lin and Huang 2011).

Example 1. The generalized Sarmanov–Lee density (3×3) with uniform marginals. Consider the partitioning of the unit square into $3 \times 3 = 9$ subsquares, induced by the lines $x, y = 1/3, 2/3$. Let the function T be defined by

$$T(x,y) = \begin{cases} 2, & 0 < x,y < \frac{1}{3} \text{ or } \frac{1}{3} < x,y < \frac{2}{3} \text{ or } \frac{2}{3} < x,y < 1, \\ -1, & \text{elsewhere.} \end{cases}$$

Then the joint density becomes

$$h(x,y)=\begin{cases}3,\ 0<x,y<\frac{1}{3}\ \text{or}\ \frac{1}{3}<x,y<\frac{2}{3}\ \text{or}\ \frac{2}{3}<x,y<1,\\ 0,\ \text{elsewhere}.\end{cases}$$

Its correlation is $\rho(h) = 8/9$, which exceeds the maximum, $\rho_{\max} = 3/4$, of the plain Sarmanov–Lee (with uniform marginals).

Example 2. The generalized Sarmanov–Lee density ($n \times n$) with uniform marginals. For $n = 2, 3, 4, \ldots$, let

$$T_n(x,y)=\begin{cases}n-1, & \frac{i-1}{n}<x,y<\frac{i}{n},\ i=1,\ldots,n,\\ -1, & \text{elsewhere},\end{cases}$$

and

$$h_n(x,y)=\begin{cases}n,\ \frac{i-1}{n}<x,y<\frac{i}{n},\ i=1,\ldots,n,\\ 0,\ \text{elsewhere}.\end{cases}$$

We have the correlation $\rho(h_n) = 1 - n^{-2} \to 1$ as $n \to \infty$.

Example 3. The generalized Sarmanov–Lee density ($n \times n$) with exponential marginals. Let

$$h_n(x,y)=\begin{cases}ne^{-x-y}, & \ln\frac{n}{n-i+1}<x,y<\ln\frac{n}{n-i},\ i=1,\ldots,n,\\ 0, & \text{elsewhere},\end{cases}$$

where $\ln(n/0) \equiv \infty$. Our calculations show that $\rho(h_n) \approx 1 - 1.08/n \to 1$ as $n \to \infty$.

The convergence to the maximal correlation is not limited to those with uniform or the exponential marginals. It holds for more general cases. The following Theorems 1 through 3 are successively stronger. Only Theorem 3 will be proved, since the others appeared in Lin and Huang (2011).

Let F and G be arbitrary distributions (not necessarily identical nor 'of the same type') with densities $F' = f$ and $G' = g$. Define the joint density of (X_n, Y_n) by

$$h_n(x,y)=\begin{cases}nf(x)g(y), & (x,y)\in\left(F^{-1}(\frac{i-1}{n}),F^{-1}(\frac{i}{n})\right)\times\left(G^{-1}(\frac{i-1}{n}),G^{-1}(\frac{i}{n})\right),\\ & i=1,2,\ldots,n,\\ 0, & \text{elsewhere},\end{cases}\tag{9}$$

where $F^{-1}(0) \equiv \lim_{t\to 0^+} F^{-1}(t)$ and $F^{-1}(1) \equiv \lim_{t\to 1^-} F^{-1}(t)$ are the left and right extremities of the distribution F, respectively. It qualifies as a generalized Sarmanov–Lee bivariate density (with marginal densities f and g).

Theorem 1. For $F = G$, (9) becomes

$$h_n(x,y)=\begin{cases}nf(x)f(y), & F^{-1}(\frac{i-1}{n})<x,y<F^{-1}(\frac{i}{n}),\ i=1,2,\ldots,n,\\ 0, & \text{elsewhere}.\end{cases}$$

If F has a finite variance σ^2, then the correlation is

$$\rho(h_n)=\frac{1}{\sigma^2}\left[n\sum_{i=1}^{n}\left(\int_{F^{-1}(\frac{i-1}{n})}^{F^{-1}(\frac{i}{n})}xf(x)dx\right)^2-(E(X))^2\right],$$

which converges to $\rho(H_+)$, which is 1 in this case, as $n \to \infty$.

Theorem 2. If in (9), X and Y satisfy any of the conditions:

(i) F^{-1} or G^{-1} is uniformly continuous on $(0, 1)$,
(ii) $a \le X, a' \le Y$ *a.s.*, (iii) $X \le b, Y \le b'$ *a.s.*,
(iv) $X \ge a,\ Y \le b$ *a.s.* and F^{-1}, G^{-1} have continuous derivatives, where $a,\ b,\ a',\ b' \in R$,

then $\rho(h_n)$ converges to $\rho(H_+)$ as $n \to \infty$.

Remarks 1. Theorem 1 follows immediately from a general result, which is interesting in its own right: For any $X \sim F$ with finite $E(X^2)$,

$$S_n \equiv n \sum_{i=1}^{n} \left(\int_{\frac{i-1}{n}}^{\frac{i}{n}} F^{-1}(t)dt \right)^2 \longrightarrow \int_0^1 \left(F^{-1}(t)\right)^2 dt = E(X^2) \text{ as } n \to \infty.$$

The proofs of Remarks 1 and Theorem 2 are based on the following two lemmas.

Lemma 1. (Chebyshev's inequality for integrals.) Let $f_1,\ f_2\ :\ (a, b)\ \to\ R$ be both increasing or both decreasing, and let $p : (a, b) \to (0, \infty)$ be an integrable function. Then

$$\int_a^b p(x)f_1(x)dx \int_a^b p(x)f_2(x)dx \le \int_a^b p(x)dx \int_a^b p(x)f_1(x)f_2(x)dx,$$

provided that all integrals exist and are finite.

Remarks 2. An extension of this inequality can be rephrased in probabilistic terms: Let X be any random variable and let f_1, f_2 be both increasing or both decreasing, then the covariance $\mathrm{Cov}(f_1(X), f_2(X)) = E[f_1(X)f_2(X)] - E[f_1(X)]\,E[f_2(X)]$ is non-negative, provided that the expectations $E[f_1(X)]$, $E[f_2(X)]$ and $E[f_1(X)f_2(X)]$ exist.

Lemma 2. (Euler–Maclaurin summation formula.) Let $m < n$ be positive integers and let f be a real-valued function on $[m, n]$. Then we have
(i) if f has a continuous derivative on $[m, n]$,

$$\sum_{k=m}^{n} f(k) = \int_m^n f(x)dx + \frac{1}{2}(f(m) + f(n)) + \int_m^n f'(x)(x - \lfloor x \rfloor - \frac{1}{2})dx;$$

(ii) if f has a continuous derivative of order 4,

$$\sum_{k=m}^{n} f(k) = \int_m^n f(x)dx + \frac{1}{2}(f(m) + f(n)) + \frac{1}{12}(f'(n) - f'(m)) + R(f),$$

where the remainder $R(f) = \frac{1}{24}\int_m^n f^{(4)}(x)(B_4 - B_4(x - \lfloor x \rfloor))dx$ in which $B_4(x) = x^4 - 2x^3 + x^2 - \frac{1}{30}$ is the Bernoulli polynomial and $B_4 = B_4(0) = -1/30$ the Bernoulli number.

Theorem 3. For arbitrary marginals F and G with densities $F' = f$ and $G' = g$, we have (i) the distribution H_n of (9) converges weakly to H_+ as $n \to \infty$, and (ii) the correlation of H_n converges to that of H_+ as $n \to \infty$, provided that F and G have finite variances.

Proof. Note that the support of (9) is contained in the region bounded by the two curves $G(y) = F(x) + \frac{1}{n}$ and $G(y) = F(x) - \frac{1}{n}$. Therefore for any (x, y) in the region $G(y) > F(x)$, we have $\Pr(X_n < x,\ Y_n > y) = 0$ for all large n ($n \ge (G(y) - F(x))^{-1}$). This implies that $H_n(x, y) = \Pr(X_n \le x,\ Y_n \le y) = \Pr(X_n \le x) - \Pr(X_n \le x,\ Y_n > y) = \Pr(X_n \le x) = F(x) = \min\{F(x), G(y)\} = H_+(x, y)$ for all $n \ge (G(y) - F(x))^{-1}$. Likewise, for each (x, y)

in the region $G(y) < F(x)$ we have $H_n(x,y) = G(y) = \min\{F(x), G(y)\} = H_+(x,y)$ for all large n. Finally, for each (x,y) with $G(y) = F(x)$, $F(x) - \frac{1}{n} \le H_n(x,y) \le F(x)$ for all n, and $\lim_{n\to\infty} H_n(x,y) = F(x) = G(y) = H_+(x,y)$. All together, we see that in all three cases, we have $\lim_{n\to\infty} H_n(x,y) = H_+(x,y)$. This proves part (i). Part (ii) follows from the next lemma which can be proved by using Hölder's inequality (see, e.g., the proof of Theorem 4 in Dou et al. 2013). □

Lemma 3. Let $(X_n, Y_n) \sim H_n$ be a sequence of bivariate random variables and $X_n \sim F$, $Y_n \sim G$ for all $n \ge 1$. Assume that (X_n, Y_n) converges in distribution to $(X_0, Y_0) \sim H_0$ as n tends to infinity. If, in addition, $E[\,|X_n|^{p+q}]$, $E[\,|Y_n|^{p+q}] < \infty$ for some positive integers p and q, then $\lim_{n\to\infty} E[\,X_n^p Y_n^q] = E[\,X_0^p Y_0^q]$.

Remarks 3. We wondered if the convergence of Theorem 3(ii) is monotone. Indeed, $\rho(H_m) \ge \rho(H_n)$ if m is a multiple of n (say, $m = kn$). To see this, write from (9)

$$E(X_nY_n) = n\sum_{i=1}^{n}\left(\int_{(i-1)/n}^{i/n} F^{-1}(t)dt\right)\left(\int_{(i-1)/n}^{i/n} G^{-1}(t)dt\right) \equiv n\sum_{i=1}^{n} A_iB_i,$$

where

$$\begin{aligned} A_iB_i &= \sum_{j=1}^{k}\left(\int_{(i-1)/n+(j-1)/(kn)}^{(i-1)/n+j/(kn)} F^{-1}(t)dt\right)\sum_{j=1}^{k}\left(\int_{(i-1)/n+(j-1)/(kn)}^{(i-1)/n+j/(kn)} G^{-1}(t)dt\right) \\ &\equiv \sum_{j=1}^{k} a_{ij}\sum_{j=1}^{k} b_{ij} \le k\sum_{j=1}^{k} a_{ij}b_{ij}, \end{aligned}$$

by Chebyshev's sum inequality (see, e.g., Dou et al. 2013, Lemma 1). This in turn implies that $E(X_mY_m) \ge E(X_nY_n)$ and hence $\rho(H_m) \ge \rho(H_n)$. That the sequence $\{\rho(H_n)\}_{n=1}^{\infty}$ fails to be monotonically increasing can be seen from the following counterexample. Let $f = g$, $f(x) = 1/6$ if $0 \le a \le |x| \le a+3$, and $f(x) = 0$ otherwise. We have $\rho(H_3) < \rho(H_2)$ for $a > (-1+\sqrt{6})/2$.

For certain marginals, the rate of convergence of Theorem 3(ii) can be determined (Lin and Huang 2011 and Theorem 4 below). The calculation is based on the following. Note that the correlation of two random variables is location-scale invariant.

Lemma 4. For positive integer $n > 1$, we have

(i) $\sum_{k=1}^{n} k\ln k = \frac{1}{2}n^2\ln n - \frac{1}{4}n^2 + \frac{1}{2}n\ln n + \frac{1}{12}\ln n + R_1(n)$, where $|R_1(n)| < \frac{5}{12}$;
(ii) $\sum_{k=1}^{n} k^2(\ln k)^2 = \frac{1}{3}n^3(\ln n)^2 - \frac{2}{9}n^3\ln n + \frac{2}{27}n^3 + \frac{1}{2}n^2(\ln n)^2 + \frac{1}{6}n(\ln n)^2 + \frac{1}{6}n\ln n + R_2(n)$, where $R_2(n) = \mathcal{O}(1)$ as $n\to\infty$;
(iii) $\sum_{k=1}^{n} k(k+1)(\ln k)\ln(k+1) = \frac{1}{3}n^3(\ln n)\ln(n+1) - \frac{1}{9}n^3(\ln n + \ln(n+1)) + \frac{2}{27}n^3 + n^2(\ln n)\ln(n+1) - \frac{1}{4}n^2\ln(n+1) - \frac{1}{12}n^2\ln n + \frac{2}{3}n(\ln n)\ln(n+1) + \frac{1}{9}n^2 + \frac{1}{12}(\ln n)\ln(n+1) + \frac{1}{4}n\ln n + \frac{1}{12}(n+1)\ln(n+1) - \frac{11}{36}n - \frac{1}{12}(\ln n)^2 + \frac{5}{36}\ln(n+1) + R_3(n)$, where $R_3(n) = \mathcal{O}(1)$ as $n\to\infty$.

Remarks 4. Recall that the Euler constant $\gamma = \lim_{n\to\infty}\left(\sum_{k=1}^{n} 1/k - \ln n\right) \approx 0.57721$. Interestingly, each of the three remainder terms in Lemma 4 also converges to a real constant as $n \to \infty$, say $\lim_{n\to\infty} R_i(n) = R_i$, $i = 1,2,3$. In fact, $R_1 = \ln A \approx$

0.24875, where $A \approx 1.28243$ is the so-called Glaisher–Kinkelin constant, and $\ln A$ can be represented as

$$\begin{aligned} \ln A &= \frac{1}{4} + \frac{1}{24}\int_1^\infty f_1^{(4)}(x)h(x)dx \\ &= \frac{1}{4} + \frac{1}{24}\sum_{k=1}^{\infty}\int_0^1 f_1^{(4)}(x+k)[-x^2(x-1)^2]\,dx \\ &= \frac{1}{4} + \frac{1}{12}\sum_{k=1}^{\infty}\{6k+3-(6k^2+6k+1)\ln(1+1/k)\}, \end{aligned}$$

where $f_1(x) = x\ln x$ and $h(x) = B_4 - B_4(x - \lfloor x \rfloor)$. Similarly,

$$\begin{aligned} R_2 &= -\frac{2}{27} + \frac{1}{24}\int_1^\infty f_2^{(4)}(x)h(x)dx \approx -0.06576, \\ R_3 &= \frac{1}{18}\ln 2 + \frac{13}{108} + \frac{\pi^2}{72} + \frac{1}{24}\int_1^\infty f_3^{(4)}(x)h(x)dx \\ &= 0.295956 + \frac{1}{24}\int_1^\infty f_3^{(4)}(x)h(x)dx \approx 0.303, \end{aligned}$$

where $f_2(x) = x^2(\ln x)^2$ and $f_3(x) = x(x+1)(\ln x)\ln(x+1)$.

Theorem 4. (i) If F is uniform and if G is a power function distribution, then the convergence rate of $\rho(h_n)$ in (9) is $1/n^2$ as $n \to \infty$.
(ii) If $F = U(0,1)$ and if G is exponential, then the convergence rate of $\rho(h_n)$ is $(\ln n)/n^2$ as $n \to \infty$. More precisely, $\rho(h_n) = \sqrt{3}/2 - (2\sqrt{3})^{-1}(\ln n)/n^2 + \mathcal{O}(n^{-2})$ as $n \to \infty$.
(iii) If $F = U(0,1)$ and if G is logistic, then the convergence rate of $\rho(h_n)$ is $(\ln n)/n^2$ as $n \to \infty$. More precisely, $\rho(h_n) = 3/\pi - \pi^{-1}(\ln n)/n^2 + \mathcal{O}(n^{-2})$ as $n \to \infty$.
(iv) If $F = G$ is exponential, then $\rho(h_n) = 1 + \mathcal{O}(n^{-1})$ as $n \to \infty$.

It is interesting to characterize the FGM and Sarmanov–Lee distributions by minimizing the χ^2 divergence. Let h be the joint density of X and Y with marginal densities $f = F'$ and $g = G'$. Define the χ^2 divergence (distance) between the joint density h and the product density fg (of independent random variables) by

$$\chi^2(h;f,g) = \int\int_{S_F\times S_G}\left[\frac{h(x,y)}{f(x)g(y)} - 1\right]^2 f(x)g(y)dxdy, \tag{10}$$

where S_F and S_G are the supports of F and G, respectively.

Nelsen (1994) obtained a characterization of the FGM distributions by minimizing the χ^2 divergence (10). Huang and Lin (2011) extended Nelsen's (1994) result to the case of Sarmanov–Lee distributions. For $i = 1,2$, consider the functions $\theta_i^* : [0,1] \longrightarrow [-1,1]$ satisfying

$$\int_0^1 \theta_i^*(u)du = 0,\ \int_0^1 (\theta_i^*(u))^2 du = c_i > 0,\quad \sup_{u\in[0,1]} \theta_i^*(u) = 1,\quad \inf_{u\in[0,1]} \theta_i^*(u) = -h_i, \tag{11}$$

where $h_i \in (0,1]$. Then we have the following.

Theorem 5. Among all absolutely continuous bivariate distributions with marginal densities $f = F'$ and $g = G'$, the one whose joint density is closest to the product density of independent random variables (in the sense of minimizing the χ^2 divergence) subject to the constraint $E\left[\theta_1^*(F(X))\theta_2^*(G(Y))\right] = c_0$, where $-1 \le \frac{c_0}{c_1c_2} \le (\max\{h_1, h_2\})^{-1}$ and θ_i^*, c_i, h_i, $i = 1, 2$, are given in (11), is the Sarmanov–Lee distribution having joint density

$$h(x, y) = f(x)g(y)\left\{1 + \frac{c_0}{c_1c_2}\theta_1^*(F(x))\theta_2^*(G(y))\right\},\ x, y \in R.$$

The Sarmanov–Lee distribution and its generalization have been used in actuarial science, financial markets, electrical engineering and quantum statistical mechanics (see the references in Lin and Huang 2011). For example, Hernández-Bastida and Fernández-Sánchez (2012) applied a Sarmanov–Lee family to the Bayes premium in a collective risk model. In the analysis of longitudinal data, Cole et al. (1995) used a Sarmanov–Lee bivariate distribution for transition probabilities in a two-state Markov model and developed an empirical Bayes estimation methodology. Recently, Pelican and Vernic (2013a; 2013b) studied the parameter estimation problems for the Sarmanov–Lee distribution.

5 Baker's distributions

Write $X_{k,n} \sim F_{k,n}$ for the kth smallest order statistic of the random sample $\{X_i\}_{i=1}^n$ from F. Likewise, let $Y_{k,n}$ and $G_{k,n}$ stand for another sequence $Y_i \sim G$, $\{Y_i\}_{i=1}^n$ independent of $\{X_i\}_{i=1}^n$. For maximal correlation, Baker (2008) proposed the bivariate distribution

$$H_+^{(n)}(x, y) = \frac{1}{n}\sum_{k=1}^{n} F_{k,n}(x)G_{k,n}(y),\ x, y \in R, \tag{12}$$

and for the minimum,

$$H_-^{(n)}(x, y) = \frac{1}{n}\sum_{k=1}^{n} F_{k,n}(x)G_{n-k+1,n}(y),\ x, y \in R. \tag{13}$$

Clearly, both $H_+^{(n)}$ and $H_-^{(n)}$ satisfy the constraint of having F and G as the marginals. The following convex combination of $H_+^{(n)}$ (or $H_-^{(n)}$), $n = 1, 2$,

$$\begin{aligned} H_{q\pm}(x, y) &= (1-q)H_\pm^{(1)}(x, y) + qH_\pm^{(2)}(x, y) \\ &= F(x)G(y)\{1 \pm q\overline{F}(x)\overline{G}(y)\},\ \ x, y \in R,\ q \in [0, 1], \end{aligned}$$

is an FGM distribution.

As a generalization of (12) and (13), Baker (2008) also introduced

$$H_{\mathbf{r}}^{(n)}(x, y) = \sum_{k=1}^{n}\sum_{\ell=1}^{n} r_{k,\ell}F_{k,n}(x)G_{\ell,n}(y),\ \ x, y \in R, \tag{14}$$

where $(r_{k,\ell}) = \mathbf{r} \in \mathcal{R}$ which consists of all $\mathbf{r}$ satisfying: $r_{k,\ell} \ge 0$ and $\sum_{k=1}^n r_{k,\ell} = \sum_{\ell=1}^n r_{k,\ell} = 1/n$ for all $k, \ell = 1, 2, \ldots, n$; namely, $n\mathbf{r}$ is a doubly stochastic matrix.

If $P\{(X_n, Y_n) = (X_{k,n}, Y_{\ell,n})\} = r_{k,\ell}$ for all k, ℓ, then $(X_n, Y_n) \sim H_{\mathbf{r}}^{(n)}$. A simple application of majorization theory leads to the inequality:

$$H_-(x, y) \le H_-^{(n)}(x, y) \le H_{\mathbf{r}}^{(n)}(x, y) \le H_+^{(n)}(x, y) \le H_+(x, y),\ \ x, y \in R,$$

and hence

$$\mathrm{Cov}(H_-) \le \mathrm{Cov}(H_-^{(n)}) \le \mathrm{Cov}(H_{\mathbf{r}}^{(n)}) \le \mathrm{Cov}(H_+^{(n)}) \le \mathrm{Cov}(H_+),$$

$$\rho(H_-) \le \rho(H_-^{(n)}) \le \rho(H_{\mathbf{r}}^{(n)}) \le \rho(H_+^{(n)}) \le \rho(H_+),$$

provided the variances of F and G are positive and finite.

Unlike the Sarmanov–Lee, all Baker's distributions retain the same support as that of $F \times G$ (and so does FGM). The former also admits discrete F and G.

For $(X, Y) \sim H$ with *continuous* marginals F and G, let $\rho_s(H)(= 12E[F(X)\ G(Y)] - 3)$ and $\tau(H)(= 4E[H(X, Y)] - 1)$ be its Spearman's rho and Kendall's tau coefficients, respectively (see, e.g., Joe 2001, pp. 31-32). Baker (2008) proved that $\lim_{n\to\infty} \rho_s(H_+^{(n)}) = 1$ and $\lim_{n\to\infty} \tau(H_+^{(n)}) = 1$. On the other hand, Lin and Huang (2010) found conditions under which Pearson's correlation $\rho(H_+^{(n)})$ converges to $\rho(H_+)$ as $n \to \infty$. It is also possible to calculate the convergence rate of $\rho(H_+^{(n)})$ for some specific marginals. For instance, we have

Theorem 6. (i) If F is uniform and if G a power function distribution, then the convergence rate of $\rho(H_+^{(n)})$ in (12) is $1/n$ as $n \to \infty$.
(ii) If $F = U(0, 1)$ and if G is exponential, then the convergence rate of $\rho(H_+^{(n)})$ is $1/n$ as $n \to \infty$. More precisely, $\rho(H_+^{(n)}) = \sqrt{3}/2 - \sqrt{3}/n + \mathcal{O}(n^{-2})$ as $n \to \infty$.
(iii) If $F = U(0, 1)$ and if G is logistic, then the convergence rate of $\rho(H_+^{(n)})$ is $1/n$ as $n \to \infty$. More precisely, $\rho(H_+^{(n)}) = 3\pi^{-1} - 6\pi^{-1}/n + \mathcal{O}(n^{-2})$ as $n \to \infty$.
(iv) If $F = G$ is exponential, then the convergence rate of $\rho(H_+^{(n)})$ is $(\ln n)/n$ as $n \to \infty$. More precisely, $\rho(H_+^{(n)}) = 1 - \frac{1}{n}\sum_{k=1}^{n} 1/k = 1 - (\ln n)/n + \mathcal{O}(n^{-1})$ as $n \to \infty$.

For $H_+^{(n)}$ in (12), Dou et al. (2013) established the weak convergence and the product-moment convergence as well as the TP_2 property described below. The results for $H_-^{(n)}$ can be derived similarly and are omitted. The next theorem is of interest in its own right.

Theorem 7. Let $(X_n, Y_n) \sim H_+^{(n)}$ in (12) with marginal densities f and g, and let $U_n = \sqrt{n}(G(Y_n) - F(X_n))$. Then, as $n \to \infty$, (X_n, U_n) converges in distribution to $(\widetilde{X}, \widetilde{U})$ having joint density

$$k(x, u) = \frac{1}{2\sqrt{\pi F(x)\overline{F}(x)}} \exp\left\{-\frac{u^2}{4F(x)\overline{F}(x)}\right\} f(x).$$

The conditional distribution of $\widetilde{U}$ given $\widetilde{X} = x$ is the normal distribution with mean 0 and variance $2F(x)\overline{F}(x)$, namely, $\widetilde{U}|_{\widetilde{X}=x} \sim N(0, 2F(x)\overline{F}(x))$.

By Theorem 7, we see that for uniform marginals $F = G = U(0, 1)$, Baker's distribution $H_+^{(n)}$ converges weakly to H_+ (with support on the diagonal line $y = x$) as $n \to \infty$. This can be further extended to the following general case by using the monotone property of distribution functions.

Theorem 8. For general marginals F and G, Baker's distribution $H_+^{(n)}$ converges weakly to the Fréchet–Hoeffding upper bound H_+ as n tends to infinity.

Theorem 9. Let $(X_n, Y_n) \sim H_+^{(n)}$ with general marginals F and G, and let $(\widetilde{X}, \widetilde{Y}) = (F^{-1}(Z), G^{-1}(Z))$, where $Z \sim U(0, 1)$. If, in addition, $E[|X_n|^{p+q}]$, $E[|Y_n|^{p+q}] < \infty$ for some positive integers p and q, then $\lim_{n\to\infty} E[X_n^p Y_n^q] = E[(\widetilde{X})^p(\widetilde{Y})^q] = \int_0^1 \left(F^{-1}(t)\right)^p \left(G^{-1}(t)\right)^q dt$.

Corollary 1. If F and G have finite non-zero variances, $\lim_{n\to\infty} \rho(H_+^{(n)}) = \rho(H_+)$.

The main tool in the proof of Theorem 7 is the following generalization of the so-called local limit theorem for binomial distribution, which is also of interest in itself.

Lemma 5. Let $0 < p, q < 1$ be constants such that $p+q = 1$, and let $u \in R$ be a constant. Let ψ be a function such that $\psi(n) = o(n^{1/6})$ as $n \to \infty$. Write $k = np + y_k\sqrt{npq} \in \{0, 1, 2, \ldots, n\}$. Then, as $n \to \infty$,

$$\binom{n}{k}\left(p + \frac{u}{\sqrt{n}}\right)^k \left(q - \frac{u}{\sqrt{n}}\right)^{n-k} \approx \frac{1}{\sqrt{2\pi npq}} \exp\left\{-\frac{1}{2}\left(y_k - \frac{u}{\sqrt{pq}}\right)^2\right\} \tag{15}$$

(asymptotically) uniformly in y_k such that $|y_k| \leq \psi(n)$. Namely, the left-hand side $a_n(k)$ and the right-hand side $b_n(y_k)$ in (15) satisfy $\sup_{\{k:|y_k|\leq\psi(n)\}} |a_n(k)/b_n(y_k) - 1| \to 0$ as $n \to \infty$.

Recall that a real-valued function k on R^2 is totally positive of order two (TP_2), a notion of strong positive dependence, if $k(x_1, y_1)k(x_2, y_2) \geq k(x_1, y_2)k(x_2, y_1)$ for all $x_1 \leq x_2$, $y_1 \leq y_2$.

Theorem 10. For general distributions F and G, Baker's distribution $H_+^{(n)}$ is TP_2.

Finally, if we allow the two sample sizes in (14) to be different, say m and n, then the resulting bivariate distribution is

$$H_{\mathbf{r}}^{(m,n)}(x, y) = \sum_{k=1}^{m}\sum_{\ell=1}^{n} r_{k,\ell} F_{k,m}(x) G_{\ell,n}(y) = C(F(x), G(y); \mathbf{r}), \tag{16}$$

where C is a Bernstein copula and the parameters $r_{k,\ell}$ satisfy

$$\sum_{\ell=1}^{n} r_{k,\ell} = \frac{1}{m}, \quad \sum_{k=1}^{m} r_{k,\ell} = \frac{1}{n}, \quad r_{k,\ell} \geq 0,\ 1 \leq k \leq m,\ 1 \leq \ell \leq n.$$

Because of the Weierstrass approximation theorem, we can use $H_{\mathbf{r}}^{(m,n)}$ in (16) to approximate smooth bivariate distributions. Moreover, since (16) is a finite mixture distribution, the EM algorithm applies to the estimation of the parameters $r_{k,\ell}$ of $H_{\mathbf{r}}^{(m,n)}$. Dou et al. (2014) took these advantages and succeeded in fitting Baker's distributions to some practical data sets including the Illinois state education data.

6 Bayramoglu's distributions

In this section, we extend Baker's distributions by an alternative approach. Starting with any joint distribution H with marginals F and G (e.g., the FGM distribution (4)), let $(X_1, Y_1), \ldots, (X_n, Y_n)$ be a random sample of size n from H. As before, sorting $\{X_\ell\}_{\ell=1}^n$ and $\{Y_\ell\}_{\ell=1}^n$, we obtain the order statistics $X_{1,n} \leq X_{2,n} \leq \cdots \leq X_{n,n}$ and $Y_{1,n} \leq Y_{2,n} \leq \cdots \leq Y_{n,n}$, respectively, and again, write $X_{r,n} \sim F_{r,n}$, $Y_{s,n} \sim G_{s,n}$.

Instead of Baker's $H_+^{(n)}$ and $H_-^{(n)}$, Bairamov and Bayramoglu (2013) proposed

$$K_+^{(n)}(x, y) = \frac{1}{n}\sum_{r=1}^{n} \Pr(X_{r,n} \leq x,\ Y_{r,n} \leq y), \quad x, y \in R, \tag{17}$$

$$K_-^{(n)}(x, y) = \frac{1}{n}\sum_{r=1}^{n} \Pr(X_{r,n} \leq x,\ Y_{n-r+1,n} \leq y), \quad x, y \in R, \tag{18}$$

both having marginals F and G. We see that if $H(x, y) = F(x)G(y)$ for all x, y, namely, if X and Y are independent, then $\Pr(X_{r,n} \leq x,\ Y_{s,n} \leq y) = F_{r,n}(x)G_{s,n}(y)$ for all x, y, and hence (17) and (18) reduce to (12) and (13), respectively. Therefore, the generalization admits a wider range of correlation.

When X and Y are not independent, the computation of the joint distribution of bivariate order statistics is complicated (see David 1981, p. 26, or David and Nagaraja 2003, p. 25):

$$K_{r,s}^{(n)}(x,y) \equiv \Pr(X_{r,n} \le x, Y_{s,n} \le y) = \sum_{i=r}^{n}\sum_{j=s}^{n}\sum_{k} f_{k,i,j}^{(n)}(x,y), \quad x,y \in R,$$

where k takes all integers such that $k,\ i-k,\ j-k,\ n-i-j+k \ge 0$, the summand

$$\begin{aligned} f_{k,i,j}^{(n)}(x,y) &= \frac{n!}{k!\,(i-k)!\,(j-k)!\,(n-i-j+k)!} H^k(x,y)(F(x)-H(x,y))^{i-k} \\ &\quad \cdot (G(y)-H(x,y))^{j-k}(\overline{H}(x,y))^{n-i-j+k}, \\ &\quad \text{if the exponents } k,\ i-k,\ j-k,\ n-i-j+k \ge 0, \end{aligned}$$

and $f_{k,i,j}^{(n)}(x,y) = 0$, otherwise. Here $\overline{H}(x,y) \equiv \Pr(X > x, Y > y) = 1-F(x)-G(y)+H(x,y)$.

Huang et al. (2013) established the following properties of $K_{r,s}^{(n)}$, $K_{+}^{(n)}$ and $K_{-}^{(n)}$. Hereafter, we shall consider x, y as fixed, and for simplicity, write F, G, H and $\overline{H}$ for $F(x)$, $G(y)$, $H(x,y)$ and $\overline{H}(x,y)$, respectively.

Theorem 11. For fixed marginals F, G and $1 \le r,\ s \le n$, the joint distribution $K_{r,s}^{(n)}$ of bivariate order statistics $(X_{r,n}, Y_{s,n})$ is increasing in H. More precisely, for given x, y, F and G, if $H_1(x,y) \le H_2(x,y)$, then $K_{r,s,1}^{(n)}(x,y) \le K_{r,s,2}^{(n)}(x,y)$, where $K_{r,s,i}^{(n)}$, $i = 1, 2$, is the distribution of $(X_{r,n}, Y_{s,n})$ generated from H_i with marginals F and G.

Theorem 12. For given H with marginals F and G, Bayramoglu's distribution $K_{+}^{(n)}$ in (17) is increasing in n. More precisely,

$$K_{+}^{(n)} = K_{+}^{(n-1)} + \frac{1}{n-1}\sum_{i=1}^{\lfloor n/2 \rfloor}\binom{n-1}{i,\, i-1,\, n-2i}(F-H)^i(G-H)^i\left(H+\overline{H}\right)^{n-2i}, \qquad (19)$$

$n \ge 2$.

Theorem 13. For given H with marginals F and G, Bayramoglu's distribution $K_{-}^{(n)}$ in (18) is decreasing in n. More precisely,

$$K_{-}^{(n)} = K_{-}^{(n-1)} - \frac{1}{n-1}\sum_{i=1}^{\lfloor n/2 \rfloor}\binom{n-1}{i,\, i-1,\, n-2i}(H\overline{H})^i(1-H-\overline{H})^{n-2i}, \quad n \ge 2. \qquad (20)$$

Recall that the bivariate distribution H with marginals F and G is positive quadrant dependent (PQD), if $H \ge FG$, that is, $H(x,y) \ge F(x)G(y)$ for all x, y, and that H is negative quadrant dependent (NQD) if $H \le FG$. As immediate consequences of Theorems 11–13, we have the following corollary (assuming $\sigma_X^2, \sigma_Y^2 \in (0,\infty)$ if necessary).

Corollary 2. (i) For fixed n and marginals F, G, both $K_{+}^{(n)}$ and $K_{-}^{(n)}$ are increasing in H;
(ii) $\rho(K_{+}^{(n)}) \ge \rho(H_{+}^{(n)})$ if H is PQD, and $\rho(K_{-}^{(n)}) \le \rho(H_{-}^{(n)})$ if H is NQD;
(iii) $K_{+}^{(2)} = FG$ if $H = H_{-}$, and in general, $0 \le \mathrm{Cov}(K_{+}^{(2)}) \le \mathrm{Cov}(K_{+}^{(3)}) \le \cdots$;
(iv) $K_{-}^{(2)} = FG$ if $H = H_{+}$, and in general, $0 \ge \mathrm{Cov}(K_{-}^{(2)}) \ge \mathrm{Cov}(K_{-}^{(3)}) \ge \cdots$.

Remarks 5. The first parts of Corollaries 2(iii) and 2(iv) can be proved directly without invoking the identities (19) and (20), but the proof is complicated. For instance, that of Corollary 2(iv) can be proceeded as follows. Let Z_1 and Z_2 be two independent copies of

$Z \sim U(0,1)$, and let $Z_{1,2} \le Z_{2,2}$ be the corresponding order statistics. Then $(X_i, Y_i) = (F^{-1}(Z_i),\ G^{-1}(Z_i))$, $i = 1,2$, are independent and have common distribution $H_+(x,y) = \min\{F(x),\ G(y)\}$, $x, y \in R$. In this case, we have

$$\begin{aligned}
K_-^{(2)}(x,y) &= \frac{1}{2}\left[\Pr(X_{1,2} \le x,\ Y_{2,2} \le y) + \Pr(X_{2,2} \le x,\ Y_{1,2} \le y)\right] \\
&= \frac{1}{2}\left[\Pr(F^{-1}(Z_{1,2}) \le x, G^{-1}(Z_{2,2}) \le y) + \Pr(F^{-1}(Z_{2,2}) \le x, G^{-1}(Z_{1,2}) \le y)\right] \\
&= \frac{1}{2}\left[\Pr(Z_{1,2} \le F(x), Z_{2,2} \le G(y)) + \Pr(Z_{2,2} \le F(x), Z_{1,2} \le G(y))\right] \\
&= \frac{1}{2}\left[\Pr(Z_1 \le F(x), Z_2 \le G(y) | Z_1 \le Z_2) + \Pr(Z_1 \le F(x), Z_2 \le G(y) | Z_1 > Z_2)\right] \\
&= \Pr(Z_1 \le F(x),\ Z_2 \le G(y)) = F(x)G(y),\quad x, y \in R,
\end{aligned}$$

the penultimate equality following from the fact that $\Pr(Z_1 \le Z_2) = \Pr(Z_1 > Z_2) = 1/2$ and the law of total probability.

In the rest of this section, we will mention some new results (Theorems 14–17) regarding (a) weak convergence, (b) product-moment convergence and (c) convergence of correlations (including Pearson's correlation, Spearman's rho and Kendall's tau) of Bayramoglu's distributions. The proofs are given in the Appendix. For proving the weak convergence, we offer an alternative approach different from that (Theorem 7) for Baker's distributions.

Theorem 14. Let H be any bivariate distribution with marginals F and G.

(i) Bayramoglu's distribution (17) converges weakly to H_+ as $n \to \infty$. Moreover, $\lim_{n\to\infty} \rho(K_+^{(n)}) = \rho(H_+)$, provided both F and G have finite nonzero variances.
(ii) Bayramoglu's distribution (18) converges weakly to H_- as $n \to \infty$. Moreover, $\lim_{n\to\infty} \rho(K_-^{(n)}) = \rho(H_-)$, provided both F and G have finite nonzero variances.

Theorem 15. Let $(X, Y) \sim H$ with marginals F and G and $E[\,|X|^{p+q}]$, $E[\,|Y|^{p+q}] < \infty$ for some positive integers p and q.

(i) If $(X_n, Y_n) \sim K_+^{(n)}$ and $(\widetilde{X}, \widetilde{Y}) = (F^{-1}(Z),\ G^{-1}(Z)) \sim H_+$, then

$$\lim_{n\to\infty} E[X_n^p Y_n^q] = E[(\widetilde{X})^p(\widetilde{Y})^q] = \int_0^1 \left(F^{-1}(t)\right)^p \left(G^{-1}(t)\right)^q dt.$$

(ii) If $(X_n^*, Y_n^*) \sim K_-^{(n)}$ and $(\widetilde{X^*}, \widetilde{Y^*}) = (F^{-1}(Z),\ G^{-1}(1-Z)) \sim H_-$, then

$$\lim_{n\to\infty} E[(X_n^*)^p(Y_n^*)^q] = E[(\widetilde{X^*})^p(\widetilde{Y^*})^q] = \int_0^1 \left(F^{-1}(t)\right)^p \left(G^{-1}(1-t)\right)^q dt.$$

For non-negative random variables X and Y, we can consider the general functions α, β of X, Y, respectively.

Theorem 16. Let $(X, Y) \sim H$, defined on $R_+^2 \equiv [0,\infty)\times[0,\infty)$, and let $X \sim F$, $Y \sim G$. Let α and β be two increasing and left-continuous functions on R_+. Assume also that $E[\,|\alpha(X)\beta(Y)|]$, $E[\,|\alpha(X)|]$ and $E[\,|\beta(Y)|]$ are finite.
(i) If $(X_n, Y_n) \sim K_+^{(n)}$ and $(\widetilde{X}, \widetilde{Y}) = (F^{-1}(Z),\ G^{-1}(Z)) \sim H_+$, then

$$\lim_{n\to\infty} E[\alpha(X_n)\beta(Y_n)] = E[\alpha(\widetilde{X})\beta(\widetilde{Y})] = \int_0^1 \alpha(F^{-1}(t))\beta(G^{-1}(t))dt.$$

(ii) If $(X_n^*, Y_n^*) \sim K_-^{(n)}$ and $(\widetilde{X^*}, \widetilde{Y^*}) = (F^{-1}(Z),\ G^{-1}(1-Z)) \sim H_-$, then

$$\lim_{n\to\infty} E[\alpha(X_n^*)\beta(Y_n^*)] = E[\alpha(\widetilde{X^*})\beta(\widetilde{Y^*})] = \int_0^1 \alpha(F^{-1}(t))\beta(G^{-1}(1-t))dt.$$

For Spearman's ρ_s and Kendall's τ, we have

Theorem 17. If H has *continuous* marginals F and G then

(i) $\lim_{n\to\infty} \rho_s(K_+^{(n)}) = 1$ and $\lim_{n\to\infty} \rho_s(K_-^{(n)}) = -1$;
(ii) $\lim_{n\to\infty} \tau(K_+^{(n)}) = 1$ and $\lim_{n\to\infty} \tau(K_-^{(n)}) = -1$.

Remarks 6. By the monotone property of $K_+^{(n)}$ and $K_-^{(n)}$ (Theorems 12 and 13), we can further conclude, in addition to Theorems 14 and 17, that all the sequences $\{\rho(K_+^{(n)})\}$, $\{\rho(K_-^{(n)})\}$, $\{\rho_s(K_+^{(n)})\}$, $\{\rho_s(K_-^{(n)})\}$, $\{\tau(K_+^{(n)})\}$ and $\{\tau(K_-^{(n)})\}$ are monotone (see, e.g., Joe 2001, p. 54).

To prove the new results, we need two more lemmas. Lemma 6 below is an extension of the classic Hoeffding's (1940) identity (2) for Cov(X, Y) to Cov$(\alpha(X), \beta(Y))$, and Lemma 7 gives an explicit form of the expectation $E[\alpha(X)\beta(Y)]$ for $(X, Y) \sim H$ on R_+^2. Our assumptions on the marginals (F, G) are weaker than those in Cuadras (2002), pp. 19-20, and the proof differs from that of Beare (2009).

Lemma 6. Let $(X, Y) \sim H, X \sim F$ and $Y \sim G$. Let $\alpha, \beta : R \to R$ be two left-continuous functions and be of bounded variation on each compact subset of R. Further, assume that the expectations $E[\,|\alpha(X)\beta(Y)|]$, $E[\,|\alpha(X)|]$, $E[\,|\beta(Y)|] < \infty$. Then

$$\begin{aligned} &\mathrm{Cov}(\alpha(X), \beta(Y)) = E[\alpha(X)\beta(Y)] - E[\alpha(X)]\,E[\beta(Y)] \\ &= \int_{-\infty}^{\infty}\int_{-\infty}^{\infty} [\overline{H}(x,y) - \overline{F}(x)\overline{G}(y)]\, d\alpha(x)d\beta(y) \\ &= \int_{-\infty}^{\infty}\int_{-\infty}^{\infty} [H(x,y) - F(x)G(y)]\, d\alpha(x)d\beta(y). \end{aligned}$$

Lemma 7. Let $(X, Y) \sim H$ defined on R_+^2 and $X \sim F,\ Y \sim G$. Let α and β be two increasing and left-continuous functions on R_+. Then

$$\begin{aligned} E[\alpha(X)\beta(Y)] = &\int_0^{\infty}\int_0^{\infty} \overline{H}(x,y)d\alpha(x)d\beta(y) - \alpha(0)\beta(0) \\ &+\alpha(0)E[\beta(Y)] + \beta(0)E[\alpha(X)], \end{aligned} \tag{21}$$

provided the expectations exist.

Corollary 3. Let $(X, Y) \sim H$ defined on R_+^2, and assume that the expectations $E[X^rY^s]$, $E[X^r]$, $E[Y^s] < \infty$ for some real numbers $r, s > 0$. Then

$$E[X^rY^s] = rs\int_0^{\infty}\int_0^{\infty} \overline{H}(x,y)x^{r-1}y^{s-1}dxdy.$$

Lemma 7 is a strengthening of Theorem 3.1 of Gupta et al. (2008). Then as a result, our corollary here extends their Corollary 2 in that no assumption of any smoothness of the bivariate distribution H is required.

7 Other related distributions

In addition to the above Sarmanov–Lee and Baker approaches, we mention some others.

For constructing bivariate distributions H with marginals F and G, Johnson and Tenenbein (1981) proposed the following steps:

(i) Let X_1' and X_2' be two i.i.d. random variables with a density k.

(ii) Define the correlated random variables

$$(X_1, X_2) = (X_1', \alpha X_1' + (1-\alpha)X_2'), \quad \alpha \in [0,1].$$

(iii) Find the copula $C(u,v)$ of (X_1, X_2), which is a function of α and k.

(iv) Define $H(x,y) = C(F(x), G(y))$, $x, y \in R$.

In general, the constructed copula C in (iii) is too complicated to have a closed form.

Takeuchi (2010) suggested using FGM distributions for low correlation, and the Gaussian copula for high correlation:

$$C_N(u,v;\rho) = \Phi_2(\Phi^{-1}(u), \Phi^{-1}(v); \rho), \quad u, v \in [0,1],$$

where Φ_2 is a two-dimensional normal distribution with correlation ρ, and Φ^{-1} is the quantile function of the standard normal distribution. For example, for constructing a bivariate (FIR-FUV) galaxy luminosity distribution, he proposed $H(x,y) = C_N(F(x), G(y); \rho)$, $x, y \in R$, where $X \sim F$ is far-infrared luminosity and $Y \sim G$ is far-ultraviolet luminosity. Note that there is no explicit form for the quantile function Φ^{-1}.

Larralde (2012) gave the maximum-entropy bivariate distribution with normal marginals. Dukic and Marić (2013) considered the convex combinations of the independent case FG and the extreme H_+ or H_-. But these are not what we really want because both H_+ and H_- are composed of singular distributions.

In conclusion, the most convenient unified approach to the above-mentioned problem is probably by way of a linear combination of the joint distributions of bivariate order statistics.

Appendix: Proof of the new results in Section 6

Proof of Lemma 6. Let (X_1, Y_1) and (X_2, Y_2) be two independent copies of (X, Y). Then

$$E[\alpha(X_i)\beta(Y_i)] = \int_{-\infty}^{\infty}\int_{-\infty}^{\infty} \alpha(x)\beta(y)dH(x,y) \quad \text{for } i = 1, 2,$$

and

$$E[\alpha(X_i)\beta(Y_j)] = \int_{-\infty}^{\infty}\int_{-\infty}^{\infty} \alpha(x)\beta(y)dF(x)G(y) \quad \text{for } i \neq j.$$

Define the function I on R^2 by

$$I(u,x) = \begin{cases} 1, & \text{if } u < x, \\ 0, & \text{otherwise} \end{cases}$$

(note that the function I here is a slight modification of those defined in Lehmann 1966, Lemma 2, and Quesada-Molina 1992, Theorem 2.1, and the references therein). Since α is left-continuous on R and is of bounded variation on each compact subset of R, we have

$$\int_{-\infty}^{\infty} [I(u,x_1) - I(u,x_2)]\, d\alpha(u) = -\int_{[x_1,x_2)} d\alpha(u) = \alpha(x_1) - \alpha(x_2) \quad \text{if } x_1 \le x_2,$$

and

$$\int_{-\infty}^{\infty} [I(u,x_1) - I(u,x_2)]\, d\alpha(u) = \int_{[x_2,x_1)} d\alpha(u) = \alpha(x_1) - \alpha(x_2) \quad \text{if } x_2 \le x_1.$$

Hence, in either case, $\int_{-\infty}^{\infty}[I(u,x_1) - I(u,x_2)]\, d\alpha(u) = \alpha(x_1) - \alpha(x_2)$, and

$$\begin{aligned}&\int_{-\infty}^{\infty}\int_{-\infty}^{\infty} [I(u,x_1) - I(u,x_2)]\,[I(v,y_1) - I(v,y_2)]\, d\alpha(u)d\beta(v)\\ &= [\alpha(x_1) - \alpha(x_2)]\,[\beta(y_1) - \beta(y_2)].\end{aligned}$$

On the other hand, note that $E[I(u,X_i)I(v,Y_j)] = \Pr(X_i > u, Y_j > v)$. This implies that

$$E[I(u,X_i)I(v,Y_j)] = \Pr(X > u, Y > v) \quad \text{if } i = j, \text{ and}$$
$$E[I(u,X_i)I(v,Y_j)] = \Pr(X > u)\Pr(Y > v) \quad \text{if } i \ne j.$$

Therefore, we have

$$\begin{aligned}&2\{E[\alpha(X)\beta(Y)] - E[\alpha(X)]\,E[\beta(Y)]\}\\
&= E[\alpha(X_1)\beta(Y_1) - \alpha(X_1)\beta(Y_2) - \alpha(X_2)\beta(Y_1) + \alpha(X_2)\beta(Y_2)]\\
&= E\left[\int_{-\infty}^{\infty}\int_{-\infty}^{\infty} \{I(u,X_1) - I(u,X_2)\}\{I(v,Y_1) - I(v,Y_2)\}d\alpha(u)d\beta(v)\right]\\
&= \int_{-\infty}^{\infty}\int_{-\infty}^{\infty} E\left[\{I(u,X_1) - I(u,X_2)\}\{I(v,Y_1) - I(v,Y_2)\}\right] d\alpha(u)d\beta(v)\\
&= \int_{-\infty}^{\infty}\int_{-\infty}^{\infty} 2\left[\Pr(X > u, Y > v) - \Pr(X > u)\Pr(Y > v)\right] d\alpha(u)d\beta(v)\\
&= 2\int_{-\infty}^{\infty}\int_{-\infty}^{\infty} [\overline{H}(x,y) - \overline{F}(x)\overline{G}(y)]\, d\alpha(x)d\beta(y)\\
&= 2\int_{-\infty}^{\infty}\int_{-\infty}^{\infty} [H(x,y) - F(x)G(y)]\, d\alpha(x)d\beta(y),\end{aligned}$$

last equality following from the fact that $H(x,y) - F(x)G(y) = \overline{H}(x,y) - \overline{F}(x)\overline{G}(y)$.

Proof of Lemma 7. By Lemma 6, we have

$$\begin{aligned}&E[\alpha(X)\beta(Y)] - E[\alpha(X)]\,E[\beta(Y)]\\
&= \int_0^{\infty}\int_0^{\infty} \left[\overline{H}(x,y) - \overline{F}(x)\overline{G}(y)\right] d\alpha(x)d\beta(y)\\
&= \int_0^{\infty}\int_0^{\infty} \overline{H}(x,y)d\alpha(x)d\beta(y) - \int_0^{\infty}\overline{F}(x)d\alpha(x)\int_0^{\infty}\overline{G}(y)d\beta(y).\end{aligned} \tag{22}$$

On the other hand, we have

$$\begin{aligned}E[\alpha(X)] &= \int_0^{\infty}\alpha(x)dF(x) = -\int_0^{\infty}\alpha(x)d\overline{F}(x) = -\lim_{b\to\infty}\int_{[0,b]}\alpha(x)d\overline{F}(x)\\
&= -\lim_{b\to\infty}\left[\alpha(b)\overline{F}(b) - \alpha(0-)\overline{F}(0-) - \int_0^b \overline{F}(x)d\alpha(x)\right]\\
&= \alpha(0) + \int_0^{\infty}\overline{F}(x)d\alpha(x).\end{aligned} \tag{23}$$

The last equality is due to the assumption $E[\,|\alpha(X)|] < \infty$ and the monotone property of the function α in case $\lim_{x\to\infty}\alpha(x) = \infty$. Similarly,

$$E[\beta(Y)] = \beta(0) + \int_0^{\infty}\overline{G}(y)d\beta(y). \tag{24}$$

Therefore, it follows from (23) and (24) that

$$\begin{aligned}
&\int_0^\infty \overline{F}(x)d\alpha(x)\int_0^\infty \overline{G}(y)d\beta(y)\\
&= \{E[\alpha(X)]-\alpha(0)\}\{E[\beta(Y)]-\beta(0)\}\\
&= E[\alpha(X)]\,E[\beta(Y)]-\alpha(0)E[\beta(Y)]-\beta(0)E[\alpha(X)]\\
&\quad+\alpha(0)\beta(0). \qquad (25)
\end{aligned}$$

Combining (22) and (25), we prove the identity (21).

Proof of Theorem 14. We first prove the following two facts.

Fact A. If $H = H_-$ with uniform marginals, $\lim_{n\to\infty}\mathrm{Cov}(K_+^{(n)}) = \frac{1}{12} = \mathrm{Cov}(H_+)$.

Fact B. If $H = H_+$ with uniform marginals, $\lim_{n\to\infty}\mathrm{Cov}(K_-^{(n)}) = -\frac{1}{12} = \mathrm{Cov}(H_-)$.

Proof of Fact A. For $F = G = U(0,1)$, and $H(x,y) = H_-(x,y) = \max\{0, x+y-1\}$, we have, by (19),

$$\begin{aligned}
K_+^{(n)} &= H+\sum_{m=2}^{n}\frac{1}{m-1}\sum_{i=1}^{\lfloor m/2\rfloor}\binom{m-1}{i,i-1,m-2i}(x-H)^i(y-H)^i(1-x-y+2H)^{m-2i}\\
&= \begin{cases}\sum_{m=2}^{n}\frac{1}{m-1}\sum_{i=1}^{\lfloor m/2\rfloor}\binom{m-1}{i,i-1,m-2i}x^iy^i(1-x-y)^{m-2i}, & \text{if } x+y\le 1,\\ x+y-1+\sum_{m=2}^{n}\frac{1}{m-1}\sum_{i=1}^{\lfloor m/2\rfloor}\binom{m-1}{i,i-1,m-2i}(1-x)^i(1-y)^i(x+y-1)^{m-2i}, & \\ \quad \text{if } x+y>1.\end{cases}
\end{aligned}$$

Then, by (2),

$$\begin{aligned}
&\mathrm{Cov}\left(K_+^{(n)}\right)\\
&= \int_0^1\int_0^{1-y}\left\{-xy+\sum_{m=2}^{n}\frac{1}{m-1}\sum_{i=1}^{\lfloor m/2\rfloor}\binom{m-1}{i,i-1,m-2i}x^iy^i(1-x-y)^{m-2i}\right\}dxdy\\
&\quad+\int_0^1\int_{1-y}^1\left\{-(1-x)(1-y)+\sum_{m=2}^{n}\frac{1}{m-1}\right.\\
&\qquad\left.\cdot\sum_{i=1}^{\lfloor m/2\rfloor}\binom{m-1}{i,i-1,m-2i}(1-x)^i(1-y)^i(x+y-1)^{m-2i}\right\}dxdy\\
&= -\int_0^1\int_0^{1-y}xydxdy-\int_0^1\int_{1-y}^1(1-x)(1-y)dxdy\\
&\quad+\int_0^1\int_0^{1-y}\sum_{m=2}^{n}\frac{1}{m-1}\sum_{i=1}^{\lfloor m/2\rfloor}\binom{m-1}{i,i-1,m-2i}x^iy^i(1-x-y)^{m-2i}dxdy\\
&\quad+\int_0^1\int_{1-y}^1\sum_{m=2}^{n}\frac{1}{m-1}\sum_{i=1}^{\lfloor m/2\rfloor}\binom{m-1}{i,i-1,m-2i}(1-x)^i(1-y)^i(x+y-1)^{m-2i}dxdy\\
&= -\frac{1}{12}+\sum_{m=2}^{n}\frac{1}{m-1}\sum_{i=1}^{\lfloor m/2\rfloor}\binom{m-1}{i,i-1,m-2i}\left\{\int_0^1\int_0^{1-y}x^iy^i(1-x-y)^{m-2i}dxdy\right.\\
&\quad\left.+\int_0^1\int_{1-y}^1(1-x)^i(1-y)^i(x+y-1)^{m-2i}dxdy\right\} \qquad (26)
\end{aligned}$$

$$= -\frac{1}{12} + \sum_{m=2}^{n} \frac{1}{m-1} \sum_{i=1}^{\lfloor m/2 \rfloor} \binom{m-1}{i, i-1, m-2i} \left(\frac{i!\, i!\, (m-2i)!}{(m+2)!} + \frac{i!\, i!\, (m-2i)!}{(m+2)!} \right) \quad (27)$$

$$= -\frac{1}{12} + 2 \sum_{m=2}^{n} \frac{1}{(m-1)m(m+1)(m+2)} \sum_{i=1}^{\lfloor m/2 \rfloor} i\,.$$

The two double integrals in (26) are equal by changing variables and in (27) we apply the representations for generalized (or multinomial) Beta function:

$$\int_0^1 \int_0^{1-y} x^i y^i (1-x-y)^{m-2i}\, dxdy = B(i+1, i+1, m-2i+1)$$
$$= \frac{\Gamma(i+1)\Gamma(i+1)\Gamma(m-2i+1)}{\Gamma(m+3)} = \frac{i!\, i!\, (m-2i)!}{(m+2)!}.$$

Note that

$$\sum_{m=2}^{n} \sum_{i=1}^{\lfloor m/2 \rfloor} = \sum_{m=2,even}^{n} \sum_{i=1}^{\lfloor m/2 \rfloor} + \sum_{m=2,odd}^{n} \sum_{i=1}^{\lfloor m/2 \rfloor} = \sum_{m=2,even}^{n} \sum_{i=1}^{m/2} + \sum_{m=3,odd}^{n} \sum_{i=1}^{(m-1)/2}$$

and that

$$\sum_{i=1}^{\lfloor m/2 \rfloor} i = \begin{cases} \frac{1}{2}\frac{m}{2}\left(\frac{m}{2}+1\right) = \frac{1}{8}m(m+2), & m = even, \\ \frac{1}{2}\frac{m-1}{2}\left(\frac{m-1}{2}+1\right) = \frac{1}{8}(m-1)(m+1), & m = odd. \end{cases}$$

We have

$$\sum_{m=2}^{n} \frac{1}{(m-1)m(m+1)(m+2)} \sum_{i=1}^{\lfloor m/2 \rfloor} i$$
$$= \sum_{m=2,even}^{n} \frac{1}{(m-1)m(m+1)(m+2)} \frac{m(m+2)}{8}$$
$$+ \sum_{m=3,odd}^{n} \frac{1}{(m-1)m(m+1)(m+2)} \frac{(m-1)(m+1)}{8}$$
$$= \frac{1}{8}\left(\sum_{m=2,even}^{n} \frac{1}{(m-1)(m+1)} + \sum_{m=3,odd}^{n} \frac{1}{m(m+2)} \right)$$
$$= \begin{cases} \frac{1}{8}\left(\sum_{m=2,even}^{n} \frac{1}{(m-1)(m+1)} + \sum_{m=3,odd}^{n-1} \frac{1}{m(m+2)} \right), & n = even, \\ \frac{1}{8}\left(\sum_{m=2,even}^{n-1} \frac{1}{(m-1)(m+1)} + \sum_{m=3,odd}^{n} \frac{1}{m(m+2)} \right), & n = odd \end{cases}$$
$$= \begin{cases} \frac{1}{8}\left(\sum_{k=1}^{n/2} \frac{1}{(2k-1)(2k+1)} + \sum_{k=1}^{n/2-1} \frac{1}{(2k+1)(2k+3)} \right), & n = even, \\ \frac{1}{8}\left(\sum_{k=1}^{(n-1)/2} \frac{1}{(2k-1)(2k+1)} + \sum_{k=1}^{(n-1)/2} \frac{1}{(2k+1)(2k+3)} \right), & n = odd \end{cases}$$
$$= \begin{cases} \frac{1}{8}\left(\frac{1}{2} - \frac{1}{2(n+1)} + \frac{1}{6} - \frac{1}{2(n+1)} \right) = \frac{1}{24}\frac{2n-1}{n+1}, & n = even, \\ \frac{1}{8}\left(\frac{1}{2} - \frac{1}{2n} + \frac{1}{6} - \frac{1}{2(n+2)} \right) = \frac{1}{24}\frac{(n-1)(2n+3)}{n(n+2)}, & n = odd. \end{cases}$$

The last equality follows from the two identities:

$$\sum_{k=1}^{N} \frac{1}{(2k-1)(2k+1)} = \frac{1}{2}\sum_{k=1}^{N}\left(\frac{1}{2k-1} - \frac{1}{2k+1} \right) = \frac{1}{2} - \frac{1}{2(2N+1)},$$
$$\sum_{k=1}^{N} \frac{1}{(2k+1)(2k+3)} = \frac{1}{2}\sum_{k=1}^{N}\left(\frac{1}{2k+1} - \frac{1}{2k+3} \right) = \frac{1}{6} - \frac{1}{2(2N+3)}.$$

Thus

$$\mathrm{Cov}\left(K_+^{(n)}\right) = \begin{cases} -\frac{1}{12} + 2(\frac{1}{24}\frac{2n-1}{n+1}) = \frac{1}{12}\frac{n-2}{n+1}, & n = even, \\ -\frac{1}{12} + 2(\frac{1}{24}\frac{(n-1)(2n+3)}{n(n+2)}) = \frac{1}{12}\frac{n^2-n-3}{n(n+2)}, & n = odd, \end{cases}$$

and in either case

$$\lim_{n\to\infty} \frac{1}{12}\frac{n-2}{n+1} = \frac{1}{12}, \quad \lim_{n\to\infty} \frac{1}{12}\frac{n^2-n-3}{n(n+2)} = \frac{1}{12}.$$

Therefore, $\mathrm{Cov}\left(K_+^{(n)}\right) \to \frac{1}{12} = \mathrm{Cov}(H_+)$ as $n \to \infty$.

Proof of Fact B. The case of $K_-^{(n)}$ from $H(x,y) = H_+(x,y) = \min\{x,y\}$ requires no new proof by noting that

$$\mathrm{Cov}\left(K_-^{(n)}\right) = \frac{1}{12} - \sum_{m=2}^{n} \frac{1}{m-1} \sum_{i=1}^{\lfloor m/2 \rfloor} \binom{m-1}{i, i-1, m-2i} \left\{ \int_0^1\!\!\int_0^y x^i (1-y)^i (y-x)^{m-2i} dxdy \right.$$
$$\left. + \int_0^1\!\!\int_y^1 y^i (1-x)^i (x-y)^{m-2i}\, dxdy \right\} \tag{28}$$

and that the two integrals in (28) are identical to those of (26) (by a simple change of variables).

Proof of Theorem 14 (continued).

(i) We divide the proof of part (i) into two steps:

(1a) Consider arbitrary H with uniform marginals, and prove that $K_+^{(n)}$ converges weakly to H_+ as $n \to \infty$ and $\lim_{n\to\infty} \sigma(K_+^{(n)}) = \sigma(H_+)$.

(1b) Extend the result (1a) to any H with general fixed marginals.

Proof of (1a). For arbitrary H with uniform marginals, since $\{K_+^{(n)}\}_{n=1}^{\infty}$ is a bounded and increasing sequence, its limit exists, say $K_{+\infty}$. Then $K_+^{(n)} \leq K_{+\infty} \leq H_+$ for all $n \geq 1$. Also, note that $K_+^{(n)} \geq K_+^{(n)}(H_-)$ due to Theorem 11. We have, since $K_+^{(n)}$ increases in n,

$$\begin{aligned}
0 &\leq \int_0^1\!\!\int_0^1 \left[H_+(x,y) - K_{+\infty}(x,y)\right] dxdy \\
&= \int_0^1\!\!\int_0^1 \left[H_+(x,y) - \lim_{n\to\infty} K_+^{(n)}(x,y)\right] dxdy \\
&= \lim_{n\to\infty} \int_0^1\!\!\int_0^1 \left[H_+(x,y) - K_+^{(n)}(x,y)\right] dxdy \\
&= \lim_{n\to\infty} \int_0^1\!\!\int_0^1 \left[(H_+(x,y) - xy) - (K_+^{(n)}(x,y) - xy)\right] dxdy \\
&= \lim_{n\to\infty} \left[\mathrm{Cov}(H_+) - \mathrm{Cov}(K_+^{(n)})\right] \leq \lim_{n\to\infty} \left[\mathrm{Cov}(H_+) - \mathrm{Cov}(K_+^{(n)}(H_-))\right] = 0.
\end{aligned}$$

The last equality follows from Fact A. Therefore $K_+^{(n)}$ converges weakly to H_+ as $n \to \infty$ and $\lim_{n\to\infty} \sigma(K_+^{(n)}) = \sigma(H_+)$.

Proof of (1b). We now extend the result of part (1a) to any H with general fixed marginals F and G. First, let C be a copula of H. We rewrite the result of part (1a) in terms of random variables. Let (Z_i, Z_i^*), $i = 1, 2, \ldots, n$, be n independent copies of $(Z, Z^*) \sim C$. And let $Z_{(k)}$ and $Z_{(k)}^*$ denote the kth smallest order statistics of $\{Z_i\}_{i=1}^n$ and $\{Z_i^*\}_{i=1}^n$, respectively. On the other hand, let K_n be a discrete random variable obeying the uniform distribution on $\{1, 2, \ldots, n\}$. Then $(Z_{(K_n)}, Z_{(K_n)}^*) \sim K_+^{(n)}(C)$ (the bivariate distribution (17) generated from C) and by part (1a), we have that as $n \to \infty$,

$$(Z_{(K_n)}, Z_{(K_n)}^*) \xrightarrow{d} (Z, Z) \sim C_+ \text{(the upper bound with uniform marginals)}.$$

Proceeding on similar lines in the proof of Theorem 3 of Dou et al. (2013), we conclude that the bivariate $(F^{-1}(Z_{(K_n)}),\ G^{-1}(Z^*_{(K_n)}))$ converges in distribution to $(F^{-1}(Z), G^{-1}(Z))$ as $n \to \infty$ (Serfling 1980, p. 24), namely, $K_+^{(n)}$ converges weakly to H_+ as $n \to \infty$. This together with Corollary 2(iii) completes the proof.

(ii) We also divide the proof of part (ii) into two steps:

(2a) Consider arbitrary H with uniform marginals, and prove that $K_-^{(n)}$ converges weakly to H_- as $n \to \infty$, and $\lim_{n\to\infty} \sigma(K_-^{(n)}) = \sigma(H_-)$.

(2b) Extend the result (2a) to any H with general fixed marginals.

Proof of (2a). For arbitrary H with uniform marginals, since $\{K_-^{(n)}\}_{n=1}^{\infty}$ is a bounded and decreasing sequence, its limit exists, say $K_{-\infty}$. Then $H_- \le K_{-\infty} \le K_-^{(n)}$ for all $n \ge 1$. Also, note that $K_-^{(n)} \le K_-^{(n)}(H_+)$ by Theorem 11. We have, since $K_-^{(n)}$ decreases in n,

$$\begin{aligned}
0 &\le \int_0^1\int_0^1 \left[K_{-\infty}(x,y) - H_-(x,y)\right] dxdy \\
&= \int_0^1\int_0^1 \left[\lim_{n\to\infty} K_-^{(n)}(x,y) - H_-(x,y)\right] dxdy \\
&= \lim_{n\to\infty} \int_0^1\int_0^1 \left[K_-^{(n)}(x,y) - H_-(x,y)\right] dxdy \\
&= \lim_{n\to\infty} \int_0^1\int_0^1 \left[(K_-^{(n)}(x,y) - xy) - (H_-(x,y) - xy)\right] dxdy \\
&= \lim_{n\to\infty} \left[\mathrm{Cov}(K_-^{(n)}) - \mathrm{Cov}(H_-)\right] \le \lim_{n\to\infty} \left[\mathrm{Cov}(K_-^{(n)}(H_+)) - \mathrm{Cov}(H_-)\right] = 0.
\end{aligned}$$

The last equality is due to Fact B. Therefore $K_-^{(n)}$ converges weakly to H_- as $n \to \infty$. This together with Corollary 2(iv) completes the proof.

Proof of (2b). This is similar to the proof of part (1b) and is omitted.

Proof of Theorem 15. Apply Lemma 3 to Theorem 14.

Proof of Theorem 16. (i) Note that as $n \to \infty$, $K_+^{(n)}$ converges weakly to H_+, the Fréchet–Hoeffding upper bound with the same marginals F and G, and that $(F^{-1}(Z),\ G^{-1}(Z)) \sim H_+$. Then the required result follows from Lemma 7 above and the Generalized Monotone Convergence Theorem (see, e.g., Royden 1988, p. 265), because

$$\begin{aligned}
E[\alpha(X_n)\beta(Y_n)] = &\int_0^\infty\int_0^\infty \overline{K}_+^{(n)}(x,y)d\alpha(x)d\beta(y) - \alpha(0)\beta(0) \\
&+\alpha(0)E[\beta(Y)] + \beta(0)E[\alpha(X)],
\end{aligned}$$

and $\overline{K}_+^{(n)}$ converges monotonically to $\overline{H}_+$ as $n \to \infty$ (see Theorems 12 and 14(i)).

(ii) The proof is similar to that of part (i), by noting that $(F^{-1}(Z),\ G^{-1}(1-Z)) \sim H_-$ and that $\overline{K}_-^{(n)}$ converges monotonically to $\overline{H}_-$ as $n \to \infty$ (see Theorems 13 and 14(ii)).

Proof of Theorem 17. It suffices to consider the case of uniform marginals. Let $(X_n, Y_n) \sim K_+^{(n)}$ with marginals $F = G = U(0,1)$, and $(X_n^*, Y_n^*) \sim K_-^{(n)}$ with marginals $F = G = U(0,1)$.

(i) From Theorem 15 it follows that

$$\lim_{n\to\infty} E[F(X_n)G(Y_n)] = \lim_{n\to\infty} E[X_nY_n] = \frac{1}{3},$$
$$\lim_{n\to\infty} E[F(X_n^*)G(Y_n^*)] = \lim_{n\to\infty} E[X_n^*Y_n^*] = \frac{1}{6}.$$

This in turn implies that $\lim_{n\to\infty} \rho_s(K_+^{(n)}) = 12\lim_{n\to\infty} E[X_nY_n] - 3 = 1$ and $\lim_{n\to\infty} \rho_s(K_-^{(n)}) = 12\lim_{n\to\infty} E[X_n^*Y_n^*] - 3 = -1$. (Note that these results also follow from Theorem 14; see, e.g., Balakrishnan and Lai 2009, p. 156.)

(ii) Let $(\widetilde{X}, \widetilde{Y}) = (Z, Z) \sim H_+$. Recall that $K_+^{(n)}$ are bounded and continuous functions, and that $K_+^{(n)}$ converges increasingly to H_+ as $n \to \infty$ (see Theorems 12 and 14). Moreover, (X_n, Y_n) converges in distribution to $(\widetilde{X}, \widetilde{Y})$ as $n \to \infty$. By the Generalized Lebesgue Dominated Convergence Theorem (see, e.g., Royden 1988, p. 270, or Hernández-Lerma and Lasserre 2000), we have

$$\lim_{n\to\infty} E[K_+^{(n)}(X_n, Y_n)] = E[H_+(\widetilde{X}, \widetilde{Y})]$$
$$= E[H_+(Z, Z)] = E[\min\{Z, Z\}] = \int_0^1 t\,dt = \frac{1}{2}.$$

Therefore, $\lim_{n\to\infty} \tau(K_+^{(n)}) = 4\lim_{n\to\infty} E[K_+^{(n)}(X_n, Y_n)] - 1 = 1$. On the other hand, letting $(\widetilde{X^*}, \widetilde{Y^*}) = (Z,\ 1 - Z) \sim H_-$, we have that (X_n^*, Y_n^*) converges in distribution to $(\widetilde{X^*}, \widetilde{Y^*})$ as $n \to \infty$, and, as before,

$$\lim_{n\to\infty} E[K_-^{(n)}(X_n^*, Y_n^*)] = E[H_-(\widetilde{X^*}, \widetilde{Y^*})]$$
$$= E[H_-(Z,\ 1 - Z)] = E[\max\{0, Z + (1 - Z) - 1\}] = 0.$$

Therefore,

$$\lim_{n\to\infty} \tau(K_-^{(n)}) = 4\lim_{n\to\infty} E[K_-^{(n)}(X_n^*, Y_n^*)] - 1 = -1.$$

This completes the proof.

Acknowledgements

The authors would like to thank the editors and referees for valuable comments and suggestions. The paper was presented at the first conference on Statistical Distributions and Applications (ICOSDA 2013) held from 10th to 12th October 2013 by Central Michigan University, USA. GD Lin would like to thank the Chairs of Organizing Committee, Professor Felix Famoye and Professor Carl Lee, for their kind invitation, and the National Science Council of Taiwan (ROC) for financial support under Grants NSC 99-2118-M-001-003-MY3 and NSC 102-2118-M-001-008-MY2.

Author details

[1]Institute of Statistical Science, Academia Sinica, Taiwan, Republic of China. [2]Waseda University, Tokyo, Japan. [3]The Institute of Statistical Mathematics, Tokyo, Japan. [4]Department of Mathematics and Statistics, University of Guelph, Guelph, Canada.

References

Bairamov, I, Bayramoglu, K: From the Huang–Kotz FGM distribution to Baker's bivariate distribution. J. Multivariate Anal. **113**, 106–115 (2013)

Bairamov, I, Kotz, S, Gebizlioglu, OL: The Sarmanov family and its generalization. S. Afr. Stat. J. **35**, 205–224 (2001)

Baker, R: An order-statistics-based method for constructing multivariate distributions with fixed marginals. J. Multivariate Anal. **99**, 2312–2327 (2008)

Balakrishnan, N, Lai, C-D: Continuous Bivariate Distributions. 2nd ed. Springer–Verlag, New York (2009)

Beare, BK: A generalization of Hoeffding's lemma, and a new class of covariance inequalities. Statist. & Probab. Lett. **79**, 637–642 (2009)

Beněs, V, Štěpan, J (eds.): Distributions With Given Marginals and Moment Problems. Kluwer Academic Publishers, Dordrecht (1997). Conference held in Prague 1996

Cole, BF, Lee, M-LT, Whitmore, GA, Zaslavsky, AM: An empirical Bayes model for Markov-dependent binary sequences with randomly missing observations. J. Amer. Stat. Assoc. **90**, 1364–1372 (1995)

Cuadras, CM: Probability distributions with given multivariate marginals and given dependence structure. J. Multivariate Anal. **42**, 51–66 (1992)
Cuadras, CM: On the covariance between functions. J. Multivariate Anal. **81**, 19–27 (2002)
Cuadras, CM, Fortiana, J, Rodriguez-Lallena, JA (eds.): Distributions With Given Marginals and Statistical Modelling. Kluwer Academic Publishers, Dordrecht (2002). Papers from the meeting held in Barcelona 2000
Dall'Aglio, G, Kotz, S, Salinetti, G (eds.): Advances in Probability Distributions with Given Marginals. Beyond the Copulas. Mathematics and its Applications 67, Kluwer Academic Publishers Group, Dordrecht (1991). Papers from the Symposium on Distributions with Given Marginals held in Rome 1990
Danaher, PJ, Smith, MS: Modeling multivariate distributions using copulas: applications in marketing. Mark. Sci. **30**, 4–21 (2011)
David, HA: Order Statistics. 2nd ed. Wiley, New York (1981)
David, HA, Nagaraja, HN: Order Statistics. 3rd ed. Wiley, New Jersey (2003)
Dolati, A, Úbeda-Flores, M: A method for constructing multivariate distributions with given bivariate margins. Brazilian Journal of Probability and Statistics. **19**, 85–92 (2005)
Dou, X, Kuriki, S, Lin, GD: Dependence structures and asymptotic properties of Baker's distributions with fixed marginals. J. Statist. Plann. Inference. **143**, 1343–1354 (2013)
Dou, X, Kuriki, S, Lin, GD, Richards, D: EM algorithms for estimating the Bernstein copula. To appear in Computational Statistics and Data Analysis (2014). Website http://dx.doi.org/10.1016/j.csda.2014.01.009.
Dukic, VM, Marić, N: Minimum correlation in construction of multivariate distributions. Phys. Rev. E. **87**(032114) (2013)
Eyraud, H: Les principes de la mesure des corrélations. Ann. Univ. Lyon Sect. A. **1**, 30–47 (1936)
Farlie, DJG: The performance of some correlation coefficients for a general bivariate distribution. Biometrika. **47**, 307–323 (1960)
Fréchet, M: On correlation matrices with fixed margins (Sur les tableaux de corrélation dont les marges sont données). Ann. Univ. Lyon Sect. A. **14**, 53–77 (1951)
Gómez-Déniz, E, Sarabia, JM, Balakrishnan, N: A multivariate discrete Poisson-Lindley distribution: extensions and actuarial applications. Astin Bulletin. **42**, 655–678 (2012)
Gumbel, EJ: Bivariate exponential distributions. J. Amer. Stat. Assoc. **55**, 698–707 (1960)
Gupta, RC, Tajdari, M, Bresinsky, H: Some general results for moments in bivariate distributions. Metrika. **68**, 173–187 (2008)
Hernández-Bastida, A, Fernández-Sánchez, MP: A Sarmanov family with beta and gamma marginal distributions: an application to the Bayes premium in a collective risk model. Stat. Methods Appl. **21**, 391–409 (2012)
Hernández-Lerma, O, Lasserre, JB: Fatou's lemma and Lebesgue's convergence theorem for measures. J. Appl. Math. Stochastic Anal. **13**, 137–146 (2000)
Hoeffding, W: Masstabinvariante Korrelations-theorie. Schr. Math. Inst. Univ. Berlin. **5**, 181–233 (1940). English translation: Scale-Invariant Correlation Theory. In: Fisher, NI, Sen, PK (eds.) The Collected Works of Wassily Hoeffding, pp. 57–107, Springer-Verlag, New York (1999)
Huang, JS, Dou, X, Kuriki, S, Lin, GD: Dependence structure of bivariate order statistics with applications to Bayramoglu's distributions. J. Multivariate Anal. **114**, 201–208 (2013)
Huang, JS, Kotz, S: Correlation structure in iterated Farlie–Gumbel–Morgenstern distributions. Biometrika. **71**, 633–636 (1984)
Huang, JS, Kotz, S: Modifications of the Farlie–Gumbel–Morgenstern distributions. A tough hill to climb. Metrika. **49**, 135–145 (1999)
Huang, JS, Lin, GD: A note on the Sarmanov bivariate distributions. Applied Mathematics and Computation. **218**, 919–923 (2011)
Joe, H: Multivariate Models and Dependence Concepts. Chapman & Hall/CRC, New York (2001)
Johnson, NL, Kotz, S: On some generalized Farlie–Gumbel–Morgenstern distributions. Commun. Statist. - Theory Meth. **4**, 415–427 (1975)
Johnson, NL, Kotz, S, Balakrishnan, B: Discrete Multivariate Analysis. Wiley, New York (1997)
Johnson, ME, Tenenbein, A: A bivariate distribution family with specified marginals. J. Amer. Statist. Assoc. **76**, 198–201 (1981)
Kotz, S, Balakrishnan, N, Johnson, NL: Continuous Multivariate Distributions, Vol. 1: Models and Applications. 2nd ed. Wiley, New York (2000)
Kotz, S, Nadarajah, S: Multivariate *t* Distributions and Their Applications. Cambridge University Press, Cambridge (2004)
Lai, CD: Constructions of continuous bivariate distributions. J. Indian Soc. Probab. Statist. **8**, 21–43 (2004)
Lai, CD: Constructions of discrete bivariate distributions. In: Balakrishnan, N, Castillo, E, Sarabia, JM (eds.) Advances in Distribution Theory, Order Statistics and Inference, pp. 29–58. Birkäuser, Boston, (2006)
Larralde, H: Maximum-entropy distributions of correlated variables with prespecified marginals. Physical Review E. **86**(061117) (2012)
Lee, M-LT: Properties and applications of the Sarmanov family of bivariate distributions. Commun. Statist. - Theory Meth. **25**, 1207–1222 (1996)
Lehmann, EL: Some concepts of dependence. Ann. Math. Statist. **37**, 1137–1153 (1966)
Li, D-Q, Jiang, S-H, Wu, S-B, Zhou, C-B, Zhang, L-M: Modeling multivariate distributions using Monte Carlo simulation for structural reliability analysis with complex performance function. Proc. Inst. Mech. Eng. Part O: J. Risk Reliabil. **227**, 109–118 (2013)
Lin, GD: Relationships between two extensions of Farlie–Gumbel–Morgenstern distribution. Ann. Inst. Statist. Math. **39**, 129–140 (1987)
Lin, GD, Huang, JS: A note on the maximum correlation for Baker's bivariate distributions with fixed marginals. J. Multivariate Anal. **101**, 2227–2233 (2010)
Lin, GD, Huang, JS: Maximum correlation for the generalized Sarmanov bivariate distributions. J. Statist. Plann. Inference. **141**, 2738–2749 (2011)
Morgenstern, D: Einfache Beispiele zweidimensionaler Verteilungen. Mitteilungsbl. Math. Statist. **8**, 234–235 (1956)

Nelsen, RB: A characterization of Farlie–Gumbel–Morgenstern distributions via Spearman's rho and chi-square divergence. Sankhyā, Ser. A. **56**, 476–479 (1994)
Pelican, E, Vernic, R: Maximum-likelihood estimation for the multivariate Sarmanov distribution: simulation study. Int. J. Comput. Math. **90**, 1958–1970 (2013a)
Pelican, E, Vernic, R: Parameters estimation for the bivariate Sarmanov distribution with normal-type marginals. ROMAI J. **9**, 155–165 (2013b)
Quesada-Molina, JJ: A generalization of an identity of Hoeffding and some applications. J. Ital. Statist. Soc. **1**, 405–411 (1992)
Royden, HL: Real Analysis. 3rd ed. Prentice-Hall, New Jersey (1988)
Rüschendorf, L, Schweizer, B, Taylor, MD (eds.): Distributions with Fixed Marginals and Related Topics. Institute of Mathematical Statistics, Lecture Notes - Monograph Series 28, Hayward, 1996
Sarabia, JM, Gómez-Déniz, E: Construction of multivariate distributions: a review of some recent results (with discussions). SORT. **32**, 3–36 (2008)
Sarabia, JM, Gómez-Déniz, E: Multivariate Poisson-Beta distributions with applications. Commun. Statist. - Theory Meth. **40**, 1093–1108 (2011)
Sarmanov, IO: Generalized normal correlation and two dimensional Fréchet classes. Soviet Math. Dokl. **7**, 596–599 (1966). [English translation; Russian original in Dokl. Akad. Nauk. SSSR **168**, 32–35 (1966)]
Schucany, WR, Parr, WC, Boyer, JE: Correlation structure in Farlie–Gumbel–Morgenstern distributions. Biometrika. **65**, 650–653 (1978)
Serfling, RJ: Approximation Theorems of Mathematical Statistics. Wiley, New York (1980)
Shubina, M, Lee, M-LT: On maximum attainable correlation and other measures of dependence for the Sarmanov family of bivariate distributions. Commun. Statist. - Theory Meth. **33**, 1031–1052 (2004)
Takeuchi, TT: Constructing a bivariate distribution function with given marginals and correlation: application to the galaxy luminosity function. Mon. Notices Roy. Astronom. Soc. **406**, 1830–1840 (2010)
Wigner, E: On the quantum correction for thermodynamic equilibrium. Phys. Rev. **40**, 749–759 (1932)

3

Efficient and adaptive rank-based fits for linear models with skew-normal errors

Joseph W McKean[1*] and John D Kloke[2]

*Correspondence:
mckean@wmich.edu
[1] Department of Statistics, Western Michigan University, Kalamazoo, USA
Full list of author information is available at the end of the article

Abstract

The rank-based fit of a linear model is based on minimizing a norm. A score function needs to be selected for the fit and the proper choice leads to asymptotically efficient regression estimators, i.e., fits equivalent to the maximum likelihood estimators (mle). In this paper, we present the family of optimal scores functions for the skew-normal family of distributions. We show the easy computation of this rank-based fit using the R package Rfit. We present the results of a small simulation study comparing the rank-based estimators and the mles in terms of efficiency and validity over skew-normal and contaminated normal distributions. We also develop and present empirical results for a Hogg-type adaptive procedure for selecting among a family of these scores based on a robust initial fit.

Keywords: Linear models; Monte Carlo; Nonparametrics; Regression rank scores; Robust; Wilcoxon procedures

1 Introduction

Rank-based fitting of linear models offers an attractive alternative to least squares (LS) and maximum likelihood (mle) fitting. The geometry of the rank-based fit is similar to that of LS. Simply replace the Euclidean squared norm used in the LS fit with another norm which, unlike LS, results in a robust fit. The accompanying robust analysis is the analogue of the LS's ANOVA and ANCOVA. The rank-based analysis offers a complete analysis including robust diagnostics to check quality of fit. This rank based analysis has recently been extended to mixed and nonlinear models; see Kloke et al. (2009) and Abebe and McKean (2013), respectively. A full development of the rank-based analysis can be found in Chapters 3-5 of the monograph by Hettmansperger and McKean (2011). The rank-based analysis is, generally, highly efficient. It can easily be optimized depending on the information available concerning the distribution of the random errors. For example, if the form of the error distribution is known, then an appropriate rank-based procedure can be selected to attain full efficiency.

In this paper, we discuss rank-based analyses which are appropriate for the skew-normal (SN) family of distributions. This is a rich family of skewed distributions developed by Azzalini (1985). The skewness of a distribution in this family is controlled by a shape parameter α, $-\infty < \alpha < \infty$. Distributions are left or right skewed, depending on whether $\alpha < 0$ or $\alpha > 0$, respectively. If $\alpha = 0$ then the distribution is normal. As we discuss in Section 3, all of these distributions are light-tailed. Such families of distributions occur frequently in accelerated failure time (AFT) models. The response of interest in such models is survival time and its log is frequently modeled in terms of a log linear model. The random errors for these log linear models are, generally, skewed. For instance, the family of distributions for log linear models when survival time follows an F-distribution contains a wide variety of skewed distributions with tail weights that range from moderate to heavy; see McKean and Sievers (1989). Hence, the skew-normal family adds a rich class of light-tailed skewed distributions which also includes the normal distribution.

In Section 2, we outline the rank-based analysis for a general linear model. The computation of these analyses is easily handled by the R package `Rfit`, developed by Kloke and McKean (2013), which can be freely downloaded at the CRAN http://cran.us.r-project.org/. We discuss this software for an example in this section and we continue such discussion in the remainder of the article. The data for the example and the R code, supplemental to `Rfit`, used in this article are available to the reader at the http://www.stat.wmich.edu/mckean/SN/.

In Section 3, we develop rank-based analyses for the skew-normal family. These analyses are efficient for this family and the appropriate analysis is fully efficient. As we discuss, these analyses are technically robust similar to the optimal rank-based analysis for normally distributed errors. In contrast, in a sensitivity analysis, we show that the maximum likelihood fit (mle) is not robust. In Section 4, we present the results of a Monte Carlo study which verify the robustness and validity of the rank-based analysis over a family of SN distributions and contaminated SN distributions. These studies confirm the nonrobustness of the mle analysis.

The rank-based analysis depends on the shape parameter α. One outcome of the Monte Carlo study is that rank-based analyses based on shape parameters in a neighborhood of the correct α had very similar behavior to that using the correct α. This suggests that a simple Hogg-type adaptive procedure would entertain excellent properties in this situation. In Section 5, we develop such an adaptive scheme for the family of SN distributions. In a simulation study, we verify the efficiency and validity of this scheme over SN situations and, further, over two contaminated situations.

2 Notation and rank-based analysis

Let $\boldsymbol{Y}$ be a $n \times 1$ vector of responses which follows the linear model given by

$$\boldsymbol{Y} = \mathbf{1}_n \beta_0 + \boldsymbol{X}\boldsymbol{\beta} + \boldsymbol{e}. \tag{1}$$

where $\mathbf{1}_n$ is a vector of n ones; $\boldsymbol{X}$ is a $n \times p$ design matrix which may contain predictors (covariates) as well as indicator (dummy) variables; β_0 is an intercept parameter; $\boldsymbol{\beta}$ is a $p \times 1$ vector of regression parameters; and $\boldsymbol{e}$ is a $n \times 1$ vector of random errors. Because we have an intercept parameter in the model, we assume without loss of generality that the design matrix $\boldsymbol{X}$ is centered, (all columns of $\boldsymbol{X}$ have mean 0). For the theory discussed below, assume that the components of $\boldsymbol{e}$ are iid with pdf $f(x)$ and cdf $F(x)$, where $F(x)$ is unknown.

The least squares (LS) estimator of $\boldsymbol{\beta}$ is the vector $\widehat{\boldsymbol{\beta}}_{LS}$ which minimizes the Euclidean distance between $\boldsymbol{Y}$ and the column space of $\boldsymbol{X}$; that is, it satisfies

$$\widehat{\boldsymbol{\beta}}_{LS} = \text{Argmin}\|\boldsymbol{Y} - \boldsymbol{X}\boldsymbol{\beta}\|_2^2., \tag{2}$$

where $\|\boldsymbol{v}\|_2^2 = \sum_{i=1}^n v_i^2$ is the Euclidean norm.

For the rank-based estimates, simply replace the Euclidean norm $\|\cdot\|_2^2$ by the pseudo-norm

$$\|\boldsymbol{v}\|_\varphi = \sum_{i=1}^{n} a[R(v_i)]\, v_i, \quad \boldsymbol{v} \in R^n, \tag{3}$$

where $R(v_i)$ is the rank of v_i among $v_1, \ldots, v_n$ and the scores $a(i)$ are generated as as $a[i] = \varphi[i/(n+1)]$, for a nondecreasing bounded square-integrable function $\varphi(u)$, satisfying, without loss of generality, the standardizing conditions $\int \varphi(u)\, du = 0$ and $\int \varphi^2(u)\, du = 1$. Then the rank-based estimator minimizes the $\|\cdot\|_\varphi$-distance between $\boldsymbol{Y}$ and the column space of $\boldsymbol{X}$; i.e.,

$$\hat{\boldsymbol{\beta}}_\varphi = \text{Argmin}\|\boldsymbol{Y} - \boldsymbol{X}\boldsymbol{\beta}\|_\varphi. \tag{4}$$

These estimators were proposed by Jaeckel (1972) and Jurečková (1971). An associated rank-based analysis, including diagnostics procedures, is discussed in Chapters 3-5 of the monograph by Hettmansperger and McKean (2011). A score function needs to be selected. Often the Wilcoxon (linear) score function is used, $\varphi[u] = \sqrt{12}[u - (1/2)]$. When Wilcoxon scores are used, we refer to the subsequent fit and analysis as the Wilcoxon analysis. Another frequent choice is the sign scores function, $\varphi[u] = \text{sgn}\,[u - (1/2)]$, which yields the l_1-fit. Score functions are discussed in terms of optimality in Section 2.2.

As with LS, the rank-based estimator of the intercept is a location estimate based on the residuals. For LS, the arithmetic mean is used while for the rank-based estimates, generally, the median is used; i.e.,

$$\hat{\beta}_0 = \text{med}_i\left\{Y_i - \boldsymbol{x}^T\hat{\boldsymbol{\beta}}_\varphi\right\}. \tag{5}$$

2.1 Theory

As shown in Hettmansperger and McKean (2011), the influence function of the rank-based estimator $\hat{\boldsymbol{\beta}}_\varphi$ is given by

$$\Omega\left(\boldsymbol{x}_0, y_0; \hat{\boldsymbol{\beta}}_\varphi\right) = \tau_\varphi(\boldsymbol{X}'\boldsymbol{X}/n)^{-1}\varphi[F(y_0)]\,\boldsymbol{x}_0, \tag{6}$$

where the point $(y_0, \boldsymbol{x}_0^T)$ represents an outlier. The parameter τ_φ is the scale parameter given by

$$\tau_\varphi^{-1} = \int \varphi(u)\varphi_f(u)\, du\,, \tag{7}$$

where

$$\varphi_f(u) = -\frac{f'(F^{-1}(u))}{f(F^{-1}(u))}\,. \tag{8}$$

Based on this influence function it is clear that $\hat{\boldsymbol{\beta}}_\varphi$ is robust in Y-space if the scores function $\varphi(u)$ is bounded. Note, though, that $\hat{\boldsymbol{\beta}}_\varphi$ is not robust in $\boldsymbol{X}$-space. A weighted version of the Wilcoxon estimator called the HBR (high breakdown rank-based) achieves 50% breakdown in both the $\boldsymbol{X}$-space and the Y-space; see Chang et al. (1999).

As can be seen from the influence function, the asymptotic distribution of the rank-based estimator is given by

$$\widehat{\boldsymbol{\beta}}_\varphi \sim N_p\left(\boldsymbol{\beta}, \tau_\varphi^2(\boldsymbol{X}'\boldsymbol{X})^{-1}\right). \tag{9}$$

Note that the only difference between the theory for the LS and rank-based estimators is that σ^2 is replaced by τ_φ^2. Hence, the asymptotic relative efficiency (ARE) between the

LS and rank-based estimator is σ^2/τ_φ^2. For Wilcoxon scores, assuming that the random errors have a normal distribution, this ARE is the familiar 0.955; that is, for normal errors using the Wilcoxon analysis instead of the LS analysis results in only a 5% loss of efficiency. Koul et al. (1987) developed an estimator of τ_φ, $\hat{\tau}_\varphi$, which is computed by `Rfit`.

Chapters 3-5 of Hettmansperger and McKean (2011) discuss rank-based analyses of linear, mixed, and nonlinear models. The examples and simulation studies of this paper make use of these confidence intervals and a robust R^2, which we briefly present. An asymptotic $(1-\alpha)100\%$ confidence interval for a linear combination of regression parameters, say, $\boldsymbol{h}^T\boldsymbol{\beta}$ is given by

$$\boldsymbol{h}^T\widehat{\boldsymbol{\beta}}_\varphi \pm t_{\alpha/2,n-p-1}\hat{\tau}_\varphi\sqrt{\boldsymbol{h}^T(\boldsymbol{X}'\boldsymbol{X})^{-1}\boldsymbol{h}}, \tag{10}$$

where $t_{\alpha/2,n-p-1}$ denotes the upper $\alpha/2$, t-critical value with $n-p-1$ degrees of freedom.

Next, consider a general linear hypothesis of the form

$$H_0 : \boldsymbol{H\beta} = \boldsymbol{0} \text{ versus } H_A : \boldsymbol{H\beta} \neq \boldsymbol{0} \tag{11}$$

where $\boldsymbol{H}$ is a specified $q \times p$ matrix. Let V_{Full} and V_{Red} denote the respective full model column space of $\boldsymbol{X}$ and the reduced model subspace of V_{Full} constrained by H_0. Denote the distance between $\boldsymbol{Y}$ and each of these subspaces respectively by $\|\boldsymbol{Y} - \boldsymbol{X}\widehat{\boldsymbol{\beta}}_\varphi\|_\varphi$ and $\|\boldsymbol{Y} - \boldsymbol{W}\widehat{\boldsymbol{\theta}}_\varphi\|_\varphi$, where $\boldsymbol{W}$ denotes a reduced model $n \times (p-q)$ design matrix and $\widehat{\boldsymbol{\theta}}_\varphi$ denotes the corresponding reduced model rank-based estimator. Then $\text{RD} = \|\boldsymbol{Y} - \boldsymbol{W}\widehat{\boldsymbol{\theta}}_\varphi\|_\varphi - \|\boldsymbol{Y} - \boldsymbol{X}\widehat{\boldsymbol{\beta}}_\varphi\|_\varphi$ denotes the reduction in distance when passing from the reduced model to the full model. This is analogous to the LS reduction in Euclidean squared-distance (reduction in sums-of-squares). The corresponding rank-based F-test is given by

$$F_\varphi = \frac{\text{RD}/q}{\hat{\tau}_\varphi/2}. \tag{12}$$

An asymptotic level α test is to reject H_0 in favor of H_A, if $F_\varphi \geq F_\alpha(q, n-p-1)$, where $F_\alpha(q, n-p-1)$ denotes the upper α-critical value of an F-distribution with q and $n-p-1$ degrees of freedom.

As an example of this test, consider the hypothesis that all regression coefficients except for the intercept parameter are 0; i.e.,

$$H_0 : \boldsymbol{\beta} = \boldsymbol{0} \text{ versus } H_A : \boldsymbol{\beta} \neq \boldsymbol{0}, \tag{13}$$

In this case the reduced model dispersion is $\|\boldsymbol{Y}\|_\varphi$. Thus the reduction in dispersion is $RD = \|\boldsymbol{Y}\|_\varphi - \|\boldsymbol{Y} - \boldsymbol{X}\widehat{\boldsymbol{\beta}}_\varphi\|_\varphi$. Using this reduction, the robust F-test statistic is given by expression (12). A nominal level α test of the hypotheses (13) is to reject H_0 if $F_\varphi \geq F_\alpha(p, n-p-1)$. This test is a robust analogue of the least squares F test statistic that all regression coefficients are 0. Recall that the traditional coefficient of determination R^2 can be expressed as a one-to-one function of the LS F-test. In the same way, a robust coefficient of determination R^2 can be formulated as

$$R_2 = \frac{RD}{RD + (n-p-1)(\hat{\tau}_\varphi/2)}; \tag{14}$$

see page 243 of Hettmansperger and McKean (2011). We refer to R_2 as a robust coefficient of determination in subsequent examples.

2.2 Optimal scores

The rank-based analysis outlined above requires the selection of a score function $\varphi(u)$. If the form of the underlying error distribution is known, we can obtain an optimal score function which minimizes the variance of the estimator. Using expressions (7) and (8), we can rewrite $1/\tau_\varphi$ as

$$\begin{aligned}
(\tau_\varphi)^{-1} &= \int_0^1 \varphi(u)\varphi_f(u)\,du \\
&= \frac{\int_0^1 \varphi(u)\varphi_f(u)\,du}{\left(\int_0^1 \varphi_f^2(u)\,du\right)^{1/2}} \left(\int_0^1 \varphi_f^2(u)\,du\right)^{1/2} \\
&= \rho\left(\int_0^1 \varphi_f^2(u)\,du\right)^{1/2} = \rho\sqrt{I(f)},
\end{aligned} \tag{15}$$

where ρ is a correlation coefficient and $\sqrt{I(f)}$ is Fisher Information. Therefore, minimizing τ_φ is equivalent to maximizing the above identity. By the last equality, this is accomplished by making $\rho = 1$; i.e., by taking $\varphi(u)$ to be $\varphi_f(u)$. So expression (8) is the score function which optimizes the rank-based analysis. Since $\widehat{\boldsymbol{\beta}}_\varphi$ is location and scale equivariant, only the form of $f(x)$ is needed. Furthermore, since in this case $\tau_\varphi = 1/\sqrt{I(f)}$, the rank-based estimator $\widehat{\boldsymbol{\beta}}_\varphi$ is asymptotically fully efficient, i.e., $\widehat{\boldsymbol{\beta}}_\varphi$ has the same asymptotic distribution as the maximum likelihood estimator (mle).

For example, if the error distribution is normal, then the optimal score function simplifies to $\varphi(u) = \Phi^{-1}(u)$, the normal scores. If the error distribution is logistic, then the linear Wilcoxon scores are obtained, while double exponential (Laplace) distributed errors produces the sign scores.

2.3 Computation of the rank-based analysis

The computation of a rank-based analysis can be obtained by using the R package `Rfit` developed by Kloke and McKean (2012), which can be downloaded at CRAN. Like R, `Rfit` is freeware and can run on all platforms (windows, linux, and mac). As we discuss in Section 3, it is easy to install new scores in `Rfit` based on a general scores function. For now, we illustrate the computation of Rfit for a Wilcoxon analysis in the following example.

Example 2.1 (Linear Model with Skew-Normal Errors). We use a simulated data set based on the model $y = \beta_0 + \beta_1 x_1 + \beta_2 x_2 + \beta_3 x_3 + e$, where $x_1 = 1, \cdots, 50$; x_2 and x_3 are variates from a standard normal distribution; and the random errors are generated from a standard skew-normal distribution with shape parameter $\alpha = -8$, as discussed in Section 3. We set $\beta_1 = 0.01$, $\beta_2 = 0.15$, and $\beta_3 = 0.0$. The sample size is $n = 50$. The data set can be downloaded at the url cited in Section 1. The code segments below assume that the R vectors `y`, `x1`, `x2`, `x3`, contain respectively the responses and values for x_1, x_2 and x_3. For this example, the following R code using the package `Rfit` computes the Wilcoxon fit of the model, prints out the table of coefficients, and saves the Studentized residuals in the vector `studw`.

```
> fitw <- rfit(y ~ x1+x2+x3)
> summary(fitw)
Call:
rfit.default(formula = y ~ x1 + x2 + x3)

Coefficients:
              Estimate Std. Error t.value  p.value
(Intercept) -0.4672019  0.1811862 -2.5786 0.013186 *
x1           0.0044180  0.0056997  0.7751 0.442228
x2           0.2435079  0.0803388  3.0310 0.003993 **
x3          -0.0012315  0.0688965 -0.0179 0.985817
---
Signif. codes:  0 '***' 0.001 '**' 0.01 '*' 0.05 '.' 0.1 ' ' 1

Multiple R-squared (Robust): 0.1811501
Reduction in Dispersion Test: 3.39212 p-value: 0.02567

> studw <- rstudent(fitw)
```

Based on the summary table, the 95% confidence intervals for β_j, $j = 1, 2, 3$, (10), trap the true parameters. The overall F_φ test that all the regression coefficients are 0 except for the intercept is significant, $p = 0.0257$. The value of the robust coefficient of determination R_2, expression (14), is 18%.

The last command stores the Studentized Wilcoxon residuals in the vector `studw`. These residuals are adjusted for both variance of the errors and location in the $\boldsymbol{X}$-space; see McKean and Sheather (2009) for a review of robust diagnostic procedures. Figure 1 displays Studentized residual and normal $q-q$ plots based on the Wilcoxon fit.

As noted in Section 3, the skew-normal distribution chosen to generate the random variates in this example is left skewed. Hence, as expected, the residuals show longer left than right tails. These plots show that scores for left-skewed error distributions are more appropriate for this data than the Wilcoxon scores. There appears to be one large outlier in the left tail, also. ■

3 Skew-normal error distributions

The family of skew-normal distributions consists of left and right skewed distributions along with the normal distribution. The pdfs in this family are of the form

$$f(x; \alpha) = 2\phi(x)\Phi(\alpha x), \tag{16}$$

where the parameter α satisfies $-\infty < \alpha < \infty$ and $\phi(x)$ and $\Phi(x)$ are the pdf and cdf of a standard normal distribution, respectively. For this paper, if a random variable X has this pdf, we say that X has a standard skew-normal distribution with parameter α and write $X \sim SN(\alpha)$. If $\alpha = 0$, then X has a standard normal distribution. Further X is distributed left skewed if $\alpha < 0$ and right skewed if $\alpha > 0$. This family of distributions was introduced by Azzalini (1985), who discussed many of its properties.

In this paper, we are interested in linear models, (1), where the random errors may have skew-normal errors. In this case, the random error can be written as $e_i = b\varepsilon_i$, where ε_i has a standard skew-normal distribution and b is a scale parameter. The rank-based

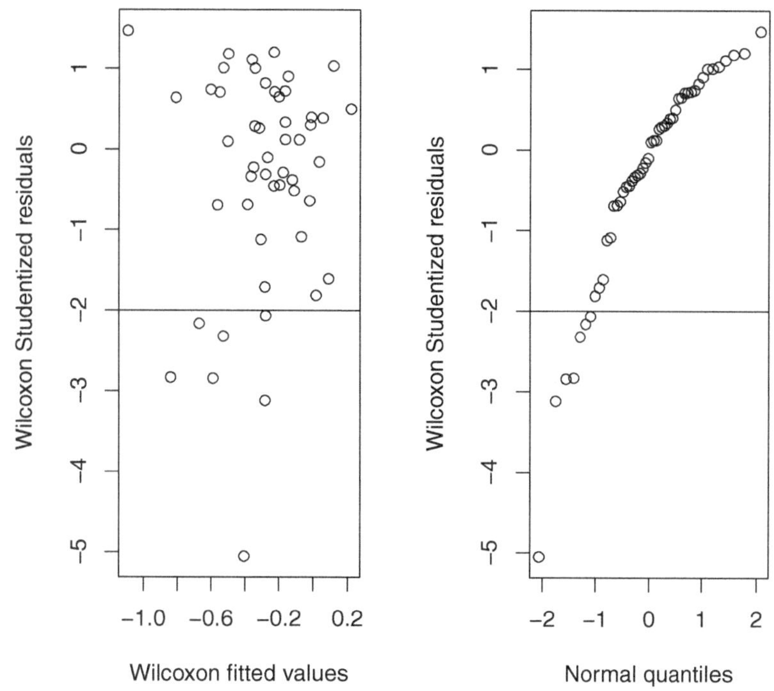

Figure 1 Studentized residual and normal $q-q$ plots based on the Wilcoxon fit of the data in Example 2.1.

estimator $\hat{\boldsymbol{\beta}}_{\varphi}$ and corresponding analysis are regression and scale equivariant, so there is no need to estimate the scale parameter b. The only scale parameter requiring estimation for standard errors is τ_{φ}. Likewise, for inference on the vector of parameters $\boldsymbol{\beta}$ there is no need to estimate the shape parameter α.

What rank scores would be best for such error distributions? To get an idea, we next discuss the optimal scores for a specified α. To obtain the optimal rank-based scores, because of equivariance, we need only the form (down to scale and location) of the pdf. So for the derivation of the scores, assume that the random variable $X \sim SN(\alpha)$ with pdf (16). It easily follows that

$$-\frac{f'(x;\alpha)}{f(x;\alpha)} = x - \frac{\alpha\phi(\alpha x)}{\Phi(\alpha x)}. \tag{17}$$

Denote the inverse of the cdf of X by $F^{-1}(u;\alpha)$. Then it follows from expression (15) that the optimal score function for X is

$$\varphi_{\alpha}(u) = F^{-1}(u;\alpha) - \frac{\alpha\phi(\alpha F^{-1}(u;\alpha))}{\Phi(\alpha F^{-1}(u;\alpha))}. \tag{18}$$

For all values of α, this score function is strictly increasing over the interval $(0, 1)$; see Azzalini (1985). As expected, for $\alpha = 0$, expression (17) simplifies to the normal scores. Due to the first term on the right-side of expression (18), all the score functions in this family are unbounded, indicating that the skew-normal family of distributions is light-tailed. Thus the influence functions of the rank-based estimators based on scores in this family are unbounded in the Y-space and, hence, are not robust. This includes the normal scores, but Huber (1981) pointed out that normal scores are technically robust and, as our simulation studies show, the family of skew-normal scores seems also to be technically robust.

Figure 2 displays the pdf's and corresponding optimal scores for three values of α: -7, $1, 5$. Note that the pdf for $\alpha = -7$ is left skewed while those for positive α values are right skewed. Unsurprisingly, the pdf for $\alpha = 1$ is closer to being symmetric than the other pdfs. The score function for the left-skewed pdf emphasizes relatively the right tails over the left tails, while the reverse is true for the right-skewed pdfs.

3.1 Computation of the rank-based analysis using skew-normal scores

The computation of the rank-based analysis can be obtained by using the R package `Rfit`. It is easy to install the family of skew-normal scores. Briefly, rank-based scores form a class in `Rfit` consisting of three parts: the score function, its derivative, and a vector of parameters used in the definition of the function. For the skew-normal scores, details are given in the appendix, but for the readers convenience the necessary R code is contained in the R function `skewns`, which we have placed at the web site cited in the introduction.

Example 3.1 (Example 2.1, Continued). We now return to Example 2.1 and show the computation of the rank-based analysis of it based on the skew-normal scores with shape parameter $\alpha = -8$. The first two lines of code define the skew-normal scores as `salp` and the third line sets the shape parameter. Details of this definition can be found in the appendix.

```
> source("skewns.R")
> salp <- skewns
> salp@param <- c(-8)
> fits <- rfit(y ~ x1+x2+x3,scores=salp)
> summary(fits)

Call:
rfit.default(formula = y ~ x1 + x2 + x3, scores = salp)

Coefficients: Estimate Std. Error t.value   p.value
(Intercept) -0.4176052  0.1443377 -2.8933 0.0058088 **
x1 0.0042469 0.0034173  1.2428 0.2202441
x2 0.1830920 0.0481674  3.8012 0.0004219 ***
x3 0.0543903 0.0413072  1.3167 0.1944525
---

Signif. codes:  0 '***' 0.001 '**' 0.01 '*' 0.05 '.' 0.1 ' ' 1

Multiple R-squared (Robust): 0.2864934 Reduction in Dispersion Test:
6.15677 p-value: 0.00131

> studs <- rstudent(fits)
```

Note that the skew-normal analysis is much more precise than the Wilcoxon analysis of the last section. The empirical ARE is $(\tau_W/\tau_{\alpha=-8})^2 = 2.78$; i.e., for this data set, the

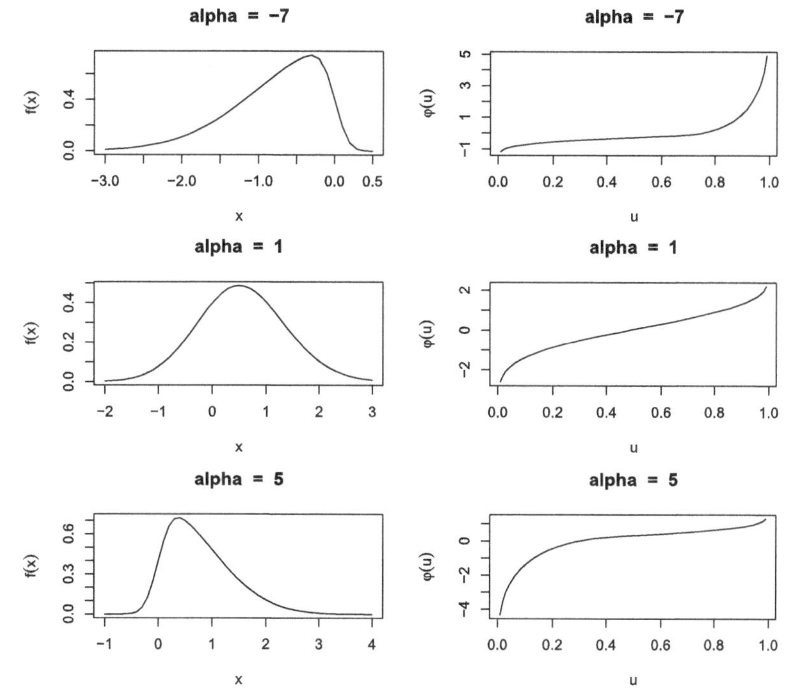

Figure 2 These plots display the pdfs of the three skew-normal distributions with shape parameter $\alpha = -7, 1$, and 5, along with the corresponding optimal scores.

skew-normal analysis is 2.8 times more efficient than the Wilcoxon analysis. Note, also, that the robust coefficient of determination, R_2, has increased from 18% to 28%.

3.2 Sensitivity analysis

For a verification of the technical robustness of the rank-based skew-normal analysis, we conducted a small sensitivity analysis. We generated $n = 50$ observations from a linear model of the form $y_i = x_i + e_i$, where x_i has a $N(0, 1)$ distribution and e_i has a $N(0, 10^2)$ distribution. The x_i's and e_i's are all independent. We added outliers of the form

$$y_{50} \leftarrow y_{50} + \Delta, \tag{19}$$

where Δ is in the set $\{0, 20, 40, 60, 80, 100, 1000, 2000\}$. The sensitivity curve for an estimator $\hat{\beta}$ is given by the function

$$S(\Delta; \hat{\beta}) = \hat{\beta} - \hat{\beta}(\Delta), \tag{20}$$

where $\hat{\beta}$ and $\hat{\beta}(\Delta)$ denote the estimates of β on the original and modified data (19), respectively. We obtained sensitivity curves for the estimators: Wilcoxon, normal scores, skew-normal ($\alpha = 3$), skew-normal ($\alpha = 5$), skew-normal ($\alpha = 7$), and maximum likelihood estimates (mle). The mles were computed by the package sn. For all values of Δ, the changes in all of the the rank-based estimates were less than 0.004. Thus the rank-based skew-normal estimators, including the normal scores estimator, exhibited technical robustness for this study. On the other hand, the mle was sensitive to the values of Δ. We show these changes in Table 1; hence, for this study, the mle was not robust.

Table 1 Values of the sensitivity function for the mle at the given values of Δ

Δ	0	20	40	60	80	100	1000	2000
mle	0.00	−0.07	−0.07	−0.00	0.12	0.30	−5.80	−6.32

3.3 Range of practical α parameters

In Sections 4 and 5, the results of simulation studies are presented. The error distributions involve families of skew-normal distributions. So a practical range of α values is needed. Pourahmadi (2007) derived properties of the skew-normal distribution including its moment generating function. In particular, he showed that if X has the $SN(\alpha)$ distribution, then X converges in distribution to $|Z|$ as $\alpha \to \infty$ where Z is $N(0,1)$; i.e., the distribution of X converges to a half-normal distribution. Likewise, X converges in distribution to $-|Z|$ as $\alpha \to -\infty$. In terms of α, the convergence is fairly fast. Table 2 serves as an illustration of this as it displays the mean (μ), median ($\tilde{\mu}$), variance (σ^2), and coefficient of skewness (ξ) for various values of α. The last column of the table shows the value of these parameters for the half-normal distribution. Positive values of α suffice because if $X \sim SN(\alpha)$ then $-X \sim SN(-\alpha)$. There is little difference between the standard skew-normal distribution and the half-normal distribution for values of α near ± 8. Based on these facts, we use skew-normal distributions with values of α between -12 and 12 for our Monte Carlo investigations.

Because we are interested in linear models, there is another practical reason for this range of α values. Note that the support of a skew-normal distribution is $(-\infty, \infty)$ making it ideal for error distributions for regression models. On the other hand, the support of a half-normal distribution is $(0, \infty)$, which is generally the support of a survival distribution. Often, the log's of such variables are modeled as accelerated failure time (AFT) models, as briefly discussed in Section 1.

4 Monte Carlo study

This section contains the results of a small simulation study concerning rank-based procedures based on skew-normal scores. The model simulated is

$$y_i = \beta_0 + \beta_1 x_i + \theta c_i + e_i, \tag{21}$$

where x_i is distributed $N(0,1)$; e_i is distributed from a selected error distribution; $i = 1, \ldots, 100$; the x_is and e_is are all independent; and the variable c_i is a treatment indicator with values of either 0 or 1. We selected two error distributions for the study. One is a skew-normal distribution with shape parameter $\alpha = 5$ while the other is a contaminated version of a skew-normal. The contaminated errors are of the form

$$e_i = (1 - I_{\epsilon,i})W_i + I_{\epsilon,i}V_i, \tag{22}$$

Table 2 Parameters (mean μ, median $\tilde{\mu}$, variance σ^2, and coefficient of skewness ξ) for $SN(\alpha)$ distributions for the given values of α

α	0.00	1.00	2.00	4.00	6.00	7.00	8.00	10.00	15.00	20.00	∞
μ	0.00	0.56	0.71	0.77	0.79	0.79	0.79	0.79	0.80	0.80	0.80
$\tilde{\mu}$	0.00	0.55	0.66	0.67	0.67	0.67	0.67	0.67	0.67	0.67	0.67
σ^2	1.00	0.68	0.49	0.40	0.38	0.38	0.37	0.37	0.37	0.36	0.36
ξ	0.00	0.14	0.45	0.78	0.89	0.92	0.93	0.96	0.98	0.98	0.99

The values in the last column (∞) are those for a half-normal distribution.

where W_i has a skew-normal distribution with shape parameter $\alpha = 5$, V_i has a $N(\mu_c = 10, \sigma_c^2 = 36)$ distribution, $I_{\epsilon,i}$ has a binomial $(1, \epsilon = 0.15)$ distribution, and W_i, V_i, and $I_{\epsilon,i}$ are all independent. Hence, this contaminated distribution is skewed with heavy right tails. The design is slightly unbalanced with $n_1 = 45$ and $n_2 = 55$. Without loss of generality β, θ, and β_0 were set to 0.

For the rank based procedures, we selected the rank-based procedure based on the score function $\varphi_5(u)$, (18), which is optimal for a skew normal distribution with $\alpha = 5$ and then three on each side of the optimal, i.e., procedures based on the score functions $\varphi_\alpha(u)$ with $\alpha = 2, 3, 4, 6, 7$, and 8. With the discussion in Section 3.3 in mind, we also selected the rank-based procedure with $\alpha = 10$. The rank-based Wilcoxon, least squares (LS) procedure, and mle procedures complete the methods investigated. The empirical results presented are the empirical AREs, which for each estimator is the ratio of the empirical mean-square error (MSE) of the mle to the empirical MSE of the estimator; hence, values of this ratio less than 1 are favorable to the mle while values greater than 1 are favorable to the estimator. Secondly, we present the empirical confidence intervals with nominal confidence 0.95. For all the procedures, we chose asymptotic confidence intervals of the form $\hat{\beta} \pm 1.96SE(\hat{\beta})$. We used a simulation size of 10,000.

The results are presented in Table 3. For the skew-normal errors, for both parameters β and θ, all the rank-based estimators except the Wilcoxon estimator are more efficient than the mle estimator. Note that the most efficient estimator for both β and θ is the rank-based estimator with $\alpha = 5$; although, empirical efficiencies are not significantly different from the empirical efficiencies for a few of the nearby (α close to 5) rank-based estimators. In terms of validity, the empirical confidences of all the procedures are close to the nominal confidence of 0.95. Not surprisingly, LS performed the worst overall.

For the contaminated error distribution, the rank-based estimators are much more efficient than the mle procedure. Further, the estimator with scores based on $\alpha = 5$ is still the most empirically powerful in the study. It has empirical efficiency of 785% relative to the mle for β and 1310% for θ. Even the Wilcoxon procedure is over 756% more efficient

Table 3 Summary of results of simulation study of rank-based procedures and the mle procedure for the skew-normal with shape $\alpha = 5$ distribution and a skew-normal contaminated distribution

	Skew normal errors				Contaminated errors			
	β		θ		β		θ	
Proced.	ARE	Conf.	ARE	Conf.	ARE	Conf.	ARE	Conf.
rb $\alpha = 2$	1.02	0.96	1.04	0.96	6.61	0.98	10.84	0.98
rb $\alpha = 3$	1.09	0.96	1.11	0.96	7.43	0.97	12.24	0.98
rb $\alpha = 4$	1.13	0.96	1.15	0.96	7.79	0.97	12.91	0.98
rb $\alpha = 5$	1.14	0.96	1.16	0.96	7.85	0.96	13.10	0.97
rb $\alpha = 6$	1.13	0.95	1.16	0.96	7.73	0.96	13.02	0.97
rb $\alpha = 7$	1.11	0.95	1.14	0.95	7.49	0.95	12.72	0.97
rb $\alpha = 8$	1.09	0.95	1.12	0.95	7.17	0.95	12.30	0.96
rb $\alpha = 10$	1.04	0.94	1.07	0.94	6.46	0.94	11.22	0.95
rb Wil.	0.78	0.95	0.79	0.95	4.70	0.96	7.56	0.97
LS	0.70	0.95	0.71	0.95	0.20	0.95	0.31	0.95
mle	1.00	0.93	1.00	0.93	1.00	0.96	1.00	0.99

than the mle for θ. On the basis of the empirical confidences, for both parameters, all procedures appear to be from slightly to moderately conservative. Least squares performed extremely poor in the contaminated part of the study. All the rank-based procedures based on skew-normal scores display technical robustness in this study.

5 Hogg-Type adaptive procedure

In Section 3, we discussed the rank-based method based on the optimal score function for a specified shape parameter α. Asymptotically, it is as efficient as the mle and, at least for the situations covered in the simulation study, the rank-based estimator appears to be more efficient than the mle for finite samples. In practice, though, the true shape parameter is not known. One could obtain the mle of α and use that score function. The mle, however, is not robust. Reconsidering the empirical study, note from Table 3 that the rank-based estimates close to the optimal rank-based estimator were more efficient than the mle and most had efficiencies that were quite close to that of the optimal. That is, in selecting a score function, perhaps close would suffice. In this section, we consider a Hogg-type adaptive scheme which has this as its goal.

Hogg et al. (1975) proposed an adaptive procedure for tests of the difference in locations for the two sample problem. The null hypothesis is that the two population distributions are the same. The selection of the test is based on a pair of selector statistics that measure respectively skewness and tail weight of the underlying error distribution. These selector statistics are functions of the order statistics of the combined samples. Several distribution-free rank tests of significance level δ comprise the tests. Under the null hypothesis, it follows from the sufficiency and completeness of the combined order statistics and the distribution-freeness of the rank test statistics that the selected test maintains the level δ. See, also, the discussion in Chapter 10 of Hogg et al. (2013).

This is fine for simple location tests where we have distribution-free rank tests, but in our case we are fitting a linear model and, hence, the adaption must be based on the residuals from an initial fit. Thus the above mentioned sufficiency result is not true for our fitting case. Shomrani (2003) developed a Hogg-type adaptive scheme for fitting a linear model based on an initial fit. In Shomrani's scheme, the selector statistics are functions of the residuals from the initial fit. While the significance level is no longer maintained, based on the results of a large simulation study, the scheme's empirical levels were generally close to the nominal value. In Chapter 6 of Kloke and McKean (2014), R software is developed for this scheme.

The adaptive scheme of Shomrani (2003) was formed for a wide range of error distributions: from left to right skewed and from light to heavy tailed distributions. We refine this scheme for the skew-normal family of distributions. As discussed above, there are two selector statistics involved. One, Q_1, selects based on skewness while the other, Q_2 selects based on tail thickness. In a preliminary study over the skew-normal family, tail thickness did not seem to be a paramount issue, so we focus on Q_1 alone.

Let $\boldsymbol{V} = (V_1, V_2, \ldots, V_n)^T$ be a random vector and define

$$Q_1(\boldsymbol{V}) = \frac{\overline{U}_{.05} - \overline{M}_{.5}}{\overline{M}_{.5} - \overline{L}_{.05}}, \tag{23}$$

where $\overline{U}_{.05}$, $\overline{M}_{.5}$, and $\overline{L}_{.05}$ are the averages of the largest 5% of the V_i's, the middle 50% of the V_i's, and the smallest 5% of the V_i's, respectively. Large values of Q_1 indicate that the right tails of the sample are longer than the left tails; i.e., indicating an underlying right skewed distribution. Likewise, small values of Q_1 indicate left-skewness. Note that as left (right) skew increases, Q_1 is likely to decrease (increase). The statistic Q_1 is not robust. One scheme under current investigation is to replace the means by medians. Keep in mind, though, that a robust diagnostic analysis is available for the initial robust fit. Hence in practice, outliers are easily flagged and modifications to the adaptive scheme can be made.

5.1 Adaptive scheme for skew normals

Our adaptive scheme consists of the 7 optimal score functions for skew-normal distributions with $\alpha = -12, -8, -4, 0, 4, 8,$ and 12. So there are three scores each for left and right skewed distributions along with the normal scores. The scheme utilizes residuals from an initial Wilcoxon fit. We chose the Wilcoxon because it is robust. Also, it is optimal for a symmetric distribution (logistic) and, hence, less likely to bias selection for left or right skewness.

We decided to set the benchmarks for the selector statistic Q_1 based on the medians of the distribution of Q_1 for the family of skew-normal distributions with $\alpha = -10, -6, -2, 2, 6,$ and 10. Then the selection part of the adaptive scheme is given by:

$$\begin{aligned}
Q_1 &< \text{med}(Q_1|\alpha = -10) \Rightarrow \text{Select score using } \alpha = -12 \\
\text{med}(Q_1|\alpha = -10) &< Q_1 < \text{med}(Q_1|\alpha = -6) \Rightarrow \text{Select score using } \alpha = -8 \\
\text{med}(Q_1|\alpha = -6) &< Q_1 < \text{med}(Q_1|\alpha = -2) \Rightarrow \text{Select score using } \alpha = -4 \\
\text{med}(Q_1|\alpha = -2) &< Q_1 < \text{med}(Q_1|\alpha = 2) \Rightarrow \text{Select score using } \alpha = 0 \\
\text{med}(Q_1|\alpha = 2) &< Q_1 < \text{med}(Q_1|\alpha = 6) \Rightarrow \text{Select score using } \alpha = 4 \\
\text{med}(Q_1|\alpha = 6) &< Q_1 < \text{med}(Q_1|\alpha = 10) \Rightarrow \text{Select score using } \alpha = 8 \\
\text{med}(Q_1|\alpha = 10) &< Q_1 \Rightarrow \text{Select score using } \alpha = 12
\end{aligned} \tag{24}$$

We estimated the medians of the sampling distributions of the statistic Q_1 based on simulations of size 10,000 drawn from the appropriate skew-normal distributions. Because Q_1 is location and scale equivariant, simulation using standard skew-normal distributions suffices. The estimated (simulated) medians used for the scheme are given in Table 4.

In summary, the algorithm for our adaptive scheme is:

1. Fit using Wilcoxon scores $\Rightarrow$ Obtain residuals $\widehat{\mathbf{e}}_W$.
2. Compute $Q_1(\widehat{\mathbf{e}}_W)$ and then select φ_α using expression (24), using the estimated medians of Q_1.
3. Fit with selected score φ_α.
4. Inference is based on the fit of Step (3).

We next try the scheme on the data of Example 2.1.

Table 4 Simulated sample median of the distribution of Q_1 drawn from the skew-normal distribution with shape parameter α

α	-10	-6	-2	2	6	10
Median Q_1	0.44	0.49	0.73	1.37	2.05	2.26

The simulation size is 10,000. These estimated medians are used in the adaptive scheme (24).

Example 5.1 (Example 2.1, continued). For the data in Example 2.1, the selector Q_1 has the value 0.344; hence the adaptive scheme (24) selects the score function with $\alpha = -12$. Table 5 summarizes the results of the fits based on skew-normal scores with α's in a neighborhood of -8.

The values of the regression coefficients are given along with the robust coefficient of determinations R_2 and estimates of τ_φ. The fits are quite similar. Notice that in terms of precision, $\hat{\tau}_\varphi$'s, that the fit with $\alpha = -8$ is the most precise. ■

5.2 Simulation study

We investigated the validity and efficiency of the adaptive scheme in a Monte Carlo study using situations similar to those in Section 4. In particular, the sample size is set at $n = 100$ and the linear model contains one predictor coefficient, β, and one indicator coefficient for the treatment, θ. We considered three sampling situations. In the first situation (I), for each simulation, α is randomly selected from the set $\{-12, -11, \dots, 11, 12\}$ where the selections are made equilikely. Then the random errors for the model are generated from this $SN(\alpha)$ distribution. For Situation II, again α is randomly selected from the same set of values but now the random errors are random variables of the form

$$e_i = (1 - I_{.15})S_i + I_{.15}C_i,$$

where $S_i \sim SN(\alpha)$, $C_i \sim N(10, 6^2)$, $I_{.15}$ is Bernoulli with proportion of success 0.15, and $S_i, C_i, I_{.15}$ are independent. Thus, Situation II is the same as the second situation of Section 3, i.e., right-skewed contamination. Situation III is the same as situation II except that $C_i \sim N(0, 6^2)$, i.e., symmetric contamination. 10,000 simulations were used for each situation.

The methods considered are: our adaptive scheme (AdSch), least squares (LS), Wilcoxon (Wil), and maximum likelihood (mle). We also considered the procedure based on the correct α; i.e., the α which is selected for the distribution of the random errors. Note that this is not a statistical method and we label it as Optrv, "rv" for random variable. Even for Situation I, the distribution of its rank-based estimate depends on the multinomial random variable involved in the selection of the simulated distribution. We only include it to serve as an yardstick for the four statistical methods.

As in the simulation study of Section 3, we considered empirical efficiency (relative to the mle) and validity of 95% confidence intervals for the parameters β and θ. The results for Situation I are summarized in Table 6. The adaptive scheme was more efficient than

Table 5 For the data of Example 2.1, the adaptive scheme chose the score function with $\alpha = -12$

α	β_1	β_2	β_3	$\hat{\tau}_\varphi$	Robust R_2
$\alpha = -12$ (Adaptive Choice)	0.0035	0.2190	0.0368	0.3675	0.2743
$\alpha = -10$	0.0030	0.2095	0.0480	0.3699	0.2732
$\alpha = -9$	0.0037	0.1923	0.0514	0.3531	0.2828
$\alpha = -8$	0.0042	0.1831	0.0544	0.3484	0.2865
$\alpha = -7$	0.0043	0.1822	0.0537	0.3533	0.2849
$\alpha = -6$	0.0044	0.1808	0.0540	0.3490	0.2884
$\alpha = -5$	0.0044	0.1805	0.0531	0.3580	0.2838
$\alpha = -4$	0.0043	0.1854	0.0449	0.3886	0.2668

This table shows the variation in the rank-based fits for score functions in a neighborhood of $\alpha = -8$.

Table 6 Empirical efficiencies and confidence coefficients for Situation I (error distributions are skew-normals)

	AdSch	Optrv	LS	Wil	mle
β, ARE	1.06	1.12	0.75	0.80	1.00
β, Conf	0.95	0.95	0.95	0.95	0.93
θ, ARE	1.05	1.11	0.73	0.79	1.00
θ, Conf	0.95	0.95	0.95	0.95	0.94

Empirical efficiencies are MSE's relative to the MSE of the mles.

all other statistical methods for both parameters. In particular, it was more efficient than the mle by 6 and 5% respectively. In terms of validity all of the methods are valid. The adaptive scheme was less efficient (by 6%) than the optimal non-statistical procedure.

For Situation I, the random errors have a skew-normal distribution with shape parameter α drawn from the set $\{-12, -11, \ldots, 12\}$, while the scheme selects scores from the set $\{-12, -8, -4, 0, 4, 8, 12\}$. These sets are different; hence, it does not make sense to consider when the scheme made the "correct" selection. We did keep track of how often the selection was within two units of the distribution simulated. For the 10,000 simulations of Situation I, the estimate of this proportion is 0.584. Note that for Situations II and III, the random errors have a contaminated skew-normal distribution. In particular, it is not a skew-normal distribution. So for Situations II and III, this proportion is irrelevant.

The results for Situation II are summarized in Table 7. For this right-skewed contaminated situation, the adaptive scheme is much more efficient than the mle, 358% and 356% respectively for β and θ. The adaptive scheme is adapting to this heavy tailed situation and the mle is not robust. As expected, the Wilcoxon performs best. All the methods are valid. We included the optimal random variable procedure since it was in Situation I. Note, though, that in Situations II and III the distribution of the random errors is not a skew-normal distribution. Besides not being a statistical method it is no longer optimal in any sense.

The results for the situation, III, with symmetric contamination can be found in Table 8. The adaptive scheme is much more efficient than the mle, 367% and 477% respectively for β and θ. As expected, the Wilcoxon performs best. All the rank-based methods are valid. The mle is slightly conservative for θ. The same comments hold for Optrv as in Situation II.

6 Conclusion

Rank-based analyses of linear models depend on the selection of a score function. In practice, often the Wilcoxon (linear) score function is chosen. These scores require no tuning constants and, further, the Wilcoxon rank-based analysis attains 95.5% efficiency relative

Table 7 Empirical efficiencies and confidence coefficients for Situation II (error distributions are skewed contaminated skew-normals)

	AdSch	Optrv	LS	Wil	mle
β, ARE	3.58	1.86	0.92	7.27	1.00
β, Conf	0.97	0.97	0.95	0.95	0.94
θ, ARE	3.56	1.56	0.94	7.36	1.00
θ, Conf	0.97	0.96	0.95	0.95	0.95

Empirical efficiencies are MSE's relative to the MSE of the mles.

Table 8 Empirical efficiencies and confidence coefficients for Situation III (error distributions are symmetrically contaminated skew-normals)

	AdSch	Optrv	LS	Wil	mle
β, ARE	3.67	1.01	0.26	6.16	1.00
β, Conf	0.94	0.97	0.95	0.97	0.95
θ, ARE	4.77	0.85	0.35	8.18	1.00
θ, Conf	0.94	0.97	0.95	0.97	0.98

Empirical efficiencies are MSE's relative to the MSE of the mles.

to the traditional least squares (LS) analysis when the random error distribution is normal. However, rank-based analyses are easily optimized if there is knowledge of the distribution of the random errors of the linear model. For example, if the random errors are normally distributed then selecting the normal scores for the rank-based analysis results in the efficiency of 100% (fully efficient) relative to the LS analysis.

In this paper, we have presented the rank-based analyses based on appropriate score functions for random errors having a distribution from the family of skew-normal distributions. In this case, the score function depends on the shape parameter α, $-\infty < \alpha < \infty$. Of course, the rank-based analysis is fully efficient if the correct α is known. The rank-based analysis is a complete analysis, including fitting, inference (rank-based ANOVA), and robust diagnostic procedures. Based on the results of our Monte Carlo, these rank-based analyses appear to be more efficient than the maximum likelihood (mle) analysis for the skew-normal distributions considered. The most efficient rank-based analysis is based on the optimal score function, but even those rank-based analyses with shape parameters within three units of the correct α were more efficient than the mle in these situations. They were much more efficient than the mle's over situations where the error distribution had a contaminated skew-normal distribution. Based on empirical confidence levels, all the methods in the study were valid.

The good efficiency results for the rank-based analyses in a neighborhood of the true α suggest that a Hogg-type adaptive scheme would have high efficiency. In Section 5, we developed such a scheme for the skew-normal family of distributions based on an initial robust Wilcoxon fit. In the Monte Carlo studies we performed, this scheme was more efficient than the mle over the family of skew-normal distributions and was much more efficient than the mle over the contaminated skew-normal situations. Furthermore, for the situations covered, this adaptive scheme appears to be valid.

Kloke and McKean (2012) developed an R package `Rfit` for these rank-based analyses, which can be freely downloaded at CRAN. The default scores are the Wilcoxon scores, but, as we discuss in Section 3 it is easy to add classes of scores including the optimal scores for skew-normal distributions. The adaptive scheme of Section 5 is also easily coded using `Rfit`. The necessary code for the scores and the adaptive scheme can be found at the web site cited in Section 1. Hence, computation of these rank-based analyses is not a problem.

The rank-based analyses using skew-normal scores are robust in Y (response) space, but not in $\boldsymbol{X}$ (factor) space. The weighted Wilcoxon fit proposed by Chang et al. (1999) yield a robust rank-based analysis which possesses 50% breakdown in $\boldsymbol{X}$ (factor) space. We are now developing such an analysis for the skew-normal scores; see Abebe et al. (2014) for discussion. This analysis could also be part of an adaptive scheme.

A Appendix: R code for the class of skew normal scores

Rank-based scores form a class in `Rfit` which consists of three principal parts. The first part is the scores function itself and the third part consists of any parameters in the function expression. Thus parts (1) and (3) for the skew-normal scores are respectively given by expression (18) and the shape parameter α. The second part is the derivative of the score function which is used in the `Rfit` function which estimates τ_φ. Let $l(x)$ denote the function defined in expression (18). The derivative of the optimal score function is given by

$$\varphi'_\alpha(u) = l'[F^{-1}(u;\alpha)]\,\frac{1}{2\phi[F^{-1}(u;\alpha)]\,\Phi[\alpha F^{-1}(u;\alpha)]}, \tag{25}$$

where

$$l'(x) = 1 + \frac{\alpha^2\phi(\alpha x)[\alpha x\Phi[\alpha F^{-1}(u;\alpha)] + \phi(\alpha x)]}{\Phi^2(\alpha x)}.$$

To complete the class statement for the skew-normal scores we need only compute the quantiles $F^{-1}(u;\alpha)$. Azzalini (2013) developed the R package `sn` (available at CRAN) which computes the quantile function $F^{-1}(u;\alpha)$ and, also, the corresponding pdf and cdf. The command `qsn(u,shape=alpha)` returns $F^{-1}(u;\alpha)$, for $0 < u < 1$. The package `sn` requires the package `mnormt`.

The following R code defines the class of skew-normal scores:

```
> library(mnormt)
> library(sn)
> source("skewns.R")
> skewns

An object of class "scores"
Slot "phi":
function (u, param)
{
    alp = param[1]
    f3 = qsn(u, shape = alp)
    top = f3 - ((alp * dnorm(alp * f3))/pnorm(alp * f3))
    top
}

Slot "Dphi":
function (u, param)
{
    alp = param[1]
    f3 = qsn(u, shape = alp)
    part1 = alp * f3 * pnorm(alp * f3) + dnorm(alp * f3)
    part2 = (alp^2 * dnorm(alp * f3) * part1)/(pnorm(alp * f3)^2)
    part3 = (1 + part2)/(2 * dnorm(f3) * pnorm(alp * f3))
    part3
}

Slot "param":
[1] 1
```

The next code segment obtains the data for a plot of the scores with shape parameter $\alpha = -7$.

```
> alp <- -7
> x <- seq(-3,.5,.1)
> y <- dsn(x,shape=alp)
> n <- 100
> ps <- (1:n)/(n+1)
> salp <- skewns
> salp@param <- c(alp)
> phialp <- getScores(salp,ps)
```

Competing interests
The authors declare that they have no competing interests.

Authors' contributions
The authors contributed equally to the manuscript. All authors read and approved the final manuscript.

Acknowledgement
We acknowledge the helpful comments of an associate editor and a referee on the original manuscript.

References

Abebe, A, McKean, JW: Weighted Wilcoxon estimators in Nonlinear Regression. Aust N Z J. Stat. **55**, 401–420 (2013)

Abebe, A, McKean, JW, Kloke, JD, Bilgic, Y: Iterated Reweighted Rank-Based Estimates for GEE Models. Technical Report (2014)

Azzalini, A: A class of distributions which includes the normal ones. Scand. J. Stat. **12**, 171–178 (1985)

Azzalini, A: R package `sn`: The skew-normal and skew-t distributions (version 0.4-18). Università di Padova, Italia (2013)

Chang, W, McKean, J, Naranjo, J, Sheather, S: High-breakdown rank regression. J. Am. Stat. Assoc. **94**, 205–219 (1999)

Hettmansperger, TP, McKean, JW: Robust Nonparametric Statistical Methods. 2nd Edition. Chapman-Hall, Boca Raton, FL (2011)

Hogg, RV, Fisher, DM, Randles, RH: A two-sample adaptive distribution-free test. J. Am. Stat. Assoc. **70**, 656–661 (1975)

Hogg, RV, McKean, JW, Craig, AT: Introduction to Mathematical Statistics. 7th Ed. Pearson, Boston (2013)

Huber, PJ: Robust Statistics. John Wiley & Son (1981)

Jaeckel, LA: Estimating regression coefficients by minimizing the dispersion of residuals. Ann. Math. Stat. **43**, 1449–1458 (1972)

Jurečková, J: Nonparametric estimate of regression coefficients. Ann. Math. Stat. **42**, 1328–1338 (1971)

Kloke, JD, McKean, JW: Rfit: Rank-based estimation for linear models. R J. **4**, 57–64 (2012)

Kloke, JD, McKean, JW: Small sample properties of JR estimators. In: JSM Proceedings. American Statistical Association, Alexandria, VA (2013)

Kloke, JD, McKean, JW: Nonparametric statistical methods using R, Chapman-Hall, Boca Raton, FL (2014)

Kloke, JD, McKean, JW, Rashid, M: Rank-based estimation and associated inferences for linear models with cluster correlated errors. J. Am. Stat. Assoc. **104**, 384–390 (2009)

Koul, HL, Sievers, GL, McKean, JW: An estimator of the scale parameter for the rank analysis of linear models under general score functions. Scand. J. Stat. **14**, 131–141 (1987)

McKean, J, Sheather, S: Diagnostic procedures. Wiley Interdiscip. S. Rev.: Comput tat. **1**(2), 221–233 (2009)

McKean, J, Sievers, G: Rank scores suitable for analysis of linear models under asymmetric error distributions. Technometrics. **31**, 207–218 (1989)

Pourahmadi, M: Construction of skew-normal random variables: Are they linear combinations of normals and half-normals. J. Stat. Theory Appl. **3**, 314–328 (2007)

Shomrani, A: A comparison of different schemes for selecting and estimating score functions based on residuals. Ph.D. thesis, Western Michigan University, Department of Statistics (2003)

4

Missing data approaches for probability regression models with missing outcomes with applications

Li Qi and Yanqing Sun*

*Correspondence: yasun@uncc.edu
Department of Mathematics and Statistics, The University of North Carolina at Charlotte, 28223 Charlotte, NC, USA

Abstract

In this paper, we investigate several well known approaches for missing data and their relationships for the parametric probability regression model $P_\beta(Y|X)$ when outcome of interest Y is subject to missingness. We explore the relationships between the mean score method, the inverse probability weighting (IPW) method and the augmented inverse probability weighted (AIPW) method with some interesting findings. The asymptotic distributions of the IPW and AIPW estimators are derived and their efficiencies are compared. Our analysis details how efficiency may be gained from the AIPW estimator over the IPW estimator through estimation of validation probability and augmentation. We show that the AIPW estimator that is based on augmentation using the full set of observed variables is more efficient than the AIPW estimator that is based on augmentation using a subset of observed variables. The developed approaches are applied to Poisson regression model with missing outcomes based on auxiliary outcomes and a validated sample for true outcomes. We show that, by stratifying based on a set of discrete variables, the proposed statistical procedure can be formulated to analyze automated records that only contain summarized information at categorical levels. The proposed methods are applied to analyze influenza vaccine efficacy for an influenza vaccine study conducted in Temple-Belton, Texas during the 2000-2001 influenza season.

Mathematics Subject Classification: Primary 62J02; Secondary 62F12

Keywords: Augmented inverse probability weighted estimator; Asymptotic results; Automated records; Auxiliary outcome; Efficiency; Inverse probability weighted estimator; Mean score estimation; Recurrent events; Vaccine efficacy; Validation sample

1 Introduction

Suppose that Y is the outcome of interest and X is a covariate vector. One is often interested in the probability regression model $P_\beta(Y|X)$ that relates Y to X. In many medical and epidemiological studies, the complete observations on Y may not be available for all study subjects because of time, cost, or ethical concerns. In some situations, an easily measured but less accurate outcome named auxiliary outcome variable, A, is supplemented. The relationship between the true outcome Y and the auxiliary outcome A in the available observations can inform about the missing values of Y. Let V be a subsample of the study subjects, termed the validation sample, for which both true and auxiliary outcomes are available. Thus observations on (X, Y, A) are available for the subjects in V and only (X, A) are observed for those not in V.

It is well known that the complete-case analysis, which uses only subjects who have all variables observed, can be biased and inefficient, cf. Little and Rubin (2002). The issues also rise when substituting auxiliary outcome for true outcome; see Ellenberg and Hamilton (1989), Prentice (1989) and Fleming (1992). Inverse probability weighting (IPW) is a statistical technique developed for surveys by Horvitz and Thompson (1952) to calculate statistics standardized to a population different from that in which the data was collected. This approach has been generalized to many aspects of statistics under various frameworks. In particular, the IPW approach is used to account for missing data through inflating the weight for subjects who are underrepresented due to missingness. The method can potentially reduce the bias of the complete-case estimator when weighting is correctly specified. However, this approach has been shown to be inefficient in several situations, see Clayton et al. (1998) and Scharfstein et al. (1999). Robins et al. (1994) developed an improved augmented inverse probability weighted (AIPW) complete-case estimation procedure. The method is more efficient and possesses double robustness property. The multiple imputation described in Rubin (1987) has been routinely used to handle missing data. Carpenter et al. (2006) compared the multiple imputation with IPW and AIPW, and found AIPW as an attractive alternative in terms of double robustness and efficiency. Using the maximum likelihood estimation (MLE) coupled with the EM-algorithm (Dempster et al. 1977), Pepe et al. (1994) proposed the mean score method for the regression model $P_\beta(Y|X)$ when both X and A are discrete.

In this paper, we investigate several well known approaches for missing data and their relationships for the parametric probability regression model $P_\beta(Y|X)$ when outcome of interest Y is subject to missingness. We explore the relationships between the mean score method, IPW and AIPW with some interesting findings. Our analysis details how efficiency is gained from the AIPW estimator over the IPW estimator through estimation of validation probability and augmentation to the IPW score function. Applying the developed missing data methods, we derive the estimation procedures for Poisson regression model with missing outcomes based on auxiliary outcomes and a validated sample for true outcomes. Further, we show that the proposed statistical procedures can be formulated to analyze automated records that only contain aggregated information at categorical levels, without using observations at individual levels.

The rest of the paper is organized as follows. Section 2 introduces several missing data approaches for the probability regression model $P_\beta(Y|X)$, where the outcome Y may be missing. Section 3 explores the relationships among these estimators. The asymptotic distributions of the IPW and AIPW estimators are derived and their efficiencies are compared. Section 3 investigates efficiency of two AIPW estimators, one is based on the augmentation using a subset of observed variables and the other is based on the augmentation using the full set of observed variables. The procedures for Poisson regression using automated data with missing outcomes are derived in Section 4. The finite-sample performances of the estimators are studied in simulations in Section 5. The proposed method is applied to analyze influenza vaccine efficacy for an influenza vaccine study conducted in Temple-Belton, Texas during the 2000-2001 influenza season. The proofs of the main results are given in the Appendix A, while the proof of a simplified variance formula in Section 4 is placed in the Appendix B.

2 Missing data approaches

Consider the probability regression model $P_\beta(Y|X)$, where Y is the outcome of interest and X is a covariate vector. Let A be the auxiliary outcome for Y and V be the validation set such that observations on (X, Y, A) are available for the subjects in V and only (X, A) are observed for those in $\bar{V}$, the complement of V. In practice, the validation sample may be selected based on the characteristics of a subset, Z, of the covariates in X. We write $X = (Z, Z^c)$. For example, Z may include exposure indicator and other discrete covariates and Z^c may be the exposure time. Let (Z_i, X_i, Y_i, A_i), $i = 1, \ldots, n$, be independent identically distributed (iid) copies of (Z, X, Y, A). Let $\xi_i = I(i \in V)$ be the selection indicator.

Most statistical methods for missing data require some assumptions on missingness mechanisms. The commonly used ones are missing completely at random (MCAR) and missing at random (MAR). MCAR assumes that the probability of missingness in a variable is independent of any characteristics of the subjects. MAR assumes that the probability that a variable is missing depends only on observed variables. In practice, if missingness is a result by design, it is often convenient to let the missing probability depend on the categorical variables only. There is also simplicity in statistical inference by modeling the missing probability based on the categorical variables. We introduce the following missing at random assumptions.

MAR I: ξ_i is independent of Y_i conditional on (X_i, A_i) and ξ_i is independent of Z_i^c conditional on (Z_i, A_i).

MAR II: ξ_i is independent of (Y_i, Z_i^c) conditional on (Z_i, A_i).

Since the conditional density $f(y, z^c|\xi, z, a) = f(z^c|\xi, z, a) f(y|z^c, \xi, z, a) = f(z^c|z, a)$ $f(y|z^c, z, a) = f(y, z^c|z, a)$, MAR I implies MAR II. It is also easy to show that MAR II implies MAR.

Let $\hat{\pi}_i$ be the estimator of the conditional probability $\pi_i = P(\xi_i = 1|X_i, A_i)$, and $\hat{\pi}_i^z$ the estimator of $\pi_i^z = P(\xi_i = 1|Z_i, A_i)$. Let $S_\beta(Y|X)$ denote the partial derivatives of $\log P_\beta(Y|X)$ with respect to β. Let $\hat{E}\left\{S_\beta(Y|X_i)|X_i, A_i\right\}$ be the estimator of the conditional expectation $E\left\{S_\beta(Y|X_i)|X_i, A_i\right\}$, and $\hat{E}\left\{S_\beta(Y|X_i)|Z_i, A_i\right\}$ the estimator of $E\left\{S_\beta(Y|X_i)|Z_i, A_i\right\}$. We investigate several estimators of β based on the following estimating equations with different choices of W_i:

$$\sum_{i=1}^{n} W_i = 0, \tag{1}$$

where W_i takes one of the following forms:

$$W_i^{I1} = \frac{\xi_i}{\hat{\pi}_i^z} S_\beta(Y_i|X_i) \tag{2}$$

$$W_i^{E1} = \xi_i S_\beta(Y_i|X_i) + (1 - \xi_i)\hat{E}\left\{S_\beta(Y|X_i)|Z_i, A_i\right\} \tag{3}$$

$$W_i^{A1} = \frac{\xi_i}{\hat{\pi}_i^z} S_\beta(Y_i|X_i) + \left(1 - \frac{\xi_i}{\hat{\pi}_i^z}\right) \hat{E}\left\{S_\beta(Y|X_i)|Z_i, A_i\right\} \tag{4}$$

$$W_i^{I2} = \frac{\xi_i}{\hat{\pi}_i} S_\beta(Y_i|X_i) \tag{5}$$

$$W_i^{E2} = \xi_i S_\beta(Y_i|X_i) + (1-\xi_i)\hat{E}\left\{S_\beta(Y|X_i)|X_i, A_i\right\} \tag{6}$$

$$W_i^{A2} = \frac{\xi_i}{\hat{\pi}_i^z} S_\beta(Y_i|X_i) + \left(1 - \frac{\xi_i}{\hat{\pi}_i^z}\right)\hat{E}\left\{S_\beta(Y|X_i)|X_i, A_i\right\}. \tag{7}$$

$$W_i^{A3} = \frac{\xi_i}{\hat{\pi}_i} S_\beta(Y_i|X_i) + \left(1 - \frac{\xi_i}{\hat{\pi}_i}\right)\hat{E}\left\{S_\beta(Y|X_i)|X_i, A_i\right\}. \tag{8}$$

The estimator $\hat{\beta}_{I1}$ obtained by using W_i^{I1} is an IPW estimator where a subject's validation probability π_i^z depends only on the category defined by (Z_i, A_i). Because $E\left\{(\pi_i^z)^{-1}\xi_i S_\beta(Y_i|X_i)\right\} = E\left\{S_\beta(Y_i|X_i)\right\} = 0$, the estimator $\hat{\beta}_{I1}$ is approximately unbiased. The estimator $\hat{\beta}_{I2}$ obtained by using W_i^{I2} is also an IPW estimator but with the validation probability π_i depending on the category defined by (Z_i, A_i) and the additional covariate Z_i^c.

The estimator $\hat{\beta}_{E1}$ obtained by using W_i^{E1} is the mean score estimator where the scores $S_\beta(Y_i|X_i)$ for those with missing outcomes are replaced by the estimated conditional expectations given (Z_i, A_i). The estimator $\hat{\beta}_{E2}$ obtained by using W_i^{E2} is the mean score estimator where the scores $S_\beta(Y_i|X_i)$ for those with missing outcomes are replaced by the estimated conditional expectations given (X_i, A_i). The estimator $\hat{\beta}_{E2}$ is the mean score estimator in Pepe et al. (1994). The mean score estimator is the MLE estimator employing the EM-algorithm (Dempster et al. 1977) under the assumption that the auxiliary outcome is noninformative in the sense that the probability model $P_\theta(A|Y, X)$ is unrelated to β.

The estimator $\hat{\beta}_{A1}$ obtained using W_i^{A1} is the AIPW estimator augmented with the estimated conditional expectation $\hat{E}\left\{S_\beta(Y|X_i)|Z_i, A_i\right\}$. The estimator $\hat{\beta}_{A2}$ obtained using W_i^{A2} is the AIPW estimator augmented with the estimated conditional expectation $\hat{E}\left\{S_\beta(Y|X_i)|X_i, A_i\right\}$. The estimator $\hat{\beta}_{A3}$ is obtained using W_i^{A3}. The W_i^{A3} differs from W_i^{A2} in that the estimated validation probability is $\hat{\pi}_i$ instead of $\hat{\pi}_i^z$.

Suppose that $\hat{\pi}_i^z$ is an asymptotically unbiased estimator of $\bar{\pi}_i^z$ and that $\hat{E}\left\{S_\beta(Y|X_i)|Z_i, A_i\right\}$ is asymptotically unbiased of $\bar{E}\left\{S_\beta(Y|X_i)|Z_i, A_i\right\}$, where both $\bar{\pi}_i^z$ and $\bar{E}\left\{S_\beta(Y|X_i)|Z_i, A_i\right\}$ are functions of (Z_i, A_i). Under MAR II, if one of the equalities, $\bar{\pi}_i^z = \pi_i^z$ and $\bar{E}\left[S_\beta(Y|X_i)|Z_i, A_i\right]\} = E\left[S_\beta(Y|X_i)|Z_i, A_i\right]\}$, holds, then

$$E\left\{\left(\bar{\pi}_i^z\right)^{-1}\xi_i S_\beta\left(Y_i|X_i\right)\right\} + E\left\{\left(1-\left(\pi_i^z\right)^{-1}\xi_i\right)\bar{E}\left[S_\beta\left(Y|X_i\right)|Z_i, A_i\right]\right\} = E\left\{S_\beta(Y_i|X_i)\right\} = 0,$$

which entails that the estimator $\hat{\beta}_{A1}$ has the double robust property in the sense that it is a consistent estimator of β if either $\hat{\pi}_i^z$ is a consistent estimator of π_i^z or $\hat{E}\left[S_\beta(Y|X_i)|Z_i, A_i\right]$ is a consistent estimator of $E\left[S_\beta(Y|X_i)|Z_i, A_i\right]$. Similarly, under MAR I, the estimator $\hat{\beta}_{A2}$ possesses the double robust property in that $\hat{\beta}_{A2}$ is a consistent estimator of β if either $\hat{\pi}_i^z$ is a consistent estimator of π_i^z or $\hat{E}\left[S_\beta(Y|X_i)|X_i, A_i\right]$ is a consistent estimator of $E\left[S_\beta(Y|X_i)|X_i, A_i\right]$. The estimator $\hat{\beta}_{A3}$ has similar double robust property as $\hat{\beta}_{A2}$.

3 Method comparisons and asymptotic results

Let $V(X_i, A_i)$ denote the subjects in V with values of (X, A) equal to (X_i, A_i), $n^V(X_i, A_i)$ the number of subjects in $V(X_i, A_i)$, and $n(X_i, A_i)$ the number of subjects with values of (X, A) equal to (X_i, A_i). When X and A are discrete and their dimensionality is reasonably small, the probability $\pi_i = P(\xi_i = 1|X_i, A_i)$ can be estimated by $\hat{\pi}_i = n^V(X_i, A_i)/n(X_i, A_i)$. The conditional expectation $E\left\{S_\beta(Y|X_i)|X_i,A_i\right\}$ can be nonparametrically estimated based on the validation sample,

$$\hat{E}\left\{S_\beta(Y|X_i)|X_i,A_i\right\} = \sum_{j\in V(X_i,A_i)} S_\beta\left(Y_j|X_j\right)/n^V(X_i,A_i), \tag{9}$$

Under MAR I, $\hat{E}\left\{S_\beta(Y|X_i)|X_i,A_i\right\}$ is an unbiased estimator of $E\left\{S_\beta(Y|X_i)|X_i,A_i\right\}$. Now we let $V(Z_i,A_i)$ denote the subjects in V with values of (Z,A) equal to (Z_i,A_i), $n^V(Z_i,A_i)$ the number of subjects in $V(Z_i,A_i)$, and $n(Z_i,A_i)$ the number of subjects in the sample with values of (Z,A) equal to (Z_i,A_i). A nonparametric estimator of $\pi_i^z = P(\xi_i = 1|Z_i,A_i)$ is given by $\hat{\pi}_i^z = n^V(Z_i,A_i)/n(Z_i,A_i)$. A nonparametric estimator of $E\{S_\beta(Y|X_i)|Z_i,A_i\}$ is given by

$$\hat{E}\left\{S_\beta(Y|X_i)|Z_i,A_i\right\} = \sum_{j\in V(Z_i,A_i)} S_\beta\left(Y_j|X_j\right)/n^V(Z_i,A_i). \tag{10}$$

Under MAR II, (Y_i,X_i) is independent of ξ_i conditional on (Z_i,A_i), then $\hat{E}\left\{S_\beta(Y|X_i)|Z_i,A_i\right\}$ is an unbiased estimator of $E\{S_\beta(Y|X_i)|Z_i,A_i\}$.

Proposition 1. *Suppose that $X = (Z,Z^c)$ and A are discrete and their dimensionality is reasonably small. Under the nonparametric estimators $\hat{\pi}_i^z = n^V(Z_i,A_i)/n(Z_i,A_i)$, $\hat{\pi}_i = n^V(X_i,A_i)/n(X_i,A_i)$ and the estimators for the conditional expectation defined in (9) and (10), the estimators $\hat{\beta}_{I1}$, $\hat{\beta}_{E1}$ and $\hat{\beta}_{A1}$ are equivalent, and the estimators $\hat{\beta}_{I2}$, $\hat{\beta}_{E2}$, $\hat{\beta}_{A2}$ and $\hat{\beta}_{A3}$ are equivalent. However, the estimator $\hat{\beta}_{A2}$ is different from $\hat{\beta}_{A1}$ unless Z_i^c is linearly related to Z_i in which case β is not identifiable.*

The results of Proposition 1 are very intriguing since research has shown that the AIPW and the mean score methods are more efficient than the IPW method. It is also intriguing that the AIPW estimators $\hat{\beta}_{A2}$ and $\hat{\beta}_{A3}$ are actually the same estimators, not affected by the validation probability. To further understand these approaches, we investigate the asymptotic properties of these methods where (X,A) are not necessarily discrete. Through the asymptotic analysis, we gain insights about what matters to the efficiency in terms of the selections of the validation sample and the augmentation function.

Suppose that $\tilde{E}\left\{S_\beta(Y|X_i)|X_i,A_i\right\}$ is a consistent parametric/nonparametric estimator of $E_a\left\{S_\beta(Y|X_i)|X_i,A_i\right\}$, where $E_a\left\{S_\beta(Y|X_i)|X_i,A_i\right\}$ is $E\left\{S_\beta(Y|X_i)|X_i,A_i\right\}$ or $E\{S_\beta(Y|X_i)|Z_i,A_i\}$. Let $\pi(X_i,A_i,\psi)$ be the parametric model for the validation probability π_i, where ψ is a q-dimensional parameter. We show in Corollary 2 that the nonparametric estimator of $\pi(X_i,A_i,\psi)$ can also be expressed in the parametric form when (X_i,A_i) are discrete. Let ψ_0 be the true value of ψ. Under MAR I, the MLE $\hat{\psi} = \left(\hat{\psi}_1,\ldots,\hat{\psi}_q\right)$ of $\psi = (\psi_1,\ldots,\psi_q)$ is obtained by maximizing the observed data likelihood,

$$\prod_{i=1}^{n}\{\pi(X_i,A_i,\psi)\}^{\xi_i}\{1-\pi(X_i,A_i,\psi)\}^{1-\xi_i}.$$

The validation probability π_i is estimated by $\tilde{\pi}_i = \pi\left(X_i,A_i,\hat{\psi}\right)$. Then by the standard likelihood based analysis, we have the approximation

$$\hat{\psi} - \psi_0 = n^{-1}\sum_{i=1}^{n}\left(I^\psi\right)^{-1}S_i^\psi + o_p\left(n^{-1/2}\right), \tag{11}$$

where S_i^{ψ} and I^{ψ} are the score vector and information matrix for $\hat{\psi}$ defined by

$$S_i^{\psi} = \frac{(\xi_i - \pi(X_i, A_i, \psi_0))}{\pi(X_i, A_i, \psi_0)(1 - \pi(X_i, A_i, \psi_0))} \frac{\partial \pi(X_i, A_i, \psi_0)}{\partial \psi},$$

$$I^{\psi} = E\left\{\frac{1}{\pi(X_i, A_i, \psi_0)(1 - \pi(X_i, A_i, \psi_0))} \left(\frac{\partial \pi(X_i, A_i, \psi_0)}{\partial \psi}\right)^{\otimes 2}\right\}, \quad (12)$$

where $a^{\otimes 2} = aa'$.

Consider the IPW estimator $\hat{\beta}_I$ obtained by solving the estimating equation

$$U_I = \sum_{i=1}^{n} \frac{\xi_i}{\tilde{\pi}_i} S_{\beta}(Y_i|X_i) \quad (13)$$

and the AIPW estimator $\hat{\beta}_A$ based on solving the estimating equation

$$U_A = \sum_{i=1}^{n} \left[\frac{\xi_i}{\tilde{\pi}_i} S_{\beta}(Y_i|X_i) + \left(1 - \frac{\xi_i}{\tilde{\pi}_i}\right) \tilde{E}\left\{S_{\beta}(Y|X_i)|X_i, A_i\right\}\right]. \quad (14)$$

Theorem 1. *Assume that $P_{\beta}(Y|X)$ and $\pi(X, A, \psi)$ have bounded third-order derivatives in a neighborhood of the true parameters and are bounded away from 0 almost surely, both $-E\left\{(\partial^2/\partial\beta^2)\left(\log P_{\beta}(Y|X)\right)\right\}$ and I^{ψ} are positive definite at the true parameters. Then, under* MAR I,

$$n^{1/2}\left(\hat{\beta}_I - \beta\right) = I^{-1}(\beta) n^{-1/2} \sum_{i=1}^{n} Q_i^I + o_p(1),$$

$$n^{1/2}\left(\hat{\beta}_A - \beta\right) = I^{-1}(\beta) n^{-1/2} \sum_{i=1}^{n} Q_i^A + o_p(1),$$

where $I(\beta) = E\left\{-(\partial^2/\partial\beta^2)\left(\log P_{\beta}(Y|X)\right)\right\} = Var\left(S_{\beta}(Y_i|X_i)\right)$,

$$Q_i^I = \xi_i/\pi_i S_{\beta}(Y_i|X_i) - E\left\{\pi_i^{-2} \xi_i S_{\beta}(Y_i|X_i)\left(\partial\pi(X_i, A_i, \psi_0)/\partial\psi\right)'\right\}(I^{\psi})^{-1} S_i^{\psi}$$

and $Q_i^A = \xi_i/\pi_i S_{\beta}(Y_i|X_i) + (1 - \xi_i/\pi_i) E_a\left\{S_{\beta}(Y|X_i)|X_i, A_i\right\}$.

Both $n^{1/2}\left(\hat{\beta}_I - \beta\right)$ and $n^{1/2}\left(\hat{\beta}_A - \beta\right)$ have asymptotically normal distributions with mean zero and covariances equal to $I^{-1}(\beta) Var\left(Q_i^I\right) I^{-1}(\beta)$ and $I^{-1}(\beta) Var\left(Q_i^A\right) I^{-1}(\beta)$, respectively. Further,

$$Var\left(Q_i^I\right) = Var\left(Q_i^A\right) + Var(B_i + O_i) \quad (15)$$

and

$$Var\left(Q_i^A\right) = I(\beta) + Var\left(\left(1 - \frac{\xi_i}{\pi_i}\right)\left\{S_{\beta}(Y_i|X_i) - E_a\left\{S_{\beta}(Y|X_i)|X_i, A_i\right\}\right\}\right), \quad (16)$$

where $O_i = E\left\{\pi_i^{-2} \xi_i S_{\beta}(Y_i|X_i)\left(\partial\pi(X_i, A_i, \psi_0)/\partial\psi\right)'\right\}(I^{\psi})^{-1} S_i^{\psi}$ *and* $B_i = (1 - \xi_i/\pi_i)$ $E_a\left\{S_{\beta}(Y|X_i)|X_i, A_i\right\}$.

Suppose that the validation probability $\pi_i = P(\xi_i = 1|X_i, A_i)$ depends only on (Z_i, A_i). That is, $\pi_i = \pi_i^z = P(\xi_i = 1|Z_i, A_i)$. Suppose that $\tilde{\pi}_i$ is the MLE of π_i^z under the parametric family $\psi(Z_i, A_i, \psi)$. Let $\hat{\beta}_{A1}$ be the estimator obtained by solving (14) where the augmented term, $\tilde{E}\left\{S_{\beta}(Y|X_i)|X_i, A_i\right\}$, is a consistent parametric/nonparametric estimator of $E\left\{S_{\beta}(Y|X_i)|Z_i, A_i\right\}$. Let $\hat{\beta}_{A2}$ be the estimator obtained by solving (14) where $\tilde{E}\left\{S_{\beta}(Y|X_i)|X_i, A_i\right\}$ is a consistent parametric/nonparametric estimator of $E\left\{S_{\beta}(Y|X_i)|X_i, A_i\right\}$. The following corollary presents the asymptotic results for two AIPW estimators of β, one that corresponds to the augmentation based on a subset,

(Z, A), of observed variables and the other that corresponds to the augmentation based on the full set, (X, A), of the observed variables.

Corollary 1. *Suppose that the validation probability $\pi_i = P(\xi_i = 1|X_i, A_i)$ depends only on (Z_i, A_i). Under the conditions of Theorem 1,*

$$n^{1/2}\left(\hat{\beta}_{A1} - \beta\right) \xrightarrow{\mathcal{D}} N\left(0, I^{-1}(\beta) + I^{-1}(\beta)\Sigma_{A1}(\beta)I^{-1}(\beta)\right), \tag{17}$$

and

$$n^{1/2}\left(\hat{\beta}_{A2} - \beta\right) \xrightarrow{\mathcal{D}} N\left(0, I^{-1}(\beta) + I^{-1}(\beta)\Sigma_{A2}(\beta)I^{-1}(\beta)\right), \tag{18}$$

where $\Sigma_{A1}(\beta) = E\left[((1 - \pi_i^z)/\pi_i^z)Var\{S_\beta(Y_i|X_i)|Z_i, A_i\}\right]$ and $\Sigma_{A2}(\beta) = E\left[((1 - \pi_i^z)/\pi_i^z)\right.$ $\left. Var\left\{S_\beta(Y_i|X_i)|X_i, A_i\right\}\right]$. The asymptotic variance of $\hat{\beta}_{A2}$ is smaller than the asymptotic variance of $\hat{\beta}_{A1}$ if the covariates Z_i are a proper subset of X_i.

Suppose that (Z, A) are discrete taking values (z, a) in a set $\mathcal{Z}$ of finite number of values. If the number of parameters in ψ equals the number of values $\psi_{z,a} = P(\xi_i = 1|Z_i = z, A_i = a)$ for all distinct pairs (z, a), then $\psi = \{\psi_{z,a}\}$ and $\pi(z, a, \psi) = \psi_{z,a}$. Further, $\frac{\partial \pi(z,a,\psi_0)}{\partial \psi}$ can be viewed as a column vector with 1 in the position for $\psi_{z,a}$ and 0 elsewhere. The information matrix I^ψ defined in (12) has the expression,

$$I^\psi = \sum_{z,a} \rho(z, a) \left\{ \frac{1}{\pi(z, a, \psi_0)(1 - \pi(z, a, \psi_0))} \frac{\partial \pi(z, a, \psi_0)}{\partial \psi} \left(\frac{\partial \pi(z, a, \psi_0)}{\partial \psi} \right)' \right\},$$

where $\rho(z, a) = P(Z_i = z, A_i = a)$. It follows that I^ψ is a diagonal matrix and its inverse matrix is also diagonal. The MLE $\hat{\psi}_{z,a} = n^V(z, a)/n(z, a)$ is in fact the nonparametric estimator for $\psi_{z,a}$ based on the proportion of validated samples in the category specified by (z, a). The equation (11) can be expressed as

$$\hat{\psi}_{z,a} - \pi(z, a, \psi_0) = n^{-1}\frac{n^V(z, a) - n(z, a)\pi(z, a, \psi_0)}{\rho(z, a)} + o_p\left(n^{-1/2}\right),$$

for $(z, a) \in \mathcal{Z}$.

By Threom 1, the possible efficiency gain of the AIPW estimator over the IPW estimator is shown through the equation (15). The AIPW estimator is more efficient unless $\mathrm{Var}(B_i + O_i) = 0$. In particular, from the proof of Theorem 1, we have

$$n^{-1/2}U_A = n^{-1/2}\sum_{i=1}^{n} Q_i^A + o_p(1) \tag{19}$$

$$n^{-1/2}U_I = n^{-1/2}\sum_{i=1}^{n} Q_i^A - n^{-1/2}\sum_{i=1}^{n}(B_i + O_i) + o_p(1), \tag{20}$$

where B_i and O_i are defined following (16). The following corollary presents the analysis of the term $n^{-1/2}\sum_{i=1}^{n}(B_i + O_i)$ when (Z_i, A_i) are discrete to understand how efficiency may be gained from the AIPW estimator over the IPW estimator.

Corollary 2. *Under the conditions of Theorem 1, suppose that $X = (Z, Z^c)$ and (Z, A) are discrete taking values (z, a) in a set $\mathcal{Z}$ of finite number of values. Suppose that the* *validation probability* $\pi_i = P(\xi_i = 1|X_i, A_i)$ *only depends on* (Z_i, A_i) *and* $\psi_{z,a} = P(\xi_i = 1|Z_i = z, A_i = a)$ *is estimated nonparametrically by* $\hat{\psi}_{z,a} = n^V(z,a)/n(z,a)$. *Then*

$$\begin{aligned} & n^{-1/2}\sum_{i=1}^{n}(O_i + B_i) \\ &= -\sum_{z,a} n^{-1/2}\sum_{j=1}^{n}\frac{\xi_j - \pi(z,a,\psi_0)}{\pi(z,a,\psi_0)} I(Z_j = z, A_j = a) \\ &\quad \left[E\{S_\beta(Y_j|X_j)|Z_j = z, A_j = a\} - E\left\{S_\beta(Y_j|X_j)|Z_j = z, Z_j^c, A_j = a\right\}\right]. \end{aligned} \tag{21}$$

By Corollary 2, (19) and (20), $\hat{\beta}_A$ is more efficient than $\hat{\beta}_I$ unless $Var\{S_\beta(Y_j|X_j)|Z_j = z, A_j = a\} = 0$ for all (z,a) for which $P(Z_i = z, A_i = a) \neq 0$. If $X = Z$ and the validation probability $\pi_i = P(\xi_i = 1|X_i, A_i)$ is nonparametrically estimated with the cell frequencies $\hat{\psi}_{z,a} = n^V(z,a)/n(z,a)$, then $\hat{\beta}_A$ and $\hat{\beta}_I$ are asymptotically equivalent.

Remark Consider the estimators of β obtained based on the estimating equation (1) corresponding to different choices of W_i given in (2) to (8). If (Z, A) are discrete and the validation probability $\pi_i^z = P(\xi_i = 1|Z_i, A_i)$ is estimated nonparametrically by the cell frequency, then by Theorem 1 and Corollary 2, $\hat{\beta}_{A1}$ and $\hat{\beta}_{I1}$ have same asymptotic normal distributions as long as $\hat{E}[S_\beta(Y|X_i)|Z_i, A_i]$ is a consistent estimator of $E[S_\beta(Y|X_i)|Z_i, A_i]$. But $\hat{\beta}_{A2}$ is more efficient than $\hat{\beta}_{I1}$ as long as $\hat{E}[S_\beta(Y|X_i)|X_i, A_i]$ is a consistent estimator of $E[S_\beta(Y|X_i)|X_i, A_i]$ since $\text{Var}(B_i + O_i)$ is not zero by (21). These results are not affected by whether $E[S_\beta(Y|X_i)|Z_i, A_i]$ and $E[S_\beta(Y|X_i)|X_i, A_i]$ are estimated nonparametrically or based on some parametric models. In addition, by Theorem 1, Corollary 1 and 2, $\hat{\beta}_{A3}$ and $\hat{\beta}_{I2}$ have the same asymptotic normal distributions as long as $\hat{E}[S_\beta(Y|X_i)|X_i, A_i]$ is a consistent estimator of $E[S_\beta(Y|X_i)|X_i, A_i]$.

4 Poisson regression using the automated data with missing outcomes

Many medical and public health data are available only in aggregated format, where the variables of interest are aggregated counts without being available at individual levels. Many existing statistical methods for missing data require observations at individual levels. Applying the missing data methods presented in Section 3, we derive some estimation procedures for the Poisson regression model with missing outcomes based on auxiliary outcomes and a validated sample for true outcomes. Further, we show that, by stratifying based on a set of discrete variables, the proposed statistical procedure can be formulated so that it can be used to analyze automated records which do not contain observations at individual levels, only summarized information at categorical levels.

Let Y represent the number of events occurring in the time-exposure interval $[0, T]$ and Z the covariates. We consider the Poisson regression model,

$$P(Y = y|Z, T) = \exp\left\{-T\exp(\beta'Z)\right\}\left\{T\exp(\beta'Z)\right\}^y/y!, \tag{22}$$

where Z is a vector of $k+1$ covariates and β a vector of $k+1$ regression coefficients. In practice, the exact number of true events may not be available for all subjects. We may instead have the number of possible events, namely, the auxiliary events. For example, in the study of vaccine adverse events associated with childhood immunizations, the number of auxiliary events A for MAARI is collected based on ICD-9 codes through hospital records. Further diagnosis may indicate that some of these events are false events. The number of true vaccine adverse events, Y, can only be confirmed for the subjects in the validation set V. Suppose that Z is the vaccination status, 1 for the vaccinated and 0 for the unvaccinated. Then, under Poisson regression, $\exp(\beta)$ is the relative rate of event occurrence per unit time of the exposed versus unexposed. We assume that the number of automated events A can be expressed as $A = Y + W$, where W is the number of false events independent of Y conditional on (Z, T). Suppose that W follows the Poisson regression model

$$P(W = w|Z, T) = \exp\left\{-T\exp(\gamma'Z)\right\}\left\{T\exp(\gamma'Z)\right\}^{w}/w!, \tag{23}$$

where $\gamma' = \left(a_0, a_1, \gamma_1, \cdots, \gamma_{k-1}\right)$.

We apply the missing data methods introduced in Section 3 on model (22). The variables (Z_i, T_i, Y_i, A_i) are observed for the validation sample V and (Z_i, T_i, A_i) are observed for the nonvalidation sample $\bar{V}$. While the covariate Z can be considered as categorical, it is natural to consider the exposure time T as a continuous variable. We assume that the validation probability depends only on the stratification of (Z, A). That is, the validation sample is a stratified random sample by the categories defined by (Z, A). Of those estimators discussed in Section 2, there are only two different estimators, $\hat{\beta}_{I1}$ and $\hat{\beta}_{A2}$. We show in Section 4.3 that the proposed method can be formulated so that it can be used to analyze the automated records with missing outcomes. First we derive the explicit estimation procedures for $\hat{\beta}_{I1}$ and $\hat{\beta}_{A2}$ and their variance estimators under model (22).

4.1 Inverse probability weighting estimation

We adopt all notations introduced in Section 3. In particular, let $\pi_i^z = P(\xi_i = 1|Z_i, A_i)$ and $\hat{\pi}_i^z = n^V(Z_i, A_i)/n(Z_i, A_i)$. Let $X = (Z, T)$ and $X_i = (Z_i, T_i)$ to be consistent with earlier notations. The score function for subject i under model (22) is $S_\beta(Y_i|X_i) = Z_i'(Y_i - T_i\exp(\beta'Z_i))$. The estimator $\hat{\beta}_{I1}$ is obtained by solving $\sum_{i=1}^n(\xi_i/\hat{\pi}_i^z)S_\beta(Y_i|X_i) = 0$, where $S_\beta(Y_i|X_i) = Z_i'(Y_i - T_i\exp(\gamma'Z_i))$. By Corollary 1, $\sqrt{n}(\hat{\beta}_{I1} - \beta)$ converges in distribution to a normal distribution with mean zero and the variance matrix $I^{-1}(\beta) + I^{-1}(\beta)\Sigma_{A1}(\beta)I^{-1}(\beta)$, where $\Sigma_{A1}(\beta) = E\left[((1-\pi_i^z)/\pi_i^z)\text{Var}\left\{S_\beta(Y_i|X_i)|Z_i, A_i\right\}\right]$.

The information matrix $I(\beta) = E(Z_iZ_i'T_i\exp(\beta'Z_i)) = \sum_z P(Z_i = z)zz'\exp(\beta'z)$ $E(T_i|Z_i = z)$ can be estimated by $\hat{I}(\hat{\beta})$ which is obtained by replacing β with $\hat{\beta}_{I1}$, $P(Z_i = z)$ by the sample proportion of the event $\{Z_i = z\}$, and $E(T_i|Z_i = z)$ with the sample average exposure time for those with covariates $Z_i = z$. The matrix $\Sigma_{A1}(\beta)$ can be estimated by

$$\hat{\Sigma}_{A1}(\hat{\beta}) = \sum_{a,z}\hat{\rho}(a,z)\frac{1-\hat{\rho}^v(a,z)}{\hat{\rho}^v(a,z)}\widehat{\text{Var}}\{S_\beta(Y|X)|A = a, Z = z\}, \tag{24}$$

where $\hat{\rho}(a,z)$ is the estimator of $P\{A_i = a, Z_i = z\}$, $\hat{\rho}^v(a,z)$ is the estimator of $P\{i \in V|A_i = a, Z_i = z\}$, and $\widehat{\text{Var}}\left\{S_\beta(Y|X)|A = a, Z = z\right\}$ is an estimator of $\text{Var}\left\{S_\beta(Y_i|X_i)|Z_i, A_i\right\}]$ which is derived in the following.

Since A is observed for all subjects, W can be determined if Y is known, and undetermined otherwise. The IPW estimator, $\hat{\gamma}_{I1}$, of γ can be estimated by solving the equation $\sum_{i=1}^n(\xi_i/\hat{\pi}_i^z)S_\gamma(W_i|X_i) = 0$, where $S_\gamma(W_i|X_i) = Z_i'(W_i - T_i\exp(\gamma'Z_i))$. The conditional distribution of Y given $A = a$, T, and $Z = z$ is Binomial (a, p_z), where $p_z = \exp(\beta'z)/(\exp(\beta'z) + \exp(\gamma'z))$. Since this conditional distribution does not depend on T, the outcome Y and T are conditionally independent given (A, Z). Therefore, $\mathrm{Var}\left\{S_\beta(Y|X)|A, Z\right\} = ZZ'\left\{\mathrm{Var}(Y|A, Z) + \exp(2\beta'Z)\mathrm{Var}(T|A, Z)\right\}$. The variance $\mathrm{Var}(Y|A = a, Z = z)$ can be estimated by $a\hat{p}_z(1 - \hat{p}_z)$, where $\hat{p}_z = \exp(\hat{\beta}'z)/\left\{\exp(\hat{\beta}'z) + \exp(\hat{\gamma}'z)\right\}$, and $\mathrm{Var}(T|A = a, Z = z) = E(T^2|A = a, Z = z) - \{E(T|A = a, Z = z)\}^2$ can be estimated nonparametrically using the first and the second sample moments conditional on each category with $A = a$ and $Z = z$.

4.2 Augmented inverse probability weighted estimation

Under the assumption that W follows the Poisson regression model (23) and is independent of Y conditional on (Z, T), $E\left\{S_\beta(Y|X)|Z, T, A\right\} = AZ'\frac{\exp(\beta'Z)}{\exp(\beta'Z)+\exp(\gamma'Z)} - TZ'\exp(\beta'Z)$. Let $\hat{E}\left\{S_\beta(Y|X_i)|X_i, A_i\right\}$ be the estimator of $E\left\{S_\beta(Y|X_i)|X_i, A_i\right\}$ for a given β by substituting γ by its estimator $\hat{\gamma}_{I1}$ of Section 4.1. Then the estimator $\hat{\beta}_{A2}$ is obtained by solving

$$\sum_{i=1}^{n} \frac{\xi_i}{\hat{\pi}_i^z} S_\beta(Y_i|X_i) + \left(1 - \frac{\xi_i}{\hat{\pi}_i^z}\right)\hat{E}\left\{S_\beta(Y|X_i)|X_i, A_i\right\} = 0. \tag{25}$$

By Corollary 1, $\sqrt{n}(\hat{\beta}_{A2} - \beta)$ converges in distribution to a normal distribution with mean zero and the variance matrix where $I^{-1}(\beta)+I^{-1}(\beta)\Sigma_{A2}(\beta)I^{-1}(\beta)$, where $\Sigma_{A2}(\beta) = E\left[((1-\pi_i^z)/\pi_i^z)\mathrm{Var}\{S_\beta(Y_i|X_i)|X_i, A_i\}\right]$. The information matrix $I(\beta)$ can be estimated by $\hat{I}(\hat{\beta})$ given in Section 4.1. The conditional variance $\mathrm{Var}\left\{S_\beta(Y|X)|Z = z, T, A = a\right\} = ap_z(1 - p_z)z^{\otimes 2}$ can be estimated by $a\hat{p}_z(1 - \hat{p}_z)$, where $\hat{p}_z = \exp(\hat{\beta}'z)/(\exp(\hat{\beta}'z) + \exp(\hat{\gamma}'z))$. It follows that $\Sigma_{A2}(\beta)$ can be consistently estimated by

$$\hat{\Sigma}_{A2}(\beta) = \sum_{a,z} \hat{\rho}(a, z)\frac{1 - \hat{\rho}^v(a, z)}{\hat{\rho}^v(a, z)} a\hat{p}_z(1 - \hat{p}_z)z^{\otimes 2},$$

where $\hat{\rho}(a, z)$ is the estimator of $P\{A_i = a, Z_i = z\}$ and $\hat{\rho}^v(a, z)$ is the estimator of $P\{i \in V|A_i = a, Z_i = z\}$.

4.3 Estimation using the automated data

This section formulates the missing data estimation procedure for (22) based on the automated (summarized) information at categorical levels defined by relevant covariates of the model. In particular, we show that $\hat{\beta}_{I1}$ and $\hat{\beta}_{A2}$ and their variance estimators can be formulated using the automated data at categorical levels.

In many applications it is convenient to write $Z = (1, Z_{(1)}, Z_{(2)})$ and $\beta = (b_0, b_1, \theta')'$, where $Z_{(1)}$ is the treatment indicator ($Z_{(1)} = 1$ for the exposed group and $Z_{(1)} = 0$ for the unexposed group) and $Z_{(2)} = (\eta_1, \cdots, \eta_{k-1})'$ as the other covariates, and $\theta = (\theta_1, \cdots, \theta_{k-1})'$. For the applications involving the automated data records, we let $\eta_1, \cdots, \eta_{k-1}$ be $k - 1$ dummy variables representing k groups. Without loss of generality, we choose the kth group as the reference group, $\eta_1 = 1$, $\eta_2 = 0$, $\cdots$, $\eta_{k-1} = 0$ for group 1, $\eta_1 = 0$, $\eta_2 = 1$, $\cdots$, $\eta_{k-1} = 0$ for group 2, so on and $\eta_1 = 0$, $\eta_2 = 0$, $\cdots$, $\eta_{k-1} = 0$ for group k. Thus each value of Z denotes a category which can be represented by (l, m) for $l = 0, 1$ and $m = 1, \cdots, k$. This correspondence is denoted by $Z \simeq (l, m)$ for convenience. For $l = 0, 1$ and $m = 1, \cdots, k - 1$, category (l, m) is defined by Z with $Z_{(1)} = l$, $\eta_m = 1$ and $\eta_j = 0$ for $j \neq m, j = 1, \ldots, k$, and category (l, k) is defined by $Z_{(1)} = l$ and $\eta_j = 0$ for $j = 1, \ldots, k - 1$. Under model (22), the expected number of events for a subject in category (l, m) with the time-exposure interval $[0, T]$ is $T\exp(b_{lm})$, for $l = 0, 1$ and

$m = 1, \cdots, k$, where $b_{1k} = b_0 + b_1$, $b_{0k} = b_0$, $b_{1m} = b_0 + b_1 + \theta_m$ and $b_{0m} = b_0 + \theta_m$ for $1 \leq m \leq k-1$. The parameter b_1 represents the log-relative rate of the exposed versus the unexposed adjusted for other factors.

The following notations are used to show that the estimators of β and their variance estimators can be calculated using the automated information at the categorical levels. Let $V(a,l,m)$ denote the set of subjects in V with $(A = a, Z \simeq (l,m))$, $V(l,m)$ for the set of subjects in V with $(Z \simeq (l,m))$, n_{alm} for the number of subjects with $(A = a, Z \simeq (l,m))$, n^v_{alm} for the number of subjects in $V(a,l,m)$, n^v_{lm} for the number of subjects in $V(l,m)$, $\lambda_{alm} = n_{alm}/n^v_{alm}$, y_{alm} for the number of events for subjects in $V(a,l,m)$, y_{lm} for the number of events for subjects in $V(l,m)$, t_{alm} for the total exposure time for subjects with $(A = a, Z \simeq (l,m))$, $t_{2,alm}$ for the total squared exposure time for subjects with $(A = a, Z \simeq (l,m))$, t_{lm} for the total exposure time for subjects with $Z \simeq (l,m)$, α_{lm} for the number of automated events for subjects with $Z \simeq (l,m)$.

Estimation with $\hat{\beta}_{I1}$ using the automated data. The validation probability π^z_i can be estimated by $1/\lambda_{alm}$ when $A_i = a$, $Z_i \simeq (l,m)$. It can be shown that the estimating equation for $\hat{\beta}_{I1}$ is equivalent to the following nonlinear equations for $\{b_{lm}$, for $l = 0, 1$, $m = 1, \cdots, k\}$,

$$\sum_{m=1}^{k} \left(\sum_{a \in A} y_{alm}\lambda_{alm} - e^{b_{lm}} \sum_{a \in A} t_{alm}\lambda_{alm} \right) = 0,$$

$$\sum_{l=0,1} \left(\sum_{a \in A} y_{alm}\lambda_{alm} - e^{b_{lm}} \sum_{a \in A} t_{alm}\lambda_{alm} \right) = 0,$$

for $l = 0, 1$ and $m = 1, \ldots, k-1$. When $k > 1$, the equations have no explicit solutions.

In the following, we show that the asymptotic variance of $\hat{\beta}_{I1}$ can be consistently estimated by only using the automated information at categorical levels. The information matrix is a $(k+1) \times (k+1)$ symmetric matrix given by

$$I(\beta) = E(Z_i Z_i' T_i \exp(\beta' Z_i))$$
$$= \begin{bmatrix} \sum_{l,m} q_{lm} & \sum_{m=1}^{k} q_{1m} & q_{11} + q_{01} & \cdots & q_{1r} + q_{0r} \\ \sum_{m=1}^{k} q_{1m} & \sum_{m=1}^{k} q_{1m} & q_{11} & \cdots & q_{1r} \\ q_{11} + q_{01} & q_{11} & q_{11} + q_{01} & \cdots & 0 \\ \vdots & \vdots & \vdots & \ddots & \vdots \\ q_{1r} + q_{0r} & q_{1r} & 0 & \cdots & q_{1r} + q_{0r} \end{bmatrix},$$

where $r = k-1$ and $q_{lm} = E(T_i e^{b_{lm}} I\{\text{individual } i \text{ in category } (l,m)\})$. The consistent estimator, $\hat{I}(\hat{\beta})$, of $I(\beta)$ is thus obtained by replacing q_{lm} with $\exp(\hat{b}_{lm}) t_{lm}/n$.

Under model (23), the expected number of false events for a subject in category (l,m) with the time-exposure interval $[0, T]$ is $T \exp(d_{lm})$, for $l = 0, 1$ and $m = 1, \cdots, k$, where $d_{1k} = a_0 + a_1$, $d_{0k} = a_0$, $d_{1m} = a_0 + a_1 + \gamma_m$ and $d_{0m} = a_0 + \gamma_m$ for $1 \leq m \leq k-1$. The conditional distribution of Y given $A = a$, T, and $Z \simeq (l,m)$ is Binomial (a, p_{lm}), where $p_{lm} = \exp(b_{lm})/(\exp(b_{lm}) + \exp(d_{lm}))$ for $a \geq 1$. Then $\text{Var}(Y|A = a, Z \simeq (l,m))$ can be estimated by $a\hat{p}_{lm}(1 - \hat{p}_{lm})$, where $\hat{p}_{lm} = e^{\hat{b}_{lm}}/(e^{\hat{b}_{lm}} + e^{\hat{d}_{lm}})$, and $\text{Var}(T|A = a, Z \simeq (l,m))$

can be estimated by $v_{a,l,m} = t_{2,a,l,m}/n_{alm} - (t_{alm}/n_{alm})^2$. By (24) and the discussion that follows, $\Sigma_{A1}(\beta)$ can be estimated by

$$\hat{\Sigma}_{A1}(\hat{\beta}) = \sum_{a,l,m} \hat{\rho}(a,l,m)\frac{1-\hat{\rho}^v(a,l,m)}{\hat{\rho}^v(a,l,m)} G_{lm}\{a\hat{p}_{lm}(1-\hat{p}_{lm}) + v_{a,l,m}\}, \tag{26}$$

where $\hat{\rho}(a,l,m) = n_{alm}/n$, $\hat{\rho}^v(a,l,m) = n^v_{alm}/n_{alm}$ and G_{lm} be the value of $G_i = z_i^{\otimes 2}$ when subject i belongs to the category (l,m). Hence the covariance matrix of $\hat{\beta}_{I1}$ can be estimated by $\hat{I}^{-1}(\hat{\beta}) + \hat{I}^{-1}(\hat{\beta})\hat{\Sigma}_{A1}(\hat{\beta})\hat{I}^{-1}(\hat{\beta})$ using the automated data.

Estimation with $\hat{\beta}_{A2}$ using the automated data. The estimating equations (25) are equivalent to the following nonlinear equations for $\{b_{lm},$ for $l = 0,1,\ m = 1,\cdots,k\}$,

$$\sum_{m=1}^{k}\left\{\sum_{a\in A} y_{alm}\lambda_{alm} - e^{b_{lm}}t_{lm} + \frac{e^{b_{lm}}}{e^{b_{lm}}+e^{\hat{d}_{lm}}}\left(\alpha_{lm} - \sum_{a\in A} a n_{alm}\lambda_{alm}\right)\right\} = 0,$$

$$\sum_{l=0,1}\left\{\sum_{a\in A} y_{alm}\lambda_{alm} - e^{b_{lm}}t_{lm} + \frac{e^{b_{lm}}}{e^{b_{lm}}+e^{\hat{d}_{lm}}}\left(\alpha_{lm} - \sum_{a\in A} a n_{alm}\lambda_{alm}\right)\right\} = 0,$$

for $l = 0,1$ and $m = 1,\ldots,k-1$.

Since $\text{Var}\left\{S_\beta(Y|X)|Z \simeq (l,m), T, A = a\right\} = ap_{lm}(1-p_{lm})G_{lm}$, $\Sigma_{A2}(\beta)$ can be consistently estimated by

$$\hat{\Sigma}_{A2}(\hat{\beta}) = \sum_{a,l,m} \hat{\rho}(a,l,m)\frac{1-\hat{\rho}^v(a,l,m)}{\hat{\rho}^v(a,l,m)} a\hat{p}_{lm}(1-\hat{p}_{lm})G_{lm}.$$

Hence the covariance matrix of $\hat{\beta}_{A2}$ can be estimated by $\hat{I}^{-1}(\hat{\beta}) + \hat{I}^{-1}(\hat{\beta})\hat{\Sigma}_{A2}(\hat{\beta})\hat{I}^{-1}(\hat{\beta})$ using the automated data.

Remark In the special case where $\rho(\alpha,l,m) \approx 0$ for $\alpha \geq 2$, a much simpler formula for the variance estimator of the log relative risk can be derived. For example in the vaccine safety study, the adverse-event rate is very small. Let

$$w_m = \frac{\alpha_{0m}\alpha_{1m}y_{0m}y_{1m}}{\alpha_{0m}y_{0m}n^v_{1m} + \alpha_{1m}y_{1m}n^v_{0m}} \Bigg/ \sum_{m=1}^{k} \frac{\alpha_{0m}\alpha_{1m}y_{0m}n^v_{1m}}{\alpha_{0m}y_{0m}n^v_{1m} + \alpha_{1m}y_{1m}n^v_{0m}}.$$

Then an estimate of variance of $\hat{b}_1$ is given by

$$\widehat{\text{Var}(\hat{b}_1)} = \sum_{m=1}^{k} w_m^2\left(\frac{1}{y_{1m}} - \frac{1}{n^v_{1m}} + \frac{1}{\alpha_{1m}} + \frac{1}{y_{0m}} - \frac{1}{n^v_{0m}} + \frac{1}{\alpha_{0m}}\right), \tag{27}$$

which is the weighted sum of the estimated variances for the estimated log relative rate of the exposed versus the unexposed over k groups. The details of deviation are given in the Appendix B.

5 A simulation study

We conduct a simulation study to examine the finite sample performance of the IPW estimator $\hat{\beta}_{I1}$ and the AIPW estimator $\hat{\beta}_{A2}$. We consider the Poisson regression model (22). The covariates Z_1 and Z_2 are generated from the Bernoulli distributions with the probability of success equals to 0.4 and 0.5 respectively. The exposure time T is generated from a uniform distribution on $[0,10]$. Given $Z = (Z_1, Z_2)$ and T, the outcome variable Y follows a Poisson distribution with mean $T \exp(b_0 + b_1 Z_1 + \theta Z_2)$ where $b_0 = -0.5$, $b_1 = -0.8$ and $\theta = -0.6$, and W follows a Poisson distribution with mean $T \exp(a_0 + a_1 Z_1 + \gamma Z_2)$ where $a_0 = -1.3$, $a_1 = -1.1$, $\gamma = -1$. We set $A = Y + W$.

Four models for the validation sample are considered. Under Model 1, the validation sample is a simple random sample with probability $\pi_i = 0.4$. Model 2 considers $\pi_i = 0.6$. In Model 3, the validation probability only depends on A through the logistic regression model $\text{logit}\{\pi_i(X, A)\} = A - 0.5$ where $X = (Z, T)$. In Model 4, the validation probability depends on A and Z_1 through the logistic regression model $\text{logit}\{\pi_i(X, A)\} = A - Z_1 - 0.5$.

Tables 1 and 2 present the simulation results for $n = 50, 100, 300, 500$ and 800. Each entry of the tables is based on 1000 simulation runs. Tables 1 and 2 summarize the bias (Bias), the empirical standard error (SSE), the average of the estimated standard error (ESE), and the empirical coverage probability (CP) of 95% confidence intervals of $\hat{\beta}_{I1}$ and $\hat{\beta}_{A2}$ for $\beta = (b_0, b_1, \theta)$. We also compare the performance of the estimators $\hat{\beta}_{I1}$ and $\hat{\beta}_{A2}$ with the complete-case (CC) estimator $\hat{\beta}_C$ obtained by simply deleting subjects with missing values of Y_i. As a gold standard, we present the estimation results for the full data where all the values of Y_i are fully observed. Table 1 presents the results under Model 1 and 2, and Table 2 shows the results under Model 3 and 4.

Table 1 shows that under Model 1 and Model 2, the bias of all estimators is very small at a level comparable with that of the full data estimator. The bias decreases with increased sample size and the increased level of the validation probability. The empirical standard errors are in good agreement with the corresponding estimated standard errors, except for the IPW estimator when $n \leq 100$ and $\pi \leq 0.6$. Among them, AIPW has the smallest standard errors for all parameters and sample sizes concerned. The coverage probabilities of the confidence intervals for b_0, b_1 and θ are close to the nominal level 95%. When the sample size and the validation probability are both small, for example, $n = 50$ and $\pi = 0.4$, the IPW has large bias and is unstable but the AIPW still performs well.

Table 2 gives the results under Model 3 and Model 4. The bias remains small for $\hat{\beta}_{I1}$ and $\hat{\beta}_{A2}$. The empirical standard errors are also close to the corresponding estimated standard errors. The coverage probabilities remain close to the nominal level 95% for all IPW and AIPW estimators. However, the complete-case estimator yields larger bias and incorrect coverage probability because of the association between the validation probability and the auxiliary variable A and/or the covariate Z_1, in which case the missing is not missing completely at random. The AIPW performs better than IPW with smaller standard errors.

6 An Application

A community-based, nonrandomized, open-label influenza vaccine (CAIV-T) study was conducted in Temple-Belton, Texas during the 2000-2001 influenza season. The total 11,606 healthy children aged 18 months - 18 years were involved in this study and about 20% of them received a single dose of CAIV-T in 2000. The primary clinical outcome was based on an nonspecific case definition called medically attended acute respiratory infection (MAARI), which included all International Classification of Diseases, Ninth Revision, Clinical Modification diagnoses codes (ICD-9 codes 381-383, 460-487) for upper and lower respiratory tract infections, otitis media and sinusitis. MAARI outcomes and demographic data were extracted from the Scott & White Health Plan administrative database. For each visit, one or two International Classification of Diseases, Ninth Revision, Clinical Modification diagnosis codes were listed. Visits for which asthma diagnosis codes alone

Table 1 Simulation comparison of the IPW estimator $\hat{\beta}_{I1}$, the AIPW estimator $\hat{\beta}_{A2}$ and the complete-case (CC) estimator $\hat{\beta}_C$ under various sample sizes and selection probabilities

		b_0				b_1				θ			
n		Bias	SSE	ESE	CP	Bias	SSE	ESE	CP	Bias	SSE	ESE	CP
						Model 1: $\pi_i = .4$							
50	IPW	-.0415	.3561	.1839	.851	-.2175	1.6737	.3354	.864	-.1610	1.2201	.2962	.847
	AIPW	-.0110	.2213	.1664	.890	-.0062	.3099	.2873	.943	-.0186	.3076	.2551	.929
	CC	-.0246	.3398	.2738	.938	-.1515	1.6082	.4730	.968	-.1038	1.1709	.4187	.959
100	IPW	-.0650	.1815	.1404	.870	-.0548	.3120	.2458	.891	-.0249	.2653	.2161	.898
	AIPW	-.0094	.1376	.1187	.914	-.0024	.2284	.1988	.926	.0027	.1994	.1780	.925
	CC	-.0240	.1728	.1685	.948	-.0086	.3086	.2981	.960	.0031	.2556	.2581	.946
300	IPW	-.0368	.0936	.0874	.931	-.0209	.1535	.1460	.946	-.0022	.1419	.1286	.929
	AIPW	-.0027	.0732	.0712	.946	-.0028	.1233	.1165	.940	.0005	.1130	.1046	.938
	CC	-.0092	.0919	.0935	.960	-.0012	.1627	.1634	.952	.0040	.1438	.1432	.952
500	IPW	-.0183	.0698	.0671	.938	-.0172	.1159	.1128	.943	-.0083	.1069	.0993	.933
	AIPW	.0022	.0566	.0550	.936	-.0022	.0956	.0902	.943	-.0068	.0867	.0811	.930
	CC	.0006	.0704	.0716	.949	-.0059	.1268	.1255	.949	-.0046	.1135	.1103	.942
800	IPW	-.0126	.0538	.0527	.942	-.0134	.0862	.0889	.950	-.0029	.0759	.0779	.947
	AIPW	.0011	.0433	.0435	.952	-.0047	.0720	.0713	.956	-.0020	.0638	.0640	.951
	CC	.0002	.0562	.0565	.948	-.0051	.0974	.0990	.958	-.0013	.0844	.0869	.958
						Model 2: $\pi_i = .6$							
50	IPW	-.0316	.2079	.1714	.926	-.0934	.8426	.3112	.944	-.0563	.3320	.2690	.937
	AIPW	-.0072	.1723	.1591	.941	-.0105	.2893	.2772	.950	-.0172	.2653	.2440	.948
	CC	-.0126	.1973	.1949	.959	-.0594	.8369	.3512	.967	-.0278	.3213	.3044	.959
100	IPW	-.0366	.1399	.1259	.926	-.0420	.2363	.2192	.944	-.0100	.2103	.1911	.925
	AIPW	-.0121	.1206	.1133	.941	-.0107	.2069	.1921	.944	.0078	.1764	.1700	.940
	CC	-.0142	.1370	.1345	.947	-.0216	.2379	.2370	.961	.0030	.2103	.2072	.949
300	IPW	-.0138	.0742	.0728	.944	-.0194	.1267	.1250	.957	-.0049	.1064	.1096	.964
	AIPW	-.0030	.0650	.0651	.948	-.0044	.1136	.1093	.949	.0005	.0960	.0974	.956
	CC	-.0017	.0763	.0759	.946	-.0118	.1345	.1328	.951	-.0035	.1147	.1169	.957
500	IPW	-.0069	.0571	.0555	.945	-.0096	.0946	.0965	.947	-.0094	.0866	.0844	.953
	AIPW	.0029	.0495	.0496	.942	-.0032	.0856	.0841	.947	-.0076	.0757	.0749	.955
	CC	.0013	.0577	.0581	.947	-.0034	.1024	.1019	.949	-.0086	.0906	.0899	.954
800	IPW	-.0072	.0437	.0438	.954	-.0069	.0754	.0763	.956	-.0025	.0692	.0664	.947
	AIPW	-.0011	.0401	.0393	.951	-.0019	.0693	.0665	.943	-.0015	.0626	.0590	.931
	CC	-.0012	.0452	.0460	.958	-.0026	.0805	.0806	.952	-.0024	.0723	.0709	.952
						Full data: $\pi_i = 1$							
50		-.0079	.1510	.1466	.952	-.0182	.2691	.2618	.948	-.0104	.2264	.2263	.957
100		-.0079	.1068	.1024	.943	-.0075	.1841	.1798	.948	-.0039	.1560	.1577	.949
300		-.0019	.0596	.0583	.950	-.0081	.1032	.1023	.936	.0001	.0934	.0898	.932
500		.0006	.0452	.0450	.951	-.0041	.0783	.0788	.950	.0014	.0656	.0693	.960
800		-.0004	.0343	.0356	.951	.0025	.0612	.0622	.938	-.0006	.0532	.0547	.955

Table 2 Simulation comparison of the IPW estimator $\hat{\beta}_{I1}$, the AIPW estimator $\hat{\beta}_{A2}$ and the complete-case (CC) estimator $\hat{\beta}_C$ under various sample sizes and selection probabilities

		b_0				b_1				θ			
n		Bias	SSE	ESE	CP	Bias	SSE	ESE	CP	Bias	SSE	ESE	CP
						Model 3							
50	IPW	.0081	.1609	.1502	.949	-.0034	.5535	.2790	.954	-.0116	.2592	.2400	.963
	AIPW	-.0070	.1543	.1486	.950	-.0134	.2715	.2690	.958	-.0185	.2364	.2330	.955
	CC	.0230	.1529	.1504	.938	.0798	.5441	.2835	.940	.0648	.2367	.2414	.952
100	IPW	-.0052	.1145	.1077	.948	-.0001	.2073	.2014	.959	.0030	.1789	.1724	.948
	AIPW	-.0124	.1077	.1041	.947	-.0085	.1869	.1840	.957	.0050	.1636	.1606	.948
	CC	.0221	.1074	.1054	.939	.1023	.1830	.1937	.924	.0828	.1625	.1664	.915
300	IPW	-.0011	.0617	.0614	.951	-.0044	.1176	.1157	.952	.0019	.1009	.0993	.953
	AIPW	-.0023	.0582	.0588	.956	-.0051	.1056	.1036	.954	.0022	.0936	.0910	.944
	CC	.0295	.0577	.0596	.924	.1051	.1070	.1095	.824	.0823	.0925	.0946	.861
500	IPW	.0018	.0451	.0473	.958	-.0037	.0853	.0895	.958	-.0069	.0765	.0767	.945
	AIPW	.0009	.0430	.0452	.957	-.0032	.0793	.0798	.947	-.0066	.0689	.0702	.951
	CC	.0317	.0429	.0459	.903	.1077	.0788	.0844	.763	.0737	.0704	.0730	.839
800	IPW	-.0006	.0374	.0375	.951	-.0030	.0671	.0708	.962	.0004	.0617	.0605	.946
	AIPW	-.0003	.0362	.0358	.949	-.0031	.0623	.0631	.954	-.0012	.0577	.0554	.935
	CC	.0315	.0353	.0364	.863	.1065	.0630	.0667	.633	.0786	.0568	.0576	.721
						Model 4							
50	IPW	.0053	.1627	.1504	.948	.0825	.3531	.2832	.913	-.0057	.2736	.2405	.948
	AIPW	-.0085	.1549	.1489	.950	-.0122	.2746	.2752	.966	-.0138	.2395	.2340	.961
	CC	.2295	.2640	.0855	.531	.4513	.3805	.1760	.517	.2954	.3285	.1409	.536
100	IPW	-.0050	.1168	.1085	.939	.0481	.2350	.2130	.922	.0016	.1884	.1761	.940
	AIPW	-.0130	.1083	.1043	.943	-.0067	.1920	.1885	.950	.0066	.1648	.1613	.949
	CC	.0196	.1077	.1063	.943	.2010	.1946	.2087	.820	.0900	.1645	.1702	.910
300	IPW	-.0001	.0630	.0624	.945	-.0001	.1323	.1311	.955	-.0011	.1043	.1038	.946
	AIPW	-.0020	.0588	.0588	.951	-.0052	.1095	.1059	.952	.0012	.0950	.0913	.931
	CC	.0271	.0582	.0601	.930	.2020	.1060	.1173	.576	.0894	.0939	.0965	.863
500	IPW	.0012	.0457	.0480	.951	-.0007	.0966	.1010	.966	-.0054	.0801	.0799	.948
	AIPW	.0006	.0433	.0453	.956	-.0010	.0821	.0813	.941	-.0059	.0697	.0704	.950
	CC	.0291	.0434	.0463	.912	.2047	.0817	.0903	.364	.0815	.0711	.0745	.820
800	IPW	-.0006	.0381	.0380	.949	.0004	.0761	.0794	.967	.0000	.0636	.0630	.947
	AIPW	-.0002	.0362	.0359	.949	-.0016	.0640	.0641	.955	-.0014	.0574	.0555	.942
	CC	.0288	.0356	.0367	.885	.2039	.0644	.0714	.166	.0864	.0570	.0588	.673

were noted, without another MAARI code, were excluded. More details about this study can be found in Halloran et al. (2003).

Any children representing with history of fever and any respiratory illness were eligible to have a throat swab for influenza virus culture. The decision to obtain specimens was made irrespective of whether a patient had received CAIV-T. The specific case definition was culture-confirmed influenza. Table 3 taken from Halloran et al. (2003) contains information on the number of children in three age groups, the number of children who are vaccinated versus unvaccinated, the number of nonspecific MAARI cases, the number of cultures performed, and the number of cultures positive for each group.

With the method developed in Section 4 for Poisson regression, we compare the risk of developing MAARI for children who received CAIV-T to the risk for children who had never received CAIV-T using the automated information provided in Table 3. The number of nonspecific MAARI cases extracted using the ICD-9 codes is the auxiliary outcome A, whereas the actual number of influenza cases Y is the outcome of interest. Let Z_1 be the treatment indicator (1=vaccine and 0=placebo). Let $Z_2 = (\eta_1, \eta_2)$ be the dummy variables indicating three age groups, where $\eta_1 = 1$ if the age is in the range 1.5–4, $\eta_1 = 0$, otherwise, and $\eta_2 = 1$ if the age is in the range 5–9, $\eta_2 = 0$, otherwise. The reference group is the age 10–18. The exposure time for all children is taken as $T = 1$ year.

Consider a Poisson regression model with mean $T \exp(b_0 + b_1 Z_1 + \theta_1 \eta_1 + \theta_2 \eta_2)$. Using the IPW estimator $\hat{\beta}_{I1}$, the estimates (standard errors) are $\hat{b}_0 = -0.7659$ ($\hat{\sigma}_{b_0} = 0.1046$), $\hat{b}_1 = -1.5830$ ($\hat{\sigma}_{b_1} = 0.5017$), $\hat{\theta}_1 = -0.5572$ ($\hat{\sigma}_{\theta_1} = 0.2111$) and $\hat{\theta}_2 = -0.0199$ ($\hat{\sigma}_{\theta_2} = 0.1472$). The age-adjusted relative rate (RR) in the vaccinated group compared with the unvaccinated group equals $\exp(\hat{b}_1) = \exp(-1.5830) = 0.2054$, which means that the rate of developing MAARI for the vaccinated group is 20% of that for the unvaccinated group. In terms of the vaccine efficacy VE $= 1 - \text{RR} = 0.7946$, this represents about 80% reduction in the risk of developing MAARI for the vaccinated group compared to the unvaccinated group. The 95% confidence interval of RR obtained by using the delta method is $(0.0768, 0.5490)$, showing clear evidence that the vaccinated children have less risk of influenza than the unvaccinated children. The 95% confidence interval for VE is $(0.4510, 0.9232)$.

Table 3 Study data for influenza epidemic season 2000-01, by age and vaccine group (from Halloran et al. 2003)

Age group (years)	Vaccine	No. of children	No. of MAARI cases	No. of MAARI cases cultured	No. of positive cultures
1.5-4	CAIV-T	537	389	16	0
	None	1844	1665	86	24
5-9	CAIV-T	807	316	17	2
	None	2232	1156	118	53
10-18	CAIV-T	937	219	19	3
	None	5249	1421	123	56
Total	CAIV-T	2281	924	52	5
	None	9325	4242	327	133

Using the AIPW estimator $\hat{\beta}_{A2}$, the estimates (standard errors) are $\hat{b}_0 = -2.0703$ ($\hat{\sigma}_{b_0} = 0.0851$), $\hat{b}_1 = -1.8072$ ($\hat{\sigma}_{b_1} = 0.3786$), $\hat{\theta}_1 = 0.6452$ ($\hat{\sigma}_{\theta_1} = 0.1966$) and $\hat{\theta}_2 = 0.6235$ ($\hat{\sigma}_{\theta_2} = 0.1265$). The age-adjusted relative rate (RR) is $\exp(\hat{b}_1) = \exp(-1.8072) = 0.1641$. The estimated VE is 0.8359 and the 95% confidence interval is $(0.6553, 0.9219)$. The estimator $\hat{\beta}_{A2}$ yields smaller standard errors and confidence intervals with more precision than using $\hat{\beta}_{I1}$.

This data was analyzed by Halloran et al. (2003) and Chu and Halloran (2004). Assuming the binary probability model for $P_\beta(Y|X)$ where X includes the vaccination status and age group indicators, and using the mean score method, Halloran et al. (2003) found that the estimated VE based on the nonspecific MAARI cases alone was 0.18 with 95% confidence interval of $(0.11, 0.24)$. The estimated VE by incorporating the surveillance cultures was 0.79 with 95% confidence interval of $(0.51, 0.91)$. Halloran et al. also reported sample-size-weighted VE$= 0.77$ with 95% confidence interval of $(0.48, 0.90)$. Chu and Halloran (2004) have developed a Bayesian method to estimate vaccine efficacy. By Chu and Halloran (2004), the estimated VE was 0.74 with 95% confidence interval $(0.50, 0.88)$ and estimated VE by the multiple imputation method was 0.71 with 95% confidence interval $(0.42, 0.86)$.

Our estimates of the vaccine efficacy are in line with the existing methods. The estimator $\hat{\beta}_{A2}$ yields smaller standard errors and therefore confidence intervals are more precise than the existing methods of Halloran et al. (2003) and Chu and Halloran (2004). Compared to the binary regression, Poisson regression model allows multiple recurrent MAARI cases for each child. Although for this particular application the exposure time is fixed at one year time interval, the proposed method is applicable to the situation where the length of exposure time may be different for different children.

7 Conclusions

In this paper, we investigated the mean score method, the IPW method and the AIPW method for the parametric probability regression model $P_\beta(Y|X)$ when outcome of interest Y is subject to missingness. The asymptotic distributions are derived for the IPW estimator and the AIPW estimator. The selection probability often needs to be estimated for the IPW estimator, and both the selection probability and the conditional expectation of the score function needs to be estimated for the AIPW estimator. We investigated the properties of the IPW estimator and the AIPW estimator when the selection probability and the conditional expectation are implemented differently.

An AIPW estimator is said to be fully augmented if the selection probability and the conditional expectation are estimated using the full set of observed variables; it is partially augmented if the selection probability and the conditional expectation are estimated using a subset of observed variables. Corollary 1 shows that the fully augmented AIPW estimator is more efficient than the partially augmented AIPW estimator. Corollary 2 shows that the AIPW estimator is more efficient than the IPW estimator. However, when the selection probability depends only on a set of discrete random variables, the IPW estimator obtained by estimating the selection probability nonparametrically with the cell frequencies is asymptotically equivalent to the AIPW estimator augmented using the same set of discrete random variables. Proposition 1 shows that the IPW estimator, the AIPW estimator and the mean score estimator are equivalent if the selection probability and the conditional expectation are estimated using same set of discrete random variables.

Applying the developed missing data methods, we derived the estimation procedures for Poisson regression model with missing outcomes based on auxiliary outcomes and a validated sample for true outcomes. By assuming the selection probability depending only on the observed discrete exposure variables, not on the continuous exposure time, we show that the IPW estimator and the AIPW estimator can be formulated to analyze data when only aggregated/summarized information are available. The simulation study shows that for a moderate sample size and selection probability, the IPW estimator and AIPW estimator perform better than the complete-case estimator. The AIPW estimator is more efficient and more stable than the IPW estimator. The proposed methods are applied to analyze a data set from for an influenza vaccine study conducted in Temple-Belton, Texas during the 2000-2001 influenza season. The data set presented in Table 3 only contains summarized information at categorical levels defined by the three age groups and vaccination status. The actual number of influenza cases (the number of positive cultures) out of the number of MAARI cases cultured, along with the number of MAARI cases, are available for each category. Our analysis using the AIPW approach shows that the age-adjusted relative rate in the vaccinated group compared to the unvaccinated group equals 0.1641, which represents about 84% reduction in the risk of developing MAARI for the vaccinated group compared to the unvaccinated group.

Appendix A

Proof of Proposition 1.

Since

$$\sum_{i=1}^{n}(1-\xi_i)\hat{E}\left\{S_\beta(Y|X_i)|Z_i,A_i\right\}=\sum_{i\in\bar{V}}\sum_{j\in V(Z_i,A_i)}S_\beta(Y_j|X_j)/n^V(Z_i,A_i)$$
$$=\sum_{i\in V}\left\{n^{\bar{V}}(Z_i,A_i)/n^V(Z_i,A_i)\right\}S_\beta(Y_i|X_i),$$

we have

$$\sum_{i=1}^{n}W_i^{E1}=\sum_{i\in V}\left(1+\frac{n^{\bar{V}}(Z_i,A_i)}{n^V(Z_i,A_i)}\right)S_\beta(Y_i|X_i)=\sum_{i=1}^{n}\frac{\xi_i}{\hat{\pi}_i^z}S_\beta(Y_i|X_i)=\sum_{i=1}^{n}W_i^{I1}. \quad \text{(A.1)}$$

This shows that the mean score estimator $\hat{\beta}_{E1}$ is the same as the IPW estimator $\hat{\beta}_{I1}$. Further, since

$$\sum_{i=1}^{n}\left(1-\frac{\xi_i}{\hat{\pi}_i^z}\right)\hat{E}\{S_\beta(Y|X_i)|Z_i,A_i\}$$
$$=\sum_{i\in\bar{V}}\sum_{j\in V(Z_i,A_i)}\frac{S_\beta(Y_j|X_j)}{n^V(Z_i,A_i)}-\sum_{i\in V}\frac{n^{\bar{V}}(Z_i,A_i)}{n^V(Z_i,A_i)}\sum_{j\in V(Z_i,A_i)}\frac{S_\beta(Y_j|X_j)}{n^V(Z_i,A_i)}$$
$$=\sum_{i\in V}\frac{n^{\bar{V}}(Z_i,A_i)}{n^V(Z_i,A_i)}S_\beta(Y_i|X_i)-\sum_{i\in V}\frac{n^{\bar{V}}(Z_i,A_i)}{n^V(Z_i,A_i)}S_\beta(Y_i|X_i)=0,$$

we have $\sum_{i=1}^{n}W_i^{A1}=\sum_{i=1}^{n}W_i^{I1}$. Thus the AIPW estimator $\hat{\beta}_{A1}$, the IPW estimator $\hat{\beta}_{I1}$ and the mean score estimator $\hat{\beta}_{E1}$ are equivalent to each other.

Note that

$$\begin{aligned}
&\sum_{i=1}^{n}\left(1-\frac{\xi_i}{\hat{\pi}_i^z}\right)\hat{E}\{S_\beta(Y|X_i)|X_i,A_i\}\\
&=\sum_{i\in\bar{V}}\sum_{j\in V(X_i,A_i)}\frac{S_\beta(Y_j|X_j)}{n^V(X_i,A_i)}-\sum_{i\in V}\frac{n^{\bar{V}}(Z_i,A_i)}{n^V(Z_i,A_i)}\sum_{j\in V(X_i,A_i)}\frac{S_\beta(Y_j|X_j)}{n^V(X_i,A_i)}\\
&=\sum_{i\in V}\frac{n^{\bar{V}}(X_i,A_i)}{n^V(X_i,A_i)}S_\beta(Y_i|X_i)-\sum_{i\in V}\frac{n^V(X_i,A_i)}{n^V(X_i,A_i)}\frac{n^{\bar{V}}(Z_i,A_i)}{n^V(Z_i,A_i)}S_\beta(Y_i|X_i)\\
&=\sum_{i\in V}\left\{\frac{n^{\bar{V}}(X_i,A_i)}{n^V(X_i,A_i)}-\frac{n^{\bar{V}}(Z_i,A_i)}{n^V(Z_i,A_i)}\right\}S_\beta(Y_i|X_i), \qquad \text{(A.2)}
\end{aligned}$$

which is not zero unless Z_i^c is linearly related to Z_i and in this case β is not identifiable. Hence the AIPW estimator $\hat{\beta}_{A2}$ is different from the AIPW estimator $\hat{\beta}_{A1}$.

By (A.1) and (A.2), we have

$$\begin{aligned}
\sum_{i=1}^{n}W_i^{A2}&=\sum_{i\in V}\left\{\frac{n(Z_i,A_i)}{n^V(Z_i,A_i)}+\frac{n^{\bar{V}}(X_i,A_i)}{n^V(X_i,A_i)}-\frac{n^{\bar{V}}(Z_i,A_i)}{n^V(Z_i,A_i)}\right\}S_\beta(Y_i|X_i)\\
&=\sum_{i\in V}\left\{1+\frac{n^{\bar{V}}(X_i,A_i)}{n^V(X_i,A_i)}\right\}S_\beta(Y_i|X_i)=\sum_{i=1}^{n}W_i^{I2}.
\end{aligned}$$

Following the same arguments leading to (A.1), we also have $\sum_{i=1}^n W_i^{E2}=\sum_{i=1}^n W_i^{I2}$. Hence, the estimators $\hat{\beta}_{I2}$, $\hat{\beta}_{E2}$ and $\hat{\beta}_{A2}$ are equivalent. By following the steps in (A.2), we also have $\sum_{i=1}^n\left(1-\frac{\xi_i}{\hat{\pi}_i}\right)\hat{E}\left\{S_\beta(Y|X_i)|X_i,A_i\right\}=0$. Hence, $\hat{\beta}_{A3}$ is the same as $\hat{\beta}_{I2}$. Therefore, these are essentially two different estimators. □

Proof of Theorem 1.

Applying the first order Taylor expansion, $\tilde{\pi}_i-\pi_i=(\partial\pi(X_i,A_i,\psi_0)/\partial\psi)'\left(\hat{\psi}-\psi_0\right)+o_p\left(n^{-1/2}\right)$. From (13), we have

$$n^{-1/2}U_I=n^{-1/2}\sum_{i=1}^{n}\frac{\xi_i}{\pi_i}S_\beta(Y_i|X_i)-n^{-1/2}\sum_{i=1}^{n}\frac{\tilde{\pi}_i-\pi_i}{\tilde{\pi}_i\pi_i}\xi_iS_\beta(Y_i|X_i) \qquad \text{(A.3)}$$

The second term of (A.3) is

$$\begin{aligned}
&n^{-1/2}\sum_{i=1}^{n}\frac{\tilde{\pi}_i-\pi_i}{\tilde{\pi}_i\pi_i}\xi_iS_\beta(Y_i|X_i)\\
&=n^{-1/2}\sum_{i=1}^{n}\pi_i^{-2}\xi_iS_\beta(Y_i|X_i)\left(\frac{\partial\pi(X_i,A_i,\psi_0)}{\partial\psi}\right)'\left(\hat{\psi}-\psi_0\right)\\
&=E\left\{\pi_i^{-2}\xi_iS_\beta(Y_i|X_i)\left(\frac{\partial\pi(X_i,A_i,\psi_0)}{\partial\psi}\right)'\right\}n^{1/2}\left(\hat{\psi}-\psi_0\right)+o_p(1) \qquad \text{(A.4)}
\end{aligned}$$

By (11), (A.3) and (A.4), we have

$$n^{-1/2}U_I=n^{-1/2}\sum_{i=1}^{n}\left(\frac{\xi_i}{\pi_i}S_\beta(Y_i|X_i)-O_i\right)+o_p(1)=n^{-1/2}\sum_{i=1}^{n}Q_i^I+o_p(1).$$

Now consider the AIPW estimator $\hat{\beta}_A$ based on solving the estimating equation (14). For simplicity, we denote $E_a\left\{S_\beta(Y|X_i)|X_i,A_i\right\}$ by E_i and $\tilde{E}\left\{S_\beta(Y|X_i)|X_i,A_i\right\}$ by $\tilde{E}_i$. We note that

$$\begin{aligned} n^{-1/2}U_A &= n^{-1/2}\sum_{i=1}^{n}\left[\frac{\xi_i}{\pi_i}S_\beta(Y_i|X_i)+\left(1-\frac{\xi_i}{\pi_i}\right)E_i\right] \\ &+ n^{-1/2}\sum_{i=1}^{n}\left[\left(\frac{\xi_i}{\tilde{\pi}_i}-\frac{\xi_i}{\pi_i}\right)\left\{S_\beta(Y_i|X_i)-E_i\right\}+\left(1-\frac{\xi_i}{\tilde{\pi}_i}\right)\left(\tilde{E}_i-E_i\right)\right]. \end{aligned}$$

Suppose that $\tilde{\pi}_i$ and $\tilde{E}_i$ are the estimates of π_i and E_i based on some parametric or non-parametric models. Then it can be shown using Taylor expansion and standard probability arguments that the second term is at the order of $o_p(1)$ under MAR I. Hence

$$n^{-1/2}U_A = n^{-1/2}\sum_{i=1}^{n}\left[\frac{\xi_i}{\pi_i}S_\beta(Y_i|X_i)+\left(1-\frac{\xi_i}{\pi_i}\right)E_i\right]+o_p(1).$$

It can be shown that under MAR I, $n^{-1}\partial U_I/\partial\beta \xrightarrow{P} I(\beta)$ and $n^{-1}\partial U_A/\partial\beta \xrightarrow{P} I(\beta)$. By routine derivations, we have

$$\begin{aligned} n^{1/2}\left(\hat{\beta}_I-\beta\right) &= I^{-1}(\beta)n^{-1/2}U_I+o_p(1)=I^{-1}(\beta)n^{-1/2}\sum_{i=1}^{n}Q_i^I+o_p(1), \\ n^{1/2}\left(\hat{\beta}_A-\beta\right) &= I^{-1}(\beta)n^{-1/2}U_A+o_p(1)=I^{-1}(\beta)n^{-1/2}\sum_{i=1}^{n}Q_i^A+o_p(1). \end{aligned}$$

By the central limit theorem, both $n^{1/2}\left(\hat{\beta}_I-\beta\right)$ and $n^{1/2}\left(\hat{\beta}_A-\beta\right)$ have asymptotically normal distributions with mean zero and covariances equal to $I^{-1}(\beta)\mathrm{Var}\left(Q_i^I\right)I^{-1}(\beta)$ and $I^{-1}(\beta)\mathrm{Var}\left(Q_i^A\right)I^{-1}(\beta)$, respectively.

Next, we examine the covariance matrices $\mathrm{Var}\left(Q_i^I\right)$ and $\mathrm{Var}\left(Q_i^A\right)$ to understand the efficiency gain of $\hat{\beta}_A$ over $\hat{\beta}_I$. Note that $Q_i^I=\xi_i/\pi_iS_\beta(Y_i|X_i)-O_i$ and $Q_i^A=\xi_i/\pi_iS_\beta(Y_i|X_i)+(1-\xi_i/\pi_i)E_i$. Denote $A_i=\xi_i/\pi_iS_\beta(Y_i|X_i)$ and $B_i=(1-\xi_i/\pi_i)E_i$. Then $Q_i^I=Q_i^A-B_i-O_i$. Under MAR I, $\mathrm{Cov}\left(Q_i^A,O_i\right)=\mathrm{E}\left(Q_i^AO_i\right)=E\left\{E(Q_i^A|\xi_i,X_i,A_i)O_i\right\}=E\{E_iO_i\}=E\{E_i\ E(O_i|X_i,A_i)\}=0$, and

$$\begin{aligned} \mathrm{Cov}(Q_i^A,B_i) &= E\left[\left\{\frac{\xi_i}{\pi_i}S_\beta(Y_i|X_i)+\left(1-\frac{\xi_i}{\pi_i}\right)E_i\right\}\left(1-\frac{\xi_i}{\pi_i}\right)E_i\right] \\ &= E\left[\left(1-\frac{\xi_i}{\pi_i}\right)^2E_i^2\right]-E\left[\left(1-\frac{\xi_i}{\pi_i}\right)^2S_\beta(Y_i|X_i)E_i\right] \\ &+E\left[\left(1-\frac{\xi_i}{\pi_i}\right)S_\beta(Y_i|X_i)E_i\right]=0. \end{aligned}$$

Hence, $\mathrm{Cov}\left(Q_i^A,B_i+O_i\right)=0$. It follows that $\mathrm{Var}(Q_i^I)=\mathrm{Var}\left(Q_i^A\right)+\mathrm{Var}(B_i+O_i)$. Since $Q_i^A=S_\beta(Y_i|X_i)-\left(1-\frac{\xi_i}{\pi_i}\right)\left\{S_\beta(Y_i|X_i)-E_i\right\}$ and the two terms are uncorrelated under MAR I, we have $\mathrm{Var}\left(Q_i^A\right)=\mathrm{Var}\left(S_\beta(Y_i|X_i)\right)+\mathrm{Var}\left((1-\xi_i/\pi_i)\left\{S_\beta(Y_i|X_i)-E_i\right\}\right)$, where the first term equals $I(\beta)$. This completes the proof of Theorem 1. □

Proof of Corollary 1.

Let $Q_i^{A1} = \xi_i/\pi_i^z S_\beta(Y_i|X_i) + (1-\xi_i/\pi_i^z)E\{S_\beta(Y|X_i)|Z_i, A_i\}$, and $Q_i^{A2} = \xi_i/\pi_i^z S_\beta(Y_i|X_i) + (1-\xi_i/\pi_i^z)E\{S_\beta(Y|X_i)|X_i, A_i\}$. By (16),

$$\begin{aligned}\text{Var}\left(Q_i^{A1}\right) &= \text{Var}\left(S_\beta(Y_i|X_i)\right) + \text{Var}\left(\left(1-\frac{\xi_i}{\pi_i^z}\right)\left[S_\beta(Y_i|X_i) - E\left\{S_\beta(Y|X_i)|Z_i, A_i\right\}\right]\right),\\ \text{Var}\left(Q_i^{A2}\right) &= \text{Var}\left(S_\beta(Y_i|X_i)\right) + \text{Var}\left(\left(1-\frac{\xi_i}{\pi_i^z}\right)\left[S_\beta(Y_i|X_i) - E\left\{S_\beta(Y|X_i)|X_i, A_i\right\}\right]\right).\end{aligned}$$

The second term of $\text{Var}\left(Q_i^{A1}\right)$ equals $\Sigma_{A1}(\beta)$ and the second term of $\text{Var}\left(Q_i^{A2}\right)$ equals $\Sigma_{A2}(\beta)$. Then it follows from the main results in Theorem 1 that (17) and (18) hold.

Also by Theorem 1, the difference in the variances of Q_i^{A1} and Q_i^{A2} contributes to the difference in the asymptotic variances of $\hat{\beta}_{A1}$ and $\hat{\beta}_{A2}$. Since $E\left\{(1-\xi_i/\pi_i^z)^2 \,|Y_i, X_i, A_i\right\} = E\left\{(1-\xi_i/\pi_i^z)^2 \,|X_i, A_i\right\} = (1-\pi_i^z)/\pi_i^z$ under MAR I,

$$\begin{aligned}\Sigma_{A2}(\beta) &= E\left(\frac{1-\pi_i^z}{\pi_i^z}\left\{E\left\{S_\beta^2(Y_i|X_i)|X_i, A_i\right\} - \left[E\left\{S_\beta(Y|X_i)|X_i, A_i\right\}\right]^2\right\}\right)\\ &= E\left(\frac{1-\pi_i^z}{\pi_i^z}\left\{E\left\{S_\beta^2(Y_i|X_i)|Z_i, A_i\right\} - \left[E\left\{S_\beta(Y|X_i)|Z_i, A_i\right\}\right]^2\right\}\right)\\ &\quad -E\left(\frac{1-\pi_i^z}{\pi_i^z}\left\{\left[E\left\{S_\beta(Y|X_i)|X_i, A_i\right\}\right]^2 - \left[E\left\{S_\beta(Y|X_i)|Z_i, A_i\right\}\right]^2\right\}\right)\\ &= \Sigma_{A1}(\beta) - E\left(\frac{1-\pi_i^z}{\pi_i^z}\text{Var}\left(\left[E\left\{S_\beta(Y|X_i)|X_i, A_i\right\}\right]|Z_i, A_i\right)\right),\end{aligned}$$

which is less than $\Sigma_{A1}(\beta)$ if the covariates Z_i is a proper subset of X_i. □

Proof of Corollary 2.

Consider the definitions of B_i and O_i given following (16). We note that

$$\begin{aligned}&E\left\{\pi_i^{-1}S_\beta(Y_i|X_i)\left(\frac{\partial\pi(Z_i, A_i, \psi_0)}{\partial\psi}\right)'\right\}\\ &= E\left[\pi_i^{-1}E\left\{S_\beta(Y_i|X_i)|Z_i, A_i\right\}\left(\frac{\partial\pi(Z_i, A_i, \psi_0)}{\partial\psi}\right)'\right]\\ &= \sum_{z,a}\rho(z,a)\pi^{-1}(z,a,\psi_0)E\left\{S_\beta(Y|X)|Z=z, A=a\right\}\left(\frac{\partial\pi(z,a,\psi_0)}{\partial\psi}\right)',\end{aligned}$$

and

$$\begin{aligned}n^{-1/2}\sum_{i=1}^n S_i^\psi &= n^{-1/2}\sum_{i=1}^n \frac{(\xi_i - \pi(Z_i, A_i, \psi_0))}{\pi(Z_i, A_i, \psi_0)(1-\pi(Z_i, A_i, \psi_0))}\frac{\partial\pi(Z_i, A_i, \psi_0)}{\partial\psi}\\ &= \sum_{z,a} n^{-1/2}\sum_{j=1}^n \frac{(\xi_j - \pi(z,a,\psi_0))\,I\left(Z_j = z, A_j = a\right)}{\pi(z,a,\psi_0)(1-\pi(z,a,\psi_0))}\frac{\partial\pi(z,a,\psi_0)}{\partial\psi}.\end{aligned}$$

From the discussions preceding Corollary 2, $\psi = \{\psi_{z,a}\}$ and $\pi(z,a,\psi) = \psi_{z,a}$, where $\psi_{z,a} = P(\xi_i = 1|Z_i = z, A_i = a)$ for all distinct pairs (z,a). Hence, $\frac{\partial\pi(z,a,\psi_0)}{\partial\psi}$ is a column vector with 1 in the position for $\psi_{z,a}$ and 0 elsewhere. And I^ψ is a diagonal matrix and its inverse matrix is also diagonal.

We have

$$n^{-1/2}\sum_{i=1}^{n} O_i = E\left\{\pi_i^{-1} S_\beta(Y_i|X_i)\left(\frac{\partial\pi(Z_i,A_i,\psi_0)}{\partial\psi}\right)'\right\}(I^\psi)^{-1}n^{-1/2}\sum_{i=1}^{n} S_i^\psi$$

$$= \sum_{z,a} E\{S_\beta(Y|X)|Z=z,A=a\}n^{-1/2}\sum_{j=1}^{n}\frac{\xi_j-\pi(z,a,\psi_0)}{\pi(z,a,\psi_0)}I\left(Z_j=z,A_j=a\right),$$

and

$$n^{-1/2}\sum_{i=1}^{n} B_i = -n^{-1/2}\sum_{i=1}^{n}\frac{(\xi_i-\pi(Z_i,A_i))}{\pi(Z_i,A_i)}E_i$$

$$= -n^{-1/2}\sum_{i=1}^{n}\frac{(\xi_i-\pi(Z_i,A_i))}{\pi(Z_i,A_i)}E\left\{S_\beta(Y_i|X_i)|X_i,A_i\right\}$$

$$= -\sum_{z,a} n^{-1/2}\sum_{j=1}^{n}\frac{\xi_j-\pi(z,a,\psi_0)}{\pi(z,a,\psi_0)}I\left(Z_j=z,A_j=a\right)E\left\{S_\beta(Y_j|X_j)|Z_j=z,Z_j^c,A_j=a\right\}.$$

Then (21) holds. It follows that $\hat{\beta}_A$ is more efficient than $\hat{\beta}_I$ unless $Var\{S_\beta(Y_j|X_j)|Z_j = z, A_j = a\} = 0$ for all (z,a) for which $P(Z_i = z, A_i = a) \neq 0$. □

Appendix B

Proof of the simplified variance formula (27)

The information matrix $I(\beta)$ is a $(k+1)\times(k+1)$ symmetric matrix given by

$$I(b_0,b_1,\theta) = \begin{bmatrix} \sum_{l,m} q_{lm} & \sum_{m=1}^{k} q_{1m} & q_{11}+q_{01} & \cdots & q_{1r}+q_{0r} \\ \sum_{m=1}^{k} q_{1m} & \sum_{m=1}^{k} q_{1m} & q_{11} & \cdots & q_{1r} \\ q_{11}+q_{01} & q_{11} & q_{11}+q_{01} & \cdots & 0 \\ \vdots & \vdots & \vdots & \ddots & \vdots \\ q_{1r}+q_{0r} & q_{1r} & 0 & \cdots & q_{1r}+q_{0r} \end{bmatrix},$$

where $r = k-1$ and $q_{lm} = E(T_i e^{b_{lm}} I\{\text{individual } i \text{ in category } (l,m)\})$. For ease of presentation in the following, we drop the augments (b_0, b_1, θ) and use I for $I(b_0, b_1, \theta)$. Let I_{ij} be the cofactor of the (i,j)th element of I.

First we need to find the elements on the second row of the information matrix $I(b_0, b_1, \theta)$. Note that for a matrix A, $((a_{ij})_{n\times n})^{-1} = (A_{ji})_{n\times n}/|A|$, where A_{ij} is the cofactor of a_{ij} in the matrix A and $|A|$ is the determinant of A. Also note that for a block matrix

$$B = \begin{pmatrix} B_{11} & B_{12} \\ B_{21} & B_{22} \end{pmatrix},$$

the determinant $|B| = |B_{22}||B_{11} - B_{12}B_{22}^{-1}B_{21}|$. We have

$$
\begin{aligned}
|I| &= \prod_{m=1}^{k-1}(q_{1m}+q_{0m}) \\
&\quad \left| \begin{pmatrix} \sum_{lm} q_{lm} & \sum_{m=1}^{k} q_{1m} \\ \sum_{m=1}^{k} q_{1m} & \sum_{m=1}^{k} q_{1m} \end{pmatrix} - \begin{pmatrix} \sum_{l=0}^{1}\sum_{m=1}^{k-1} q_{lm} & \sum_{m=1}^{k-1} q_{1m} \\ \sum_{m=1}^{k-1} q_{1m} & \sum_{m=1}^{k-1} \frac{(q_{1m})^2}{q_{1m}+q_{0m}} \end{pmatrix} \right| \\
&= \prod_{m=1}^{k-1}(q_{1m}+q_{0m}) \begin{vmatrix} q_{1k}+q_{0k} & q_{1k} \\ q_{1k} & q_{1k}+\sum_{m=1}^{k-1}\frac{q_{0m}q_{1m}}{q_{1m}+q_{0m}} \end{vmatrix} \\
&= \prod_{m=1}^{k}(q_{1m}+q_{0m}) \sum_{m=1}^{k} \frac{q_{0m}q_{1m}}{q_{1m}+q_{0m}},
\end{aligned}
$$

and

$$
\begin{aligned}
I_{21} &= - \begin{vmatrix} \sum_{m=1}^{k} q_{1m} & q_{11}+q_{01} & \cdots & q_{1r}+q_{0r} \\ q_{11} & q_{11}+q_{01} & \cdots & 0 \\ \vdots & \vdots & \ddots & \vdots \\ q_{1r} & 0 & \cdots & q_{1r}+q_{0r} \end{vmatrix} \\
&= -\prod_{m=1}^{k-1}(q_{1m}+q_{0m}) \left[\sum_{m=1}^{k} q_{1m} - \sum_{m=1}^{k-1} q_{1m} \right] \\
&= -q_{1k} \prod_{m=1}^{k-1}(q_{1m}+q_{0m}).
\end{aligned}
$$

Hence, the (2, 1)th element of I^{-1} is

$$
\left(I^{-1}\right)_{21} = \frac{I_{21}}{|I|} = -\frac{q_{1k}}{W(q_{1k}+q_{0k})},
$$

where $W = \sum_{m=1}^{k} q_{0m}q_{1m}/(q_{1m}+q_{0m})$.

To calculate (2, 2)th element of I^{-1}, we have

$$
\begin{aligned}
I_{22} &= \begin{vmatrix} \sum_{l,m}^{k} q_{lm} & q_{11}+q_{01} & \cdots & q_{1r}+q_{0r} \\ q_{11}+q_{01} & q_{11}+q_{01} & \cdots & 0 \\ \vdots & \vdots & \ddots & \vdots \\ q_{1r}+q_{0r} & 0 & \cdots & q_{1r}+q_{0r} \end{vmatrix} \\
&= \prod_{m=1}^{k-1}(q_{1m}+q_{0m}) \left[\sum_{l,m} q_{lm} - \sum_{l=0}^{1}\sum_{m=1}^{k-1} q_{lm} \right] = \prod_{m=1}^{k}(q_{1m}+q_{0m}).
\end{aligned}
$$

Hence, the $(2,2)$th element of I^{-1} is $\left(I^{-1}\right)_{22} = I_{22}/|I| = 1/W$.

To calculate $(2,3)$th element of I^{-1}, we have

$$\begin{aligned}
I_{23} &= -\begin{vmatrix} \sum_{l,m} q_{lm} & \sum_{m=1}^{k} q_{1m} & q_{12}+q02 & \cdots & q_{1r}+q_{0r} \\ q_{11}+q_{01} & q_{11} & 0 & \cdots & 0 \\ q_{12}+q_{02} & q_{12} & q_{12}+q_{02} & \cdots & 0 \\ \vdots & \vdots & \vdots & \ddots & \vdots \\ q_{1r}+q_{0r} & q_{1r} & 0 & \cdots & q_{1r}+q_{0r} \end{vmatrix} \\
&= -\prod_{m=2}^{k-1}(q_{1m}+q_{0m}) \left| \begin{pmatrix} \sum_{lm} q_{lm} & \sum_{m=1}^{k} q_{1m} \\ q_{11}+q_{01} & q_{11} \end{pmatrix} \right. \\
&\quad \left. - \begin{pmatrix} \sum_{m=2}^{k-1}(q_{1m}+q_{0m}) & \sum_{m=2}^{k-1} q_{1m} \\ 0 & 0 \end{pmatrix} \right| \\
&= -\prod_{m=2}^{k-1}(q_{1m}+q_{0m})(q_{0k}q_{11}-q_{1k}q_{01}).
\end{aligned}$$

Hence

$$(I^{-1})_{23} = \frac{I_{23}}{|I|} = -\frac{q_{11}}{W(q_{11}+q_{01})} + \frac{q_{1k}}{W(q_{1k}+q_{0k})}.$$

To calculate $(2,4)$th element of I^{-1}, we have

$$I_{24} = \begin{vmatrix} \sum_{l,m} q_{lm} & \sum_{m=1}^{k} q_{1m} & q_{11}+q_{01} & q_{13}+q_{03} & \cdots & q_{1r}+q_{0r} \\ q_{11}+q_{01} & q_{11} & q_{11}+q_{01} & 0 & \cdots & 0 \\ q_{12}+q_{02} & q_{12} & 0 & 0 & \cdots & 0 \\ q_{13}+q_{03} & q_{13} & 0 & q_{13}+q_{03} & \cdots & 0 \\ \vdots & \vdots & \vdots & \vdots & \ddots & \vdots \\ q_{1r}+q_{0r} & q_{1r} & 0 & 0 & \cdots & q_{1r}+q_{0r} \end{vmatrix}.$$

By switching the 2nd and the 3rd row, we have

$$I_{24} = -n^k \begin{vmatrix} \sum_{l,m} q_{lm} & \sum_{m=1}^{k} q_{1m} & q_{11}+q_{01} & q_{13}+q_{03} & \cdots & q_{1r}+q_{0r} \\ q_{12}+q_{02} & q_{12} & 0 & 0 & \cdots & 0 \\ q_{11}+q_{01} & q_{11} & q_{11}+q_{01} & 0 & \cdots & 0 \\ q_{13}+q_{03} & q_{13} & 0 & q_{13}+q_{03} & \cdots & 0 \\ \vdots & \vdots & \vdots & \vdots & \ddots & \vdots \\ q_{1r}+q_{0r} & q_{1r} & 0 & 0 & \cdots & q_{1r}+q_{0r} \end{vmatrix}.$$

Hence

$$\left(I^{-1}\right)_{24} = \frac{I_{24}}{|I|} = -\frac{q_{12}}{W(q_{12}+q_{02})} + \frac{q_{1k}}{W(q_{1k}+q_{0k})}.$$

In general, to calculate $I_{2(m+2)}$ for $m = 1, \cdots, k-1$, we can obtain a matrix with a $(k-2) \times (k-2)$ diagonal right lower block by switching rows of $I_{2(m+2)}$ even number of times when m is odd and by switching rows odd number of times when m is even. Then similar to calculating $\left(I^{-1}\right)_{24}$, we have

$$\left(I^{-1}\right)_{2(m+2)} = -\frac{q_{1m}}{W\,(q_{1m}+q_{0m})} + \frac{q_{1k}}{W(q_{1k}+q_{0k})}.$$

For $l = 1$ and $1 \le m \le k$, the (i,j)th element of G_{lm} is $g_{ij} = 1$ for $(i,j) = (1,1), (1,2), (1,m+2), (2,1), (2,2), (2,m+2), (m+2,1), (m+2,2), (m+2,m+2)$, and $g_{ij} = 0$ elsewhere. We have the $(2,2)$th element of $I^{-1}G_{lm}I^{-1}$ as

$$\begin{aligned}&\left(I^{-1}\right)_{21}^2 + \left(I^{-1}\right)_{22}^2 + \left(I^{-1}\right)_{2(m+2)}^2 + 2\left(I^{-1}\right)_{21}\left(I^{-1}\right)_{2(m+2)}\\ &+2\left(I^{-1}\right)_{22}\left(I^{-1}\right)_{2(m+2)} + 2\left(I^{-1}\right)_{21}\left(I^{-1}\right)_{22}\\ &= \frac{1}{W^2}\left(\frac{q_{0m}}{q_{1m}+q_{0m}}\right)^2.\end{aligned}$$

Since for $l = 0$ and $1 \le m \le k$, $g_{ij} = 1$ for $(i,j) = (1,1), (1,m+2), (m+2,1), (m+2,m+2)$, and $g_{ij} = 0$ elsewhere, in this case the $(2,2)$th element of $I^{-1}G_{lm}I^{-1}$ is

$$\begin{aligned}&(I^{-1})_{21}^2 + (I^{-1})_{2(m+2)}^2 + 2(I^{-1})_{21}(I^{-1})_{2(m+2)}\\ &= ((I^{-1})_{21} + (I^{-1})_{2(m+2)})^2\\ &= \frac{1}{W^2}\left(\frac{q_{1m}}{q_{1m}+q_{0m}}\right)^2.\end{aligned}$$

Hence the $(2,2)$th element of the asymptotic covariance matrix of $\hat{\beta}_{A2}$ is given by $\sigma_{b_1}^2 = 1/W + U/W^2$, where

$$U = \sum_{\alpha,l,m} \rho(\alpha,l,m)\frac{1-\rho^V(\alpha,l,m)}{\rho^V(\alpha,l,m)}\alpha p_{lm}(1-p_{lm})\left(\frac{q_{(1-l)m}}{q_{0m}+q_{1m}}\right)^2.$$

Note that $P(Y_i = 1|A_i = 1, i \in \text{category }(l,m))$ can be estimated by y_{lm}/n_{lm}^v and $\rho(l,m)$ by α_{lm}/n. Thus q_{lm} can be estimated by $(\alpha_{lm}/n)(y_{lm}/n_{lm}^v)$. Then we can estimate W by $\hat{W} = n^{-1}\sum_{m=1}^k \alpha_{0m}\alpha_{1m}y_{0m}y_{1m}/(\alpha_{0m}y_{0m}n_{1m}^v + \alpha_{1m}y_{1m}n_{0m}^v)$. By replacing $\rho_{l,m}^v$ with n_{lm}^v/α_{lm} and p_{lm} with y_{lm}/n_{lm}^v, we can estimate U by

$$\begin{aligned}\hat{U} &= n^{-1}\sum_{l,m}\alpha_{lm}\frac{\alpha_{lm}-n_{lm}^v}{n_{lm}^v}\frac{y_{lm}}{n_{lm}^v}\frac{n_{lm}^v - y_{lm}}{n_{lm}^v}\left(\alpha_{(1-l)m}\frac{y_{(1-l)m}}{n_{(1-l)m}^v}\right)\left(\alpha_{1m}\frac{y_{1m}}{n_{1m}^v}+\alpha_{0m}\frac{y_{0m}}{n_{0m}^v}\right)^{-2}\\ &= n^{-1}\sum_{m=1}^k\left[\frac{\alpha_{1m}-n_{1m}^v}{n_{1m}^v}\frac{n_{1m}^v-y_{1m}}{n_{1m}^v}\alpha_{0m}\frac{y_{0m}}{n_{0m}^v} + \alpha_{0m} - n_{0m}^v n_{0m}^v\frac{n_{0m}^v - y_{0m}}{n_{0m}^v}\alpha_{1m}\frac{y_{1m}}{n_{1m}^v}\right]\\ &\times\left(\alpha_{1m}\frac{y_{1m}}{n_{1m}^v}\alpha_{0m}\frac{y_{0m}}{n_{0m}^v}\right)\left(\alpha_{1m}\frac{y_{1m}}{n_{1m}^v}+\alpha_{0m}\frac{y_{0m}}{n_{0m}^v}\right)^{-2}\\ &= n^{-1}\sum_{m=1}^k\left[\left(\frac{1}{y_{1m}}-\frac{1}{n_{1m}^v}+\frac{1}{\alpha_{1m}}-\frac{n_{1m}^v}{\alpha_{1m}y_{1m}}\right)+\left(\frac{1}{y_{0m}}-\frac{1}{n_{0m}^v}+\frac{1}{\alpha_{0m}}-\frac{n_{0m}^v}{\alpha_{0m}y_{0m}}\right)\right]\\ &\times\left(\frac{\alpha_{1m}y_{1m}\alpha_{0m}y_{0m}}{\alpha_{1m}y_{1m}n_{0m}^v+\alpha_{0m}y_{0m}n_{1m}^v}\right)^2.\end{aligned}$$

From $\hat{W}$ and $\hat{U}$, we obtain an estimate of $\sigma_{b_1}^2$ as follows

$$\begin{aligned}&\hat{\sigma}_{b_1}^2\\ &= \frac{1}{\hat{W}^2}\sum_{m=1}^k\left[\left(\frac{1}{y_{1m}}-\frac{1}{n_{1m}^v}+\frac{1}{\alpha_{1m}}\right)+\left(\frac{1}{y_{0m}}-\frac{1}{n_{0m}^v}+\frac{1}{\alpha_{0m}}\right)\right]\left(\frac{\alpha_{1m}y_{1m}\alpha_{0m}y_{0m}}{\alpha_{1m}y_{1m}n_{0m}^v+\alpha_{0m}y_{0m}n_{1m}^v}\right)^2\\ &+\frac{1}{\hat{W}}+\frac{1}{\hat{W}^2}\sum_{m=1}^k\left[-\frac{n_{1m}^v}{\alpha_{1m}y_{1m}}-\frac{n_{0m}^v}{\alpha_{0m}y_{0m}}\right]\left(\frac{\alpha_{1m}y_{1m}\alpha_{0m}y_{0m}}{\alpha_{1m}y_{1m}n_{0m}^v+\alpha_{0m}y_{0m}n_{1m}^v}\right)^2\\ &= \frac{1}{\hat{W}^2}\sum_{m=1}^k\left[\left(\frac{1}{y_{1m}}-\frac{1}{n_{1m}^v}+\frac{1}{\alpha_{1m}}+\frac{1}{y_{0m}}-\frac{1}{n_{0m}^v}+\frac{1}{\alpha_{0m}}\right)\right]\left(\frac{\alpha_{1m}y_{1m}\alpha_{0m}y_{0m}}{\alpha_{1m}y_{1m}n_{0m}^v+\alpha_{0m}y_{0m}n_{1m}^v}\right)^2.\end{aligned}$$

The variance of $\hat{b}_1$ is estimated by $\widehat{\mathrm{Var}}\left(\hat{b}_1\right) = n^{-1}\hat{\sigma}^2_{\hat{b}_1}$. □

Competing interests
The authors declare that they have no competing interests.

Authors' contributions
YS proved Proposition 1, Theorem 1, Corollary 1 and 2, and wrote the manuscript. LQ developed the Poisson regression using the automated data with missing outcomes, performed the computations for the simulation study and for the application, and wrote the manuscript. Both authors read and approved the final manuscript.

Acknowledgements
The authors thank the two referees for their valuable comments. The authors also thank Yuriko Nagano for some help with the data analysis. This research was partially supported by the National Science Foundation grant DMS-1208978, the National Institute of Health NIAID grant R37 AI054165, and a fund provided by The University of North Carolina at Charlotte.

References
Carpenter, JR, Kenward, MG, Vansteelandt, S: A comparison of multiple imputation and doubly robust estimation for analyses with missing data. J. R. Stat. Soc. A. **169**, 571–584 (2006)
Clayton, D, Spiegelhalter, D, Dunn, G, Pickles, A: Analysis of longitudinal binary data from multiphase sampling (with discussion). J. R. Stat. Soc. B. **60**, 71–87 (1998)
Chu, H, Halloran, EM: Estimating vaccine efficacy using auxiliary outcome data and a small validation sample. Stat. Med. **23**, 2697–2711 (2004)
Dempster, AP, Laird, NM, Rubin, DB: Maximum likelihood from incomplete data via the EM algorithm. J. R. Stat. Soc. B. **39**, 1–38 (1977)
Ellenberg, SS, Hamilton, JM: Surrogate endpoints in clinical trials: cancer. Stat. Med. **8**, 405–413 (1989)
Fleming, TR: Evaluating therapeutic interventions: some issues and experiences. Stat. Sci. **7**, 428–456 (1992)
Halloran, EM, Longini, IM, Gaglani, MJ, Piedra, PA, Chu, H, Herschler, GB, Glezed, WP: Estimating efficacy of trivalent, cold-adapted, influenza virus vaccine (CAIV-T) against influenza A (H1N1) and B Using Surveillance Cultures. Am. J. Epid. **158**, 305–311 (2003)
Horvitz, DG, Thompson, DJ: A generalization of sampling without replacement from a finite universe. J. Am. Stat. Assoc. **47**, 663–685 (1952)
Little, RJA, Rubin, DB: Statistical analysis of missing data. Wiley, New York (2002)
Pepe, MS, Reilly, M, Fleming, TR: Auxiliary outcome data and the mean score method. J. Stat. Plan. Inference. **42**, 137–160 (1994)
Prentice, RL: Surrogate endpoints in clinical trials: definition and operational criteria. Stat. Med. **8**, 431–440 (1989)
Rubin, DB: Multiple imputation for nonresponse in surveys. Wiley, New York (1987)
Robins, JM, Rotnitzky, A, Zhao, LP: Estimation of regression coefficients when some regressors are not always observed. J. Am. Stat. Assoc. **89**, 846–866 (1994)
Scharfstein, DO, Rotnitzky, A, Robins, JM: Adjusting for nonignorable drop-out using semiparametric nonresponse models: rejoinder. J. Am. Stat. Assoc. **94**, 1135–1146 (1999)

Exponentiated Kumaraswamy-Dagum distribution with applications to income and lifetime data

Shujiao Huang and Broderick O Oluyede*

*Correspondence:
boluyede@georgiaSouthern.edu
Department of Mathematical Sciences, Georgia Southern University, Statesboro, GA 30458, USA

Abstract

A new family of distributions called exponentiated Kumaraswamy-Dagum (EKD) distribution is proposed and studied. This family includes several well known sub-models, such as Dagum (D), Burr III (BIII), Fisk or Log-logistic (F or LLog), and new sub-models, namely, Kumaraswamy-Dagum (KD), Kumaraswamy-Burr III (KBIII), Kumaraswamy-Fisk or Kumaraswamy-Log-logistic (KF or KLLog), exponentiated Kumaraswamy-Burr III (EKBIII), and exponentiated Kumaraswamy-Fisk or exponentiated Kumaraswamy-Log-logistic (EKF or EKLLog) distributions. Statistical properties including series representation of the probability density function, hazard and reverse hazard functions, moments, mean and median deviations, reliability, Bonferroni and Lorenz curves, as well as entropy measures for this class of distributions and the sub-models are presented. Maximum likelihood estimates of the model parameters are obtained. Simulation studies are conducted. Examples and applications as well as comparisons of the EKD and its sub-distributions with other distributions are given.

Mathematics Subject Classification (2000): 62E10; 62F30

Keywords: Dagum distribution; Exponentiated Kumaraswamy-Dagum distribution; Maximum likelihood estimation

1 Introduction

Camilo Dagum proposed the distribution which is referred to as Dagum distribution in 1977. This proposal enable the development of statistical distributions used to fit empirical income and wealth data, that could accommodate heavy tails in income and wealth distributions. Dagum's proposed distribution has both Type-I and Type-II specification, where Type-I is the three parameter specification and Type-II deals with four parameter specification. This distribution is a special case of generalized beta distribution of the second kind (GB2), McDonald (1984), McDonald and Xu (1995), when the parameter $q = 1$, where the probability density function (pdf) of the GB2 distribution is given by:

$$f_{GB2}(y; a, b, p, q) = \frac{a y^{ap-1}}{b^{ap} B(p,q)[1 + (\frac{y}{b})^{a}]^{p+q}}, \quad \text{for } y > 0.$$

See Kleiber and Kotz (2003) for details. Note that $a > 0, p > 0, q > 0$, are the shape parameters and b is the scale parameter and $B(p, q) = \frac{\Gamma(p)\Gamma(q)}{\Gamma(p+q)}$ is the beta function. Kleiber (2008) traced the genesis of Dagum distribution and summarized several statistical properties of this distribution. Domma et al. (2011) obtained the maximum likelihood estimates of the parameters of Dagum distribution for censored data. Domma and Condino (2013) presented the beta-Dagum distribution. Cordeiro et al. (2013) proposed the beta exponentiated Weibull distribution. Cordeiro et al. (2010) introduced and studied some mathematical properties of the Kumaraswamy Weibull distribution. Oluyede and Rajasooriya (2013) developed the Mc-Dagum distribution and presented its statistical properties. See references therein for additional results.

The pdf and cumulative distribution function (cdf) of Dagum distribution are given by:

$$g_D(x;\lambda,\beta,\delta) = \beta\lambda\delta x^{-\delta-1}\left(1+\lambda x^{-\delta}\right)^{-\beta-1} \tag{1}$$

and

$$G_D(x;\lambda,\beta,\delta) = \left(1+\lambda x^{-\delta}\right)^{-\beta}, \tag{2}$$

for $x > 0$, where λ is a scale parameter, δ and β are shape parameters. Dagum (1977) refers to his model as the generalized logistic-Burr distribution. The kth raw or non central moments are given by

$$E\left(X^k\right) = \beta\lambda^{\frac{k}{\delta}}B\left(\beta+\frac{k}{\delta},1-\frac{k}{\delta}\right),$$

for $k < \delta$, and $\lambda, \beta > 0$, where $B(\cdot,\cdot)$ is the beta function. The qth percentile is

$$x_q = \lambda^{\frac{1}{\delta}}\left(q^{-\frac{1}{\beta}}-1\right)^{-\frac{1}{\delta}}.$$

In this paper, we present generalizations of the Dagum distribution via Kumaraswamy distribution and its exponentiated version. This leads to the exponentiated Kumaraswamy Dagum distribution.

The motivation for the development of this distribution is the modeling of size distribution of personal income and lifetime data with a diverse model that takes into consideration not only shape, and scale but also skewness, kurtosis and tail variation. Also, the EKD distribution and its sub-models has desirable features of exhibiting a non-monotone failure rate, thereby accommodating different shapes for the hazard rate function and should be an attractive choice for survival and reliability data analysis.

This paper is organized as follows. In section 3, we present the exponentiated Kumaraswamy-Dagum distribution and its sub models, as well as series expansion, hazard and reverse hazard functions. Moments, moment generating function, Lorenz and Bonferroni curves, mean and median deviations, and reliability are obtained in section 4. Section 5 contains results on the distribution of the order statistics and Renyi entropy. Estimation of model parameters via the method of maximum likelihood is presented in section 6. In section 7, various simulations are conducted for different sample sizes. Section 8 contains examples and applications of the EKD distribution and its sub-models, followed by concluding remarks.

2 Methods, results and discussions

Methods, results and discussions for the class of EKD distributions are presented in sections 3 to 8. These sections include the sub-models, series expansion of the pdf, closed form expressions for the hazard and reverse hazard functions, moments, moment generating function, Bonferroni and Lorenz curves, reliability, mean and median deviations, distribution of order statistics and entropy, as well as estimation of model parameters and applications.

3 The exponentiated Kumaraswamy-Dagum distribution

In this section, we present the proposed distribution and its sub-models. Series expansion, hazard and reverse hazard functions are also studied in this section.

3.1 Kumaraswamy-Dagum distribution

Kumaraswamy (1980) introduced a two-parameter distribution on $(0, 1)$. Its cdf is given by

$$G(x) = 1 - \left(1 - x^{\psi}\right)^{\phi}, x \in (0, 1),$$

for $\psi > 0$ and $\phi > 0$.

For an arbitrary cdf $F(x)$ with pdf $f(x) = \frac{dF(x)}{dx}$, the family of Kumaraswamy-G distributions with cdf $G_k(x)$ is given by

$$G_K(x) = 1 - \left(1 - F^{\psi}(x)\right)^{\phi},$$

for $\psi > 0$ and $\phi > 0$. By letting $F(x) = G_D(x)$, we obtain the Kumaraswamy-Dagum (KD) distribution, with cdf

$$G_{KD}(x) = 1 - \left(1 - G_D^{\psi}(x)\right)^{\phi}.$$

3.2 The EKD distribution

In general, the EKD distribution is $G_{EKD}(x) = [F_{KD}(x)]^{\theta}$, where $F_{KD}(x)$ is a baseline (Kum-Dagum) cdf, $\theta > 0$, with the corresponding pdf given by $g_{EKD}(x) = \theta[F_{KD}(x)]^{\theta-1} f_{KD}(x)$. For large values of x, and for $\theta > 1(< 1)$, the multiplicative factor $\theta[F_{KD}(x)]^{\theta-1} > 1(< 1)$, respectively. The reverse statement holds for smaller values of x. Consequently, this implies that the ordinary moments of $g_{EKD}(x)$ are larger (smaller) than those of $f_{KD}(x)$ when $\theta > 1(< 1)$.

Replacing the dependent parameter $\beta\psi$ by α, the cdf and pdf of the EKD distribution are given by

$$G_{EKD}(x; \alpha, \lambda, \delta, \phi, \theta) = \left\{1 - \left[1 - \left(1 + \lambda x^{-\delta}\right)^{-\alpha}\right]^{\phi}\right\}^{\theta}, \tag{3}$$

and

$$g_{EKD}(x; \alpha, \lambda, \delta, \phi, \theta) = \alpha\lambda\delta\phi\theta x^{-\delta-1}\left(1 + \lambda x^{-\delta}\right)^{-\alpha-1}\left[1 - \left(1 + \lambda x^{-\delta}\right)^{-\alpha}\right]^{\phi-1} \times \left\{1 - \left[1 - \left(1 + \lambda x^{-\delta}\right)^{-\alpha}\right]^{\phi}\right\}^{\theta-1}, \tag{4}$$

for $\alpha, \lambda, \delta, \phi, \theta > 0$, and $x > 0$, respectively. The quantile function of the EKD distribution is in closed form,

$$G_{EKD}^{-1}(q) = x_q = \lambda^{\frac{1}{\delta}}\left\{\left[1 - \left(1 - q^{\frac{1}{\theta}}\right)^{\frac{1}{\phi}}\right]^{-\frac{1}{\alpha}} - 1\right\}^{-\frac{1}{\delta}}. \tag{5}$$

Plots of the pdf for some combinations of values of the model parameters are given in Figure 1. The plots indicate that the EKD pdf can be decreasing or right skewed. The EKD distribution has a positive asymmetry.

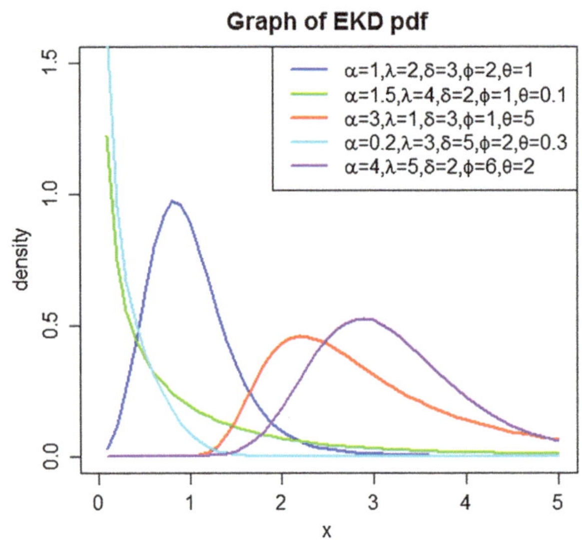

Figure 1 Graph of pdfs.

3.3 Some sub-models

Sub-models of EKD distribution for selected values of the parameters are presented in this section.

1. When $\theta = 1$, we obtain Kumaraswamy-Dagum distribution with cdf:

$$G(x; \alpha, \lambda, \delta, \phi) = 1 - \left[1 - \left(1 + \lambda x^{-\delta}\right)^{-\alpha}\right]^{\phi},$$

for $\alpha, \lambda, \delta, \phi > 0$ and $x > 0$.

2. When $\phi = \theta = 1$, we obtain Dagum distribution with cdf:

$$G(x; \alpha, \lambda, \delta) = \left(1 + \lambda x^{-\delta}\right)^{-\alpha},$$

for $\alpha, \lambda, \delta > 0$ and $x > 0$.

3. When $\lambda = 1$, we obtain exponentiated Kumaraswamy-Burr III distribution with cdf:

$$G(x; \alpha, \delta, \phi, \theta) = \left\{1 - \left[1 - \left(1 + x^{-\delta}\right)^{-\alpha}\right]^{\phi}\right\}^{\theta},$$

for $\alpha, \delta, \phi, \theta > 0$ and $x > 0$.

4. When $\lambda = \theta = 1$, we obtain Kumaraswamy-Burr III distribution with cdf:

$$G(x; \alpha, \delta, \phi) = 1 - \left[1 - \left(1 + x^{-\delta}\right)^{-\alpha}\right]^{\phi},$$

for $\alpha, \delta, \phi > 0$ and $x > 0$.

5. When $\lambda = \phi = \theta = 1$, we obtain Burr III distribution with cdf:

$$G(x; \alpha, \delta) = \left(1 + x^{-\delta}\right)^{-\alpha},$$

for $\alpha, \delta > 0$ and $x > 0$.

6. When $\alpha = 1$, we obtain exponentiated Kumaraswamy-Fisk or Kumaraswamy-Log-logistic distribution with cdf:

$$G(x; \lambda, \delta, \phi, \theta) = \left\{1 - \left[1 - \left(1 + \lambda x^{-\delta}\right)^{-1}\right]^{\phi}\right\}^{\theta},$$

for $\lambda, \delta, \phi, \theta > 0$ and $x > 0$.

7. When $\alpha = \theta = 1$, we obtain Kumaraswamy-Fisk or Kumaraswamy-Log-logistic distribution with cdf:

$$G(x; \lambda, \delta, \phi) = 1 - \left[1 - \left(1 + \lambda x^{-\delta}\right)^{-1}\right]^{\phi},$$

for $\lambda, \delta, \phi > 0$ and $x > 0$.

8. When $\alpha = \phi = \theta = 1$, we obtain Fisk or Log-logistic distribution with cdf:

$$G(x; \lambda, \delta) = \left(1 + \lambda x^{-\delta}\right)^{-1},$$

for $\lambda, \delta > 0$ and $x > 0$.

3.4 Series expansion of EKD distribution

We apply the series expansion

$$(1-z)^{b-1} = \sum_{j=0}^{\infty} \frac{(-1)^j \Gamma(b)}{\Gamma(b-j) j!} z^j, \tag{6}$$

for $b > 0$ and $|z| < 1$, to obtain the series expansion of the EKD distribution.

By using equation (6),

$$g_{EKD}(x) = \sum_{i=0}^{\infty} \sum_{j=0}^{\infty} \omega(i,j) x^{-\delta-1} \left(1 + \lambda x^{-\delta}\right)^{-\alpha(j+1)-1}, \tag{7}$$

where $\omega(i,j) = \alpha\lambda\delta\phi\theta \frac{(-1)^{i+j}\Gamma(\theta)\Gamma(\phi i+\phi)}{\Gamma(\theta-i)\Gamma(\phi i+\phi-j) i! j!}$.

Note that in the Dagum$(\alpha, \delta, \lambda)$ distribution, α and δ are shape parameters, and λ is a scale parameter. In the Exponentiated-Kumaraswamy(ψ, ϕ, θ) distribution, ψ is a skewness parameter, ϕ is a tail variation parameter, and the parameter θ characterizes the skewness, kurtosis, and tail of the distribution.

Consequently, for the EKD$(\alpha, \lambda, \delta, \phi, \theta)$ distribution, α is shape and skewness parameter, δ is shape parameter, λ is a scale parameter, ϕ is a tail variation parameter, and the parameter θ characterizes the skewness, kurtosis, and tail of the distribution.

3.5 Hazard and reverse hazard function

The hazard function of the EKD distribution is

$$\begin{aligned} h_{EKD}(x) &= \frac{g_{EKD}(x)}{1 - G_{EKD}(x)} \\ &= \alpha\lambda\delta\phi\theta x^{-\delta-1} \left(1 + \lambda x^{-\delta}\right)^{-\alpha-1} \left[1 - \left(1 + \lambda x^{-\delta}\right)^{-\alpha}\right]^{\phi-1} \\ &\quad \times \{1 - \left[1 - \left(1 + \lambda x^{-\delta}\right)^{-\alpha}\right]^{\phi}\}^{\theta-1} \\ &\quad \times \left(1 - \left\{1 - \left[1 - \left(1 + \lambda x^{-\delta}\right)^{-\alpha}\right]^{\phi}\right\}^{\theta}\right)^{-1}. \end{aligned} \tag{8}$$

Plots of the hazard function are presented in Figure 2. The plots show various shapes including monotonically decreasing, unimodal, and bathtub followed by upside down bathtub shapes with five combinations of the values of the parameters. This attractive flexibility makes the EKD hazard rate function useful and suitable for non-monotone empirical hazard behaviors which are more likely to be encountered or observed in real life situations. Unfortunately, the analytical analysis of the shape of both the density (except for zero modal when $\alpha\delta \leq 1$, and unimodal if $\alpha\delta > 1$, both for $\phi = \theta = 1$,) and

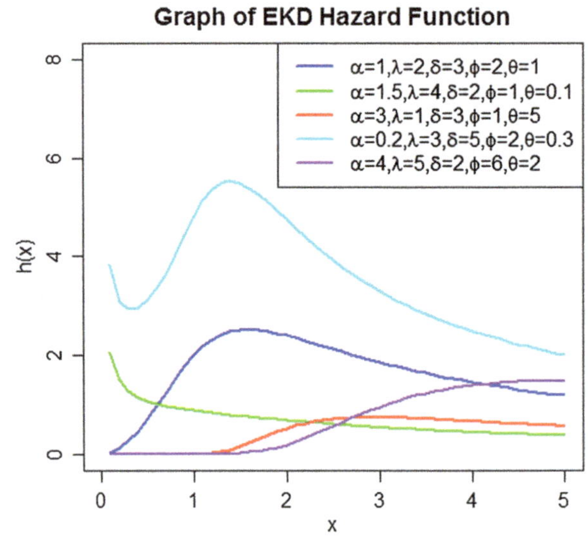

Figure 2 Graphs of hazard functions.

hazard rate function seems to be very complicated. We could not determine any specific rules for the shapes of the hazard rate function.

The reverse hazard function of the EKD distribution is

$$\begin{aligned}\tau_{EKD}(x) &= \frac{g_{EKD}(x)}{G_{EKD}(x)} \\ &= \alpha\lambda\delta\phi\theta x^{-\delta-1}\left(1+\lambda x^{-\delta}\right)^{-\alpha-1}\left[1-\left(1+\lambda x^{-\delta}\right)^{-\alpha}\right]^{\phi-1} \\ &\times\left\{1-\left[1-\left(1+\lambda x^{-\delta}\right)^{-\alpha}\right]^{\phi}\right\}^{-1}. \end{aligned} \tag{9}$$

4 Moments, moment generating function, Bonferroni and Lorenz curves, mean and median deviations, and reliability

In this section, we present the moments, moment generating function, Bonferroni and Lorenz curves, mean and median deviations as well as the reliability of the EKD distribution. The moments of the sub-models can be readily obtained from the general results.

4.1 Moments and moment generating function

Let $t = (1+\lambda x^{-\delta})^{-1}$ in equation (7), then the sth raw moment of the EKD distribution is given by

$$\begin{aligned}E(X^s) &= \int_0^\infty x^s \cdot g_{EKD}(x)dx \\ &= \sum_{i=0}^{\infty}\sum_{j=0}^{\infty}\omega(i,j)\lambda^{\frac{s}{\delta}-1}\cdot\frac{1}{\delta}\cdot B\left(\alpha(j+1)+\frac{s}{\delta},1-\frac{s}{\delta}\right) \\ &= \sum_{i=0}^{\infty}\sum_{j=0}^{\infty}\omega(i,j,s)B\left(\alpha(j+1)+\frac{s}{\delta},1-\frac{s}{\delta}\right), \end{aligned} \tag{10}$$

where $\omega(i,j,s) = \alpha\lambda^{\frac{s}{\delta}}\phi\theta\frac{(-1)^{i+j}\Gamma(\theta)\Gamma(\phi i+\phi)}{\Gamma(\theta-i)\Gamma(\phi i+\phi-j)i!j!}$, and $s < \delta$.

The moment generating function of the EKD distribution is given by

$$M(t) = \sum_{r=0}^{\infty}\sum_{i=0}^{\infty}\sum_{j=0}^{\infty} \omega(i,j,r)\frac{t^r}{r!}B\left(\alpha(j+1)+\frac{r}{\delta}, 1-\frac{r}{\delta}\right),$$

for $r < \delta$.

4.2 Bonferroni and Lorenz curves

Bonferroni and Lorenz curves are widely used tool for analyzing and visualizing income inequality. Lorenz curve, $L(p)$ can be regarded as the proportion of total income volume accumulated by those units with income lower than or equal to the volume a, and Bonferroni curve, $B(p)$ is the scaled conditional mean curve, that is, ratio of group mean income of the population. Plots of Bonferroni and Lorenz curves are given in Figure 3.

Let $I(a) = \int_0^a x \cdot g_{EKD}(x)dx$ and $\mu = E(X)$, then Bonferroni and Lorenz curves are given by

$$B(p) = \frac{I(q)}{p\mu} \quad \text{and} \quad L(p) = \frac{I(q)}{\mu},$$

respectively, for $0 \leq p \leq 1$, and $q = G_{EKD}^{-1}(p)$. The mean of the EKD distribution is obtained from equation (10) with $s = 1$ and the quantile function is given in equation (5). Consequently,

$$I(a) = \sum_{i=0}^{\infty}\sum_{j=0}^{\infty} \omega(i,j,1)B_{t(a)}\left(\alpha(j+1)+\frac{1}{\delta}, 1-\frac{1}{\delta}\right), \tag{11}$$

for $\delta > 1$, where $t(a) = (1+\lambda a^{-\delta})^{-1}$, and $B_{G(x)}(c,d) = \int_0^{G(x)} t^{c-1}(1-t)^{d-1}dt$ for $|G(x)| < 1$ is incomplete Beta function.

4.3 Mean and median deviations

If X has the EKD distribution, we can derive the mean deviation about the mean $\mu = E(X)$ and the median deviation about the median M from

$$\delta_1 = \int_0^{\infty} |x-\mu|\, g_{EKD}(x)dx \quad \text{and} \quad \delta_2 = \int_0^{\infty} |x-M|\, g_{EKD}(x)dx,$$

respectively. The mean μ is obtained from equation (10) with $s = 1$, and the median M is given by equation (5) when $q = \frac{1}{2}$.

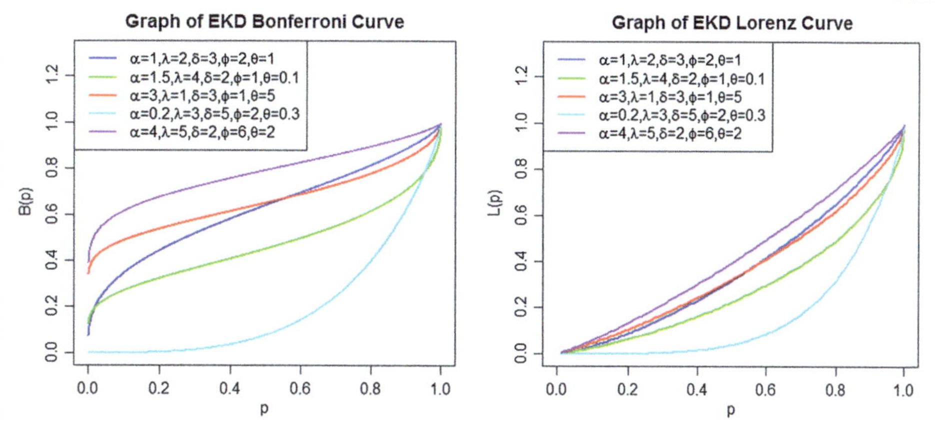

Figure 3 Graphs of Bonferroni and Lorenz curves.

The measure δ_1 and δ_2 can be calculated by the following relationships:

$$\delta_1 = 2\mu\, G_{EKD}(\mu) - 2\mu + 2T(\mu) \quad \text{and} \quad \delta_2 = 2T(M) - \mu,$$

where $T(a) = \int_a^{\infty} x \cdot g_{EKD}(x)dx$ follows from equation (11), that is

$$T(a) = \sum_{i=0}^{\infty}\sum_{j=0}^{\infty} \omega(i,j,1)\left[B\left(\alpha(j+1)+\frac{1}{\delta}, 1-\frac{1}{\delta}\right) - B_{t(a)}\left(\alpha(j+1)+\frac{1}{\delta}, 1-\frac{1}{\delta}\right)\right].$$

4.4 Reliability

The reliability $R = P(X_1 > X_2)$ when X_1 and X_2 have independent $EKD(\alpha_1, \lambda_1, \delta_1, \phi_1, \theta_1)$ and $EKD(\alpha_2, \lambda_2, \delta_2, \phi_2, \theta_2)$ distributions is given by

$$\begin{aligned} R &= \int_0^{\infty} g_1(x)G_2(x)dx \\ &= \sum_{i=0}^{\infty}\sum_{j=0}^{\infty}\sum_{k=0}^{\infty}\sum_{l=0}^{\infty} \zeta(i,j,k,l) \int_0^{\infty} x^{-\delta_1-1}\left(1+\lambda_1 x^{-\delta_1}\right)^{-\alpha_1(j+1)-1}\left(1+\lambda_2 x^{-\delta_2}\right)^{-\alpha_2 l} dx, \end{aligned}$$

where $\zeta(i,j,k,l) = \alpha_1\lambda_1\delta_1\phi_1\theta_1 \frac{(-1)^{i+j+k+l}\Gamma(\theta_1)\Gamma(\phi_1 i+\phi_1)\Gamma(\theta_2+1)\Gamma(\phi_2 k+1)}{\Gamma(\theta_1-i)\Gamma(\phi_1 i+\phi_1-j)\Gamma(\theta_2+1-k)\Gamma(\phi_2 k+1-l)i!j!k!l!}$.

If $\lambda = \lambda_1 = \lambda_2$ and $\delta = \delta_1 = \delta_2$, then reliability can be reduced to

$$R = \sum_{i=1}^{\infty}\sum_{j=1}^{\infty}\sum_{k=1}^{\infty}\sum_{l=1}^{\infty} \frac{\zeta(i,j,k,l)}{\lambda\delta[\alpha_1(j+1)+\alpha_2 l]}.$$

5 Order statistics and entropy

In this section, the distribution of the kth order statistic and Renyi entropy (Renyi 1960) for the EKD distribution are presented. The entropy of a random variable is a measure of variation of the uncertainty.

5.1 Order statistics

The pdf of the kth order statistics from a pdf $f(x)$ is

$$\begin{aligned} f_{k:n}(x) &= \frac{f(x)}{B(k, n-k+1)} F^{k-1}(x)[1-F(x)]^{n-k} \\ &= k\binom{n}{k} f(x)F^{k-1}(x)[1-F(x)]^{n-k}. \end{aligned} \tag{12}$$

Using equation (6), the pdf of the kth order statistic from EKD distribution is given by

$$g_{k:n}(x) = \sum_{i=0}^{\infty}\sum_{j=0}^{\infty}\sum_{p=0}^{\infty} K(i,j,p,k)\cdot x^{-\delta-1}\left(1+\lambda x^{-\delta}\right)^{-\alpha-\alpha p-1},$$

where $K(i,j,p,k) = \frac{(-1)^{i+j+p}\Gamma(n-k+1)\Gamma(\theta k+\theta i)\Gamma(\phi j+\phi)}{\Gamma(n-k+1-i)\Gamma(\theta k+\theta i-j)\Gamma(\phi j+\phi-p)i!j!p!} k\binom{n}{k}\alpha\lambda\delta\phi\theta$.

5.2 Entropy

Renyi entropy of a distribution with pdf $f(x)$ is defined as

$$I_R(\tau) = (1-\tau)^{-1}\log\left\{\int_{\mathbb{R}} f^{\tau}(x)dx\right\}, \tau > 0, \tau \neq 1.$$

Using equation (6), Renyi entropy of EKD distribution is given by

$$\begin{aligned} I_R(\tau) = (1-\tau)^{-1}\log\Bigg[\sum_{i=0}^{\infty}\sum_{j=0}^{\infty} &\frac{(-1)^{i+j}\Gamma(\theta\tau-\tau+1)\Gamma(\phi\tau-\tau+\phi i+1)}{\Gamma(\theta\tau-\tau+1-i)\Gamma(\phi\tau-\tau+\phi i+1-j)i!j!} \\ &\times \alpha^{\tau}\lambda^{-\frac{\tau}{\delta}+\frac{1}{\delta}}\delta^{\tau-1}\phi^{\tau}\theta^{\tau} B\left(\alpha\tau+\alpha j+\frac{1-\tau}{\delta}, \tau+\frac{\tau-1}{\delta}\right)\Bigg]. \end{aligned}$$

for $\alpha\tau+\alpha j+\frac{1-\tau}{\delta} > 0$ and $\tau+\frac{\tau-1}{\delta} > 0$. Renyi entropy for the sub-models can be readily obtained.

6 Estimation of model parameters

In this section, we present estimates of the parameters of the EKD distribution via method of maximum likelihood estimation. The elements of the score function are presented. There are no closed form solutions to the nonlinear equations obtained by setting the elements of the score function to zero. Thus, the estimates of the model parameters must be obtained via numerical methods.

6.1 Maximum likelihood estimation

Let $\mathbf{x} = (x_1, \cdots, x_n)^T$ be a random sample of the EKD distribution with unknown parameter vector $\Theta = (\alpha, \lambda, \delta, \phi, \theta)^T$. The log-likelihood function for Θ is

$$\begin{aligned} l(\Theta) &= n\left(\ln\alpha + \ln\lambda + \ln\delta + \ln\phi + \ln\theta\right) - (\delta+1)\sum_{i=1}^{n}\ln x_i \\ &\quad - (\alpha+1)\sum_{i=1}^{n}\ln\left(1+\lambda x_i^{-\delta}\right) + (\phi-1)\sum_{i=1}^{n}\ln\left[1-\left(1+\lambda x_i^{-\delta}\right)^{-\alpha}\right] \\ &\quad + (\theta-1)\sum_{i=1}^{n}\ln\left\{1-\left[1-\left(1+\lambda x_i^{-\delta}\right)^{-\alpha}\right]^{\phi}\right\}. \end{aligned} \tag{13}$$

The partial derivatives of $l(\Theta)$ with respect to the parameters are

$$\begin{aligned} \frac{\partial l}{\partial \alpha} &= \frac{n}{\alpha} - \sum_{i=1}^{n}\ln\left(1+\lambda x_i^{-\delta}\right) + (\phi-1)\sum_{i=1}^{n}\frac{\left(1+\lambda x_i^{-\delta}\right)^{-\alpha}\ln\left(1+\lambda x_i^{-\delta}\right)}{1-\left(1+\lambda x_i^{-\delta}\right)^{-\alpha}} \\ &\quad - (\theta-1)\phi\sum_{i=1}^{n}\frac{\left[1-\left(1+\lambda x_i^{-\delta}\right)^{-\alpha}\right]^{\phi-1}\left(1+\lambda x_i^{-\delta}\right)^{-\alpha}\ln\left(1+\lambda x_i^{-\delta}\right)}{1-\left[1-\left(1+\lambda x_i^{-\delta}\right)^{-\alpha}\right]^{\phi}}, \end{aligned}$$

$$\begin{aligned} \frac{\partial l}{\partial \lambda} &= \frac{n}{\lambda} - (\alpha+1)\sum_{i=1}^{n}\frac{x_i^{-\delta}}{1+\lambda x_i^{-\delta}} + (\phi-1)\alpha\sum_{i=1}^{n}\frac{\left(1+\lambda x_i^{-\delta}\right)^{-\alpha-1}x_i^{-\delta}}{1-\left(1+\lambda x_i^{-\delta}\right)^{-\alpha}} \\ &\quad - (\theta-1)\phi\alpha\sum_{i=1}^{n}\frac{\left[1-\left(1+\lambda x_i^{-\delta}\right)^{-\alpha}\right]^{\phi-1}\left(1+\lambda x_i^{-\delta}\right)^{-\alpha-1}x_i^{-\delta}}{1-\left[1-\left(1+\lambda x_i^{-\delta}\right)^{-\alpha}\right]^{\phi}}, \end{aligned}$$

$$\begin{aligned} \frac{\partial l}{\partial \delta} &= \frac{n}{\delta} - \sum_{i=1}^{n}\ln x_i + (\alpha+1)\lambda\sum_{i=1}^{n}\frac{x_i^{-\delta}\ln x_i}{1+\lambda x_i^{-\delta}} \\ &\quad - (\phi-1)\alpha\lambda\sum_{i=1}^{n}\frac{\left(1+\lambda x_i^{-\delta}\right)^{-\alpha-1}x_i^{-\delta}\ln x_i}{1-\left(1+\lambda x_i^{-\delta}\right)^{-\alpha}} \\ &\quad + (\theta-1)\phi\alpha\lambda\sum_{i=1}^{n}\frac{\left[1-\left(1+\lambda x_i^{-\delta}\right)^{-\alpha}\right]^{\phi-1}\left(1+\lambda x_i^{-\delta}\right)^{-\alpha-1}x_i^{-\delta}\ln x_i}{1-\left[1-\left(1+\lambda x_i^{-\delta}\right)^{-\alpha}\right]^{\phi}}, \end{aligned}$$

$$\frac{\partial l}{\partial \phi} = \frac{n}{\phi} + \sum_{i}^{n}\ln\left[1-\left(1+\lambda x_i^{-\delta}\right)^{-\alpha}\right] - (\theta-1)\sum_{i=1}^{n}\frac{\left[1-\left(1+\lambda x_i^{-\delta}\right)^{-\alpha}\right]^{\phi}\ln\left[1-\left(1+\lambda x_i^{-\delta}\right)^{-\alpha}\right]}{1-\left[1-\left(1+\lambda x_i^{-\delta}\right)^{-\alpha}\right]^{\phi}},$$

and

$$\frac{\partial l}{\partial \theta} = \frac{n}{\theta} + \sum_{i=1}^{n}\ln\left\{1-\left[1-\left(1+\lambda x_i^{-\delta}\right)^{-\alpha}\right]^{\phi}\right\},$$

respectively. The MLE of the parameters $\alpha, \lambda, \delta, \phi$, and θ, say $\hat{\alpha}, \hat{\lambda}, \hat{\delta}, \hat{\phi}$, and $\hat{\theta}$, must be obtained by numerical methods.

6.2 Asymptotic confidence intervals

In this section, we present the asymptotic confidence intervals for the parameters of the EKD distribution. The expectations in the Fisher Information Matrix (FIM) can be obtained numerically. Let $\hat{\Theta} = (\hat{\alpha}, \hat{\lambda}, \hat{\delta}, \hat{\phi}, \hat{\theta})$ be the maximum likelihood estimate of $\Theta = (\alpha, \lambda, \delta, \phi, \theta)$. Under the usual regularity conditions and that the parameters are in the interior of the parameter space, but not on the boundary, we have: $\sqrt{n}(\hat{\Theta} - \Theta) \xrightarrow{d} N_5(\underline{0}, I^{-1}(\Theta))$, where $I(\Theta)$ is the expected Fisher information matrix. The asymptotic behavior is still valid if $I(\Theta)$ is replaced by the observed information matrix evaluated at $\hat{\Theta}$, that is $J(\hat{\Theta})$. The multivariate normal distribution $N_5\left(\underline{0}, J(\hat{\Theta})^{-1}\right)$, where the mean vector $\underline{0} = (0,0,0,0,0)^T$, can be used to construct confidence intervals and confidence regions for the individual model parameters and for the survival and hazard rate functions.

The approximate $100(1-\eta)\%$ two-sided confidence intervals for α, λ, δ, ϕ and θ are given by:

$$\widehat{\alpha} \pm Z_{\frac{\eta}{2}}\sqrt{I_{\alpha\alpha}^{-1}(\widehat{\Theta})}, \quad \widehat{\lambda} \pm Z_{\frac{\eta}{2}}\sqrt{I_{\lambda\lambda}^{-1}(\widehat{\Theta})}, \quad \widehat{\delta} \pm Z_{\frac{\eta}{2}}\sqrt{I_{\delta\delta}^{-1}(\widehat{\Theta})}$$

$$\widehat{\phi} \pm Z_{\frac{\eta}{2}}\sqrt{I_{\phi\phi}^{-1}(\widehat{\Theta})}, \quad \widehat{\theta} \pm Z_{\frac{\eta}{2}}\sqrt{I_{\theta\theta}^{-1}(\widehat{\Theta})}$$

respectively, where $Z_{\frac{\eta}{2}}$ is the upper $\frac{\eta}{2}^{th}$ percentile of a standard normal distribution.

We can use the likelihood ratio (LR) test to compare the fit of the EKD distribution with its sub-models for a given data set. For example, to test $\theta = 1$, the LR statistic is

$$\omega = 2\left[\ln\left(L\left(\hat{\alpha}, \hat{\lambda}, \hat{\delta}, \hat{\phi}, \hat{\theta}\right)\right) - \ln\left(L\left(\tilde{\alpha}, \tilde{\lambda}, \tilde{\delta}, \tilde{\phi}, 1\right)\right)\right],$$

where $\hat{\alpha}$, $\hat{\lambda}$, $\hat{\delta}$, $\hat{\phi}$ and $\hat{\theta}$ are the unrestricted estimates, and $\tilde{\alpha}$, $\tilde{\lambda}$, $\tilde{\delta}$ and $\tilde{\phi}$ are the restricted estimates. The LR test rejects the null hypothesis if $\omega > \chi_d^2$, where χ_d^2 denote the upper $100d\%$ point of the χ^2 distribution with 1 degrees of freedom.

7 Simulation study

In this section, we examine the performance of the EKD distribution by conducting various simulations for different sizes (n=200, 400, 800, 1200) via the subroutine NLP in SAS. We simulate 2000 samples for the true parameters values $I: \alpha = 2, \lambda = 1, \delta = 3, \phi = 2, \theta = 2$ and $II: \alpha = 1, \lambda = 1, \delta = 1, \phi = 1, \theta = 1$. Table 1 lists the means MLEs of the five model parameters along with the respective root mean squared errors (RMSE). From the results, we can verify that as the sample size n increases, the mean estimates of the parameters tend to be closer to the true parameter values, since RMSEs decay toward zero.

8 Application: EKD and sub-distributions

In this section, applications based on real data, as well as comparison of the EKD distribution with its sub-models are presented. We provide examples to illustrate the flexibility of the EKD distribution in contrast to other models, including the exponentiated

Table 1 Monte Carlo simulation results: mean estimates and RMSEs

		I		II	
n	Parameter	Mean	RMSE	Mean	RMSE
200	α	4.41621	3.979304324	1.7899006	1.992043574
	λ	1.3580866	2.642335804	1.4287071	1.528578784
	δ	3.1167852	2.601663026	1.0337146	0.5898521
	ϕ	5.7270324	6.535452517	2.4702434	3.712081559
	θ	4.5560563	4.306946865	2.8884959	3.689669972
400	α	3.5972873	3.071770841	1.5456974	1.513782811
	λ	1.1196079	0.900800533	1.1382897	0.732002869
	δ	2.9333424	1.821450521	1.0064105	0.377302664
	ϕ	4.6989703	5.277876069	1.5488732	1.872088246
	θ	4.1188983	3.616692978	2.4213684	2.969367761
800	α	3.1040595	2.417025941	1.4359333	1.278449373
	λ	1.0626388	0.609066006	1.0432761	0.346996974
	δ	2.8960167	1.36814261	1.0017278	0.250650155
	ϕ	3.7437056	3.919777583	1.176675	0.766203302
	θ	3.4890255	2.748229594	1.9733522	2.197844717
1200	α	2.8399564	2.058703427	1.3884174	1.169251427
	λ	1.0429655	0.501712467	1.021836	0.258884917
	δ	2.9152476	1.133666485	1.0014919	0.193825437
	ϕ	3.1751818	3.043071803	1.083574	0.392293513
	θ	3.164176	2.346236284	1.731924	1.788360814

Kumaraswamy-Weibull (EKW), and beta-Kumaraswamy-Weibull (BKW) distributions for data modeling. The pdfs of EKW and BKW distributions are

$$f_{EKW}(x) = \theta abc\lambda^c x^{c-1} e^{-(\lambda x)^c}\left[1-e^{-(\lambda x)^c}\right]^{a-1}\left\{1-\left[1-e^{-(\lambda x)^c}\right]^a\right\}^{b-1}$$
$$\times\left[1-\left\{1-\left[1-e^{-(\lambda x)^c}\right]^a\right\}^b\right]^{\theta-1},$$

and

$$f_{BKW}(x) = \frac{1}{B(a,b)}\alpha\beta c\lambda^c x^{c-1} e^{-(\lambda x)^c}\left[1-e^{-(\lambda x)^c}\right]^{\alpha-1}$$
$$\times\left\{1-\left[1-e^{-(\lambda x)^c}\right]^\alpha\right\}^{\beta b-1}\left[1-\left\{1-\left[1-e^{-(\lambda x)^c}\right]^\alpha\right\}^\beta\right]^{a-1},$$

respectively.

The first data set consists of the number of successive failures for the air conditioning system of each member in a fleet of 13 Boeing 720 jet airplanes (Proschan 1963). The data is presented in Table 2. The second data set consists of the salaries of 818 professional baseball players for the year 2009 (USA TODAY).

The third data set represents the poverty rate of 533 districts with more than 15,000 students in 2009 (Digest of Education Statistics "http://nces.ed.gov/programs/digest/d11/tables/dt11_096.asp"). These data sets are modeled by the EKD distribution and compared with the corresponding sub-models, the Kumaraswamy-Dagum and Dagum

Table 2 Air conditioning system data

194	413	90	74	55	23	97	50	359	50	130	487	57
102	15	14	10	57	320	261	51	44	9	254	493	33
18	209	41	58	60	48	56	87	11	102	12	5	14
14	29	37	186	29	104	7	4	72	270	283	7	61
100	61	502	220	120	141	22	603	35	98	54	100	11
181	65	49	12	239	14	18	39	3	12	5	32	9
438	43	134	184	20	386	182	71	80	188	230	152	5
36	79	59	33	246	1	79	3	27	201	84	27	156
21	16	88	130	14	118	44	15	42	106	46	230	26
59	153	104	20	206	5	66	34	29	26	35	5	82
31	118	326	12	54	36	34	18	25	120	31	22	18
216	139	67	310	3	46	210	57	76	14	111	97	62
39	30	7	44	11	63	23	22	23	14	18	13	34
16	18	130	90	163	208	1	24	70	16	101	52	208
95	62	11	191	14	71							

distributions, and as well as EKW, BKW distributions. Table 3 gives a descriptive summary of each sample. The air conditioning system sample has far more variability and the baseball player salary sample has the lowest variability.

The maximum likelihood estimates (MLEs) of the parameters are computed by maximizing the objective function via the subroutine NLMIXED in SAS. The estimated values of the parameters (standard error in parenthesis), -2 Log-likelihood statistic, Akaike Information Criterion, $AIC = 2p - 2\ln(L)$, Bayesian Information Criterion, $BIC = p\ln(n) - 2\ln(L)$, and Consistent Akaike Information Criterion, $AICC = AIC + 2\frac{p(p+1)}{n-p-1}$, where $L = L(\hat{\Theta})$ is the value of the likelihood function evaluated at the parameter estimates, n is the number of observations, and p is the number of estimated parameters for the EKD distribution and its sub-distributions are tabulated. See Table 4, Table 5 and Table 6.

Plots of the fitted EKD, KD, D and the histogram of the data are given in Figure 4. The probability plots (Chambers et al. 1983) consists of plots of the observed probabilities, against the probabilities predicted by the fitted model are also presented in Figure 5. For the EKD distribution, we plotted for example,

$$G(x_{(j)}) = \left\{1 - \left[1 - \left(1 + \hat{\lambda}x_{(j)}^{-\hat{\delta}}\right)^{-\hat{\alpha}}\right]^{\hat{\phi}}\right\}^{\hat{\theta}}$$

against $\frac{j-0.375}{n+0.25}, j = 1, 2, \ldots, n$, where $x_{(j)}$ are the ordered values of the observed data. A measure of closeness of the plot to the diagonal line given by the sum of squares

$$SS = \sum_{j=1}^{n}\left[G(x_{(j)}) - \left(\frac{j-0.375}{n+0.25}\right)\right]^2$$

Table 3 Descriptive statistics

Data	Mean	Median	Mode	SD	Variance	Skewness	Kurtosis	Min.	Max.
I	92.07	54.00	14.00	107.92	11646	2.16	5.19	1.0	603.0
II	3.26	1.15	0.40	4.36	19.05	2.10	5.13	0.4	33.0
III	17.71	16.80	9.30	8.80	77.38	0.80	0.73	2.7	53.6

Table 4 Estimation of models for air conditioning system data

Distribution	Estimates						Statistics				
	α	λ	δ	ϕ	θ		-2 log likelihood	AIC	AICC	BIC	SS
EKD	20.6164	4.7323	0.6192	18.1616	0.1657		2065.0	2075.0	2075.3	2091.2	0.0309
	(1.2347)	(0.4174)	(0.0459)	(5.8028)	(0.0089)						
KD	5.0354	4.3846	0.3762	21.7047	1		2066.9	2074.9	2075.2	2087.9	0.0368
	(2.1177)	(3.0727)	(0.1253)	(27.9167)	-						
D	1.2390	94.1526	1.2626	1	1		2078.4	2084.4	2084.5	2094.1	0.1344
	(0.1749)	(33.7549)	(0.0663)	-	-						
	a	b	c	λ	θ						
EKW	3.7234	0.1219	1.0595	0.0495	0.3784		2063.7	2073.7	2074.0	2089.8	0.0254
	(0.8783)	(0.0183)	(0.1448)	(0.0224)	(0.1136)						
	a	b	α	β	c	λ					
BKW	1.4342	0.0830	2.0054	1.9100	0.7412	0.1809	2064.6	2076.6	2077.1	2096.1	0.0338
	(1.2507)	(0.0875)	(1.6573)	(1.9807)	(0.0343)	(0.0388)					

Table 5 Estimation of models for baseball player salary data

Distribution	Estimates						Statistics				
	α	λ	δ	ϕ	θ		-2 log likelihood	AIC	AICC	BIC	SS
EKD	69.1586	0.000043	7.6321	0.0591	0.4075		2864.1	2874.1	2874.2	2897.7	7.8153
	(0.000036)	(0.0000058)	(0.0557)	(0.0044)	(0.0327)						
KD	69.0839	0.000011	7.2375	0.0996	1		2957.2	2965.2	2965.2	2984.0	7.7095
	(0.000061)	(0.00000133)	(0.037)	(0.0036)	-						
D	70.0780	0.0116	1.0312	1	1		3225.6	3231.6	3231.6	3245.7	6.4568
	(34.4988)	(0.0058)	(0.0301)	-	-						
	a	b	c	λ	θ						
EKW	15.0514	0.1368	0.6376	8.8903	0.5419		3209.8	3219.8	3219.9	3243.3	5.3289
	(2.0692)	(0.0266)	(0.0756)	(4.9198)	(0.2098)						
	a	b	α	β	c	λ					
BKW	24.0047	0.03783	14.4799	4.6029	0.5168	32.1184	3088.4	3100.4	3100.5	3128.7	18.0516
	(0.6879)	(0.0039)	(0.2069)	(0.4549)	(0.006)	(2.4559)					

Table 6 Estimation of models for poverty rate data

Distribution	Estimates						Statistics				
	α	λ	δ	ϕ	θ		-2 log likelihood	AIC	AICC	BIC	SS
EKD	75.5803	0.851500	0.8183	60.9069	0.3091		3750.7	3760.7	3760.8	3782.1	0.1305
	(11.1276)	(0.32)	(0.0714)	(29.1324)	(0.02229)						
KD	60.8898	0.304000	0.4666	90.2889	1		3758.9	3766.9	3767.0	3784.0	0.2604
	(17.5714)	(0.0963)	(0.0555)	(54.8283)	-						
D	1.7954	350.0100	2.4175	1	1		3831.8	3837.8	3837.9	3850.7	0.9210
	(0.2034)	(105.94)	(0.0784)	-	-						
	a	b	c	λ	θ						
EKW	0.1013	2.2289	2.741	0.02545	20.0336		3752.8	3762.8	3762.9	3784.2	0.1071
	(0.0944)	(1.8026)	(2.2276)	(0.0199)	(30.0233)						
	a	b	α	β	c	λ					
BKW	0.9985	1.0006	1.9999	0.03989	2.0006	0.1141	4727.5	4739.5	4739.7	4765.2	80.9942
	(0.0069)	(0.0431)	(0.0584)	(0.0017)	(0.2564)	(0.0075)					

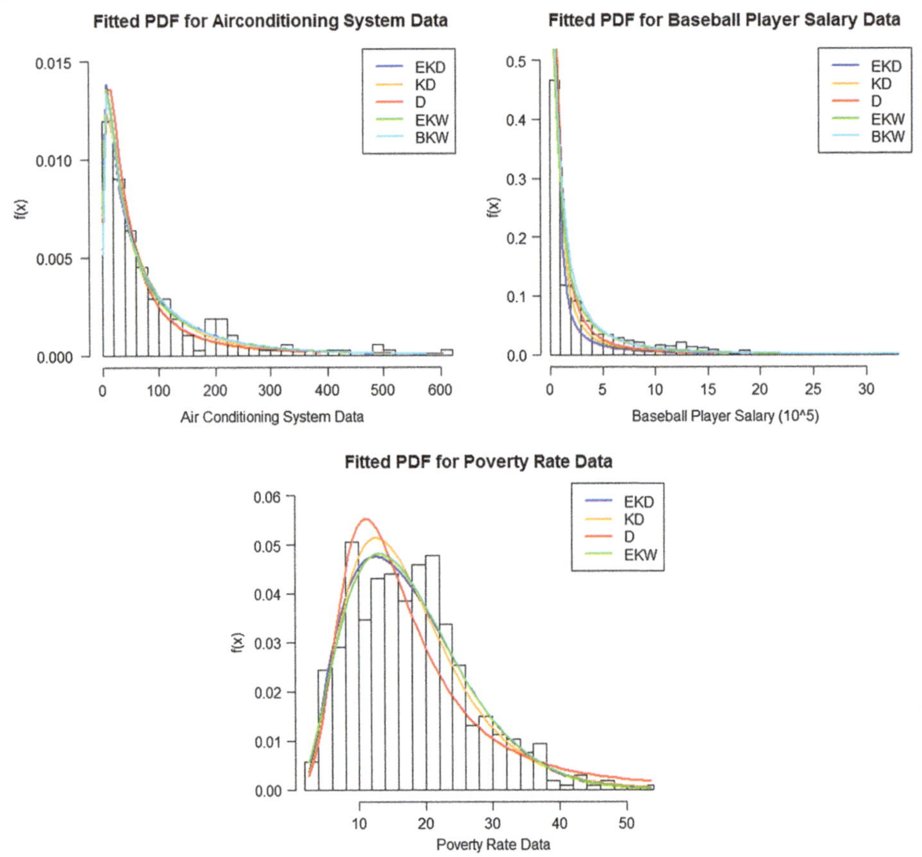

Figure 4 Fitted PDF for data sets.

was calculated for each plot. The plot with the smallest SS corresponds to the model with points that are closer to the diagonal line. Plots of the empirical and estimated survival functions for the models are also presented in Figure 6.

For the air conditioning system data, initial values $\alpha = 1, \lambda = 2, \delta = 0.6, \phi = 3, \theta = 1$ are used in SAS code for EKD model. The LR statistics for the test of the hypothesis $H_0 : KD$ against $H_a : EKD$ and $H_0 : D$ against $H_a : EKD$ are 1.9 (p-value $= 0.17$) and 13.4 (p-value $= 0.0012$). Consequently, KD distribution is the best distribution based on the LR statistic. The KD distribution gives smaller SS value than Dagum distribution and slightly bigger than EKD. For the non nested models, the values of AIC and AICC for KD and EKW models are very close, however the BIC value for KD distribution is slightly smaller than the corresponding value for the EKW distribution. We conclude that KD model compares favorably with the EKW distribution and thus provides a good fit for the air conditioning system data.

For the baseball player salary data set, initial values for EKD model in SAS code are $\alpha = 70, \lambda = 0.01, \delta = 1.026, \phi = 0.1, \theta = 1$. The EKD distribution is a better fit than KD and Dagum distributions for this data, as well as the other distributions. The values of the statistics AIC, AICC and BIC for KD distribution are smaller compared to the non nested distributions. The LR statistics for the test of the hypotheses $H_0 : KD$ against $H_a : EKD$ and $H_0 : D$ against $H_a : EKD$ are 93.1 (p-value< 0.0001) and 361.5

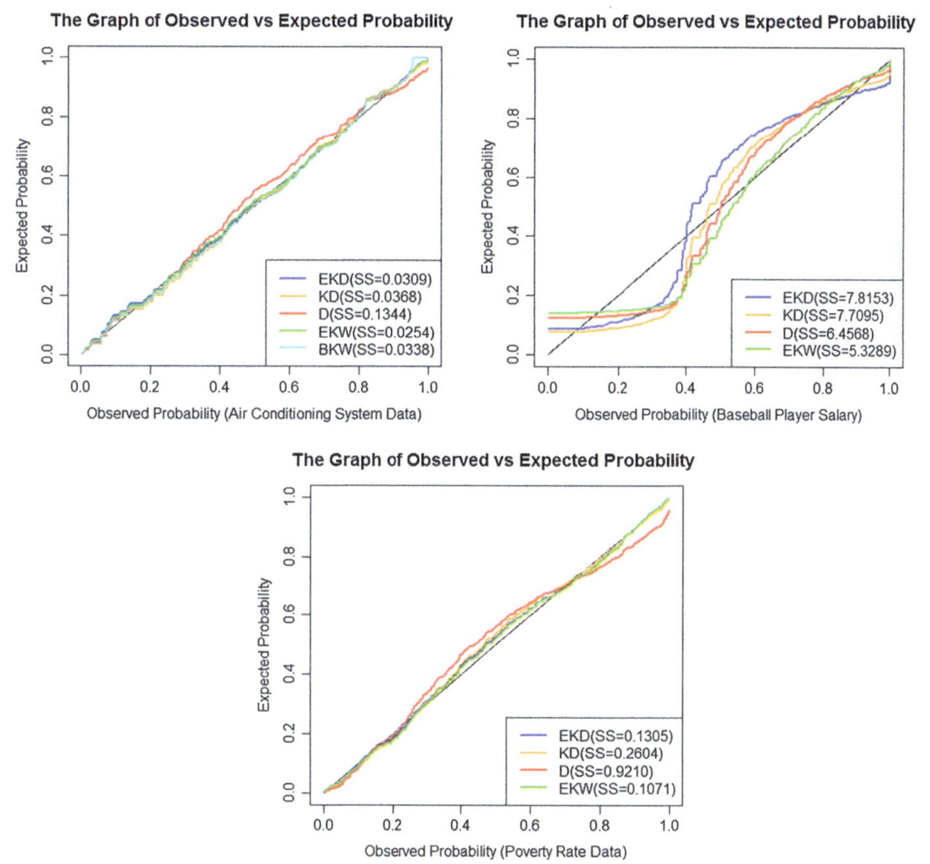

Figure 5 Observed probability vs predicted probability for data sets.

(p-value < 0.0001). Consequently, we reject the null hypothesis in favor of the EKD distribution and conclude that the EKD distribution is significantly better than the KD and Dagum distributions based on the LR statistic. The value of AIC, AICC and BIC statistics are lower for the EKD distribution when compared to those for the EKW and BKW distributions.

For poverty rate data, initial values for EKD model are $\alpha = 73, \lambda = 0.1, \delta = 0.15, \phi = 60, \theta = 0.33$. The LR statistic for the test of the hypotheses $H_0 : KD$ against $H_a : EKD$ and $H_0 : D$ against $H_a : EKD$ are 8.2 (p-value $= 0.0042$) and 81.1 (p-value < 0.0001), respectively. The values of AIC, AICC and BIC statistics shows EKD distributions is a better model and the SS value of EKD model is comparatively smaller than the corresponding values for the KD and D distributions. Consequently, we conclude that EKD distribution is the best fit for the poverty rate data.

9 Conclusions

We have proposed and presented results on a new class of distributions called the EKD distribution. This class of distributions have applications in income and lifetime data analysis. Properties of this class of distributions including the series expansion of pdfs, cdfs, moments, hazard function, reverse hazard function, income inequality measures such as Lorenz and Bonferroni curves are derived. Renyi entropy, order statistics, reliability, mean

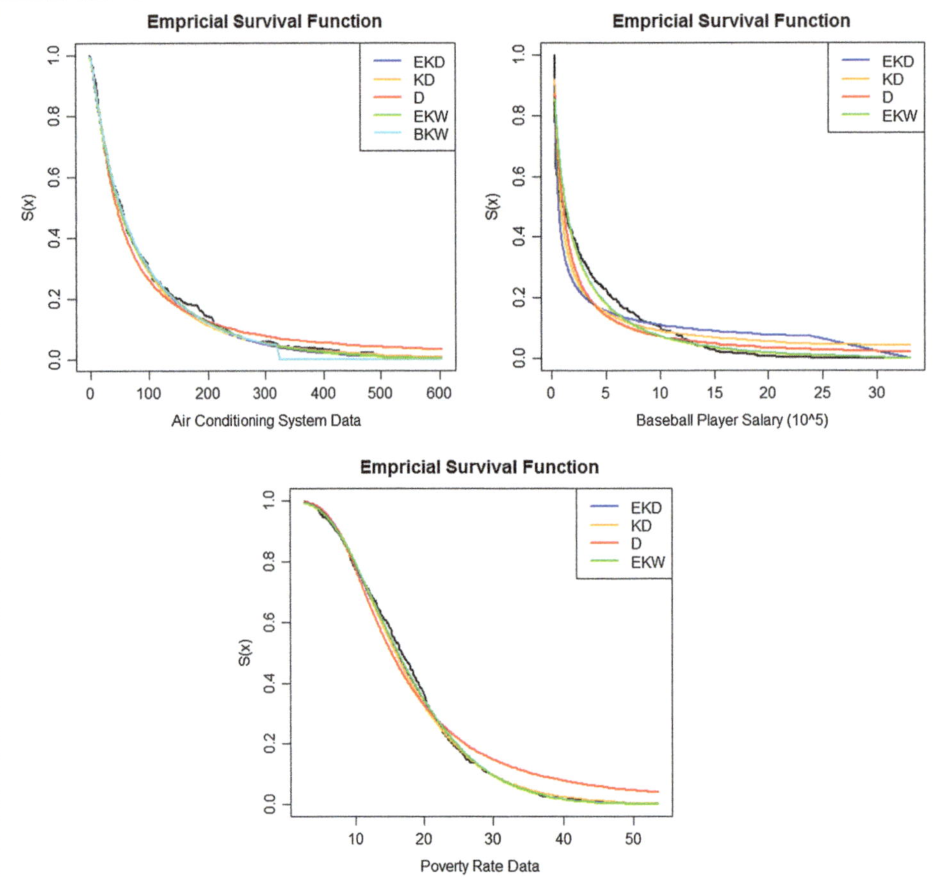

Figure 6 Empirical survival function for data sets.

and median deviations are presented. Estimation of the parameters of the models and applications are also given. Future work include MCMC methods with censored data and regression problems with concomitant information.

Appendix

R Codes

```
install.packages('stats4')
install.packages('bbmle')
library(stats4)
library(bbmle)
library(stats)
# Define Function
# define EKD pdf
g=function(alpha,lambda,delta,phi,theta,x){
y=alpha*lambda*delta*phi*theta*((x)^(-delta-1))*((1+lambda*(x^(-delta)))^(-alpha-1))*((1-((1+lambda*(x^(-delta)))^(-
alpha)))^(phi-1))*((1-((1-((1+lambda*(x^(-delta)))^(-alpha)))^(phi)))^(theta-1))
return(y)
}
# define EKD cdf
```

```
G=function(alpha,lambda,delta,phi,theta,x){
y=(1-((1-((1+lambda*(x^(-delta)))^(-alpha)))^(phi)))^(theta)
return(y)
}
# define EKD hazard
h=function(alpha,lambda,delta,phi,theta,x){
y=g(alpha,lambda,delta,phi,theta,x)/(1-G(alpha,lambda,delta,phi,theta,x))
return(y)
}
# define EKD quantile
quantile=function(alpha,lambda,delta,phi,theta,q){
((lambda)^(1/delta))*((((1-((1-((q)^(1/theta)))^(1/phi)))^(-1/alpha))-1)^(-1/delta))
}
# define EKD moments.
# note: k<delta
moments=function(alpha,lambda,delta,phi,theta,k){
f=function(alpha,lambda,delta,phi,theta,k,x){(x^k)*(g(alpha,lambda,delta,phi,theta,x))}
y=integrate(f,lower=0,upper=Inf,subdivisions=100000,alpha=alpha,lambda=lambda,delta=delta,phi=phi,theta=theta,k=k)
return(y)
}
# define EKD I(a)
Ia=function(alpha,lambda,delta,phi,theta,a){
n=length(a)
y=0
for(i in 1:n){
y[i]=integrate(function(alpha,lambda,delta,phi,theta,x){x*g(alpha,lambda,delta,phi,theta,x)},lower=0,upper=a[i],
subdivisions=100000,alpha=alpha,lambda=lambda,delta=delta,phi=phi,theta=theta)$value
}
return(y)
}
# define EKD bonferroni
# note: p is between (0,1)
bonferroni=function(alpha,lambda,delta,phi,theta,p){
q=quantile(alpha,lambda,delta,phi,theta,p)
mu=moments(alpha,lambda,delta,phi,theta,1)$value
y=(Ia(alpha,lambda,delta,phi,theta,q))/(p*mu)
return(y)
}
# define EKD lorenz
# note: p is between (0,1)
lorenz=function(alpha,lambda,delta,phi,theta,p){
q=quantile(alpha,lambda,delta,phi,theta,p)
mu=moments(alpha,lambda,delta,phi,theta,1)$value
y=(Ia(alpha,lambda,delta,phi,theta,q))/(mu)
return(y)
}
```

Competing interests

The authors declare that they have no competing interests.

Authors' contributions

SH provided the R codes, conducted the simulations and the applications. BOO proposed, developed the statistical properties of the new family of distributions and drafted the manuscript. All authors read and approved the final manuscript.

Authors' information

Shujiao Huang is a graduate student at Georgia Southern University and Broderick O. Oluyede is Professor of Mathematics and Statistics at Georgia Southern University.

References

Chambers, J, Cleveland, W, Kleiner, B, Tukey, P: Graphical Methods for Data Analysis. Chapman and Hall, London (1983)

Cordeiro, GM, Ortega, EMM, Nadarajah, S: The Kumaraswamy Weibull distribution with application to failure data. J. Franklin Inst. **347**, 1399–1429 (2010)

Cordeiro, GM, Gomes, AE, de-Silva, CQ, Ortega, EMM: The Beta Exponentiated Weibull Distribution. J. Stat. Comput. Simulat. **83**(1), 114–138 (2013)

Dagum, CA: New model of personal income distribution: specification and estimation. Economie Applique'e. **30**, 413–437 (1977)

Domma, F, Condino, F: The Beta-Dagum distribution: definition and properties. Communications in Statistics-Theory and Methods. **44**(22), 4070–4090 (2013)

Domma, F, Giordano, S, Zenga, M: Maximum likelihood estimation in Dagum distribution with censored samples. J. Appl. Stat. **38**(21), 2971–2985 (2011)

Kleiber, C: A Guide to the Dagum Distributions. In: Duangkamon, C (ed.) *Modeling Income Distributions and Lorenz Curve Series: Economic Studies in Inequality, Social Exclusion and Well-Being 5*. Springer, New York, (2008)

Kleiber, C, Kotz, S: Statistical size distributions in economics and actuarial sciences. Wiley, New York (2003)

Kumaraswamy, P: Generalized probability density function for double-bounded random process. J. Hydrol. **46**, 79–88 (1980)

McDonald, B: Some generalized functions for the size distribution of income. Econometrica. **52**(3), 647–663 (1984)

McDonald, B, Xu, J: A generalization of the beta distribution with application. J. Econometrics. **69**(2), 133–152 (1995)

Oluyede, BO, Rajasooriya, S: The Mc-Dagum Distribution and Its Statistical Properties with Applications. Asian J. Mathematics and Applications. **2013**(85) (2013). http://scienceasia.asia/index.php/ama/article/view/85/44, 1–16

Proschan, F: Theoretical explanation of observed decreasing failure rate. Technometrics. **5**, 375–383 (1963)

Renyi, A: On measures of entropy and information. Berkeley Symp. Math. Stat. Probability. **1**(1), 547–561 (1960)

6

Failure time regression with continuous informative auxiliary covariates

Lipika Ghosh[1], Jiancheng Jiang[1*], Yanqing Sun[1] and Haibo Zhou[2]

*Correspondence: jjiang1@uncc.edu
[1] Department of Mathematics and Statistics, University of North Carolina at Charlotte, Charlotte, NC 28223, USA
Full list of author information is available at the end of the article

Abstract

In this paper we use Cox's regression model to fit failure time data with continuous informative auxiliary variables in the presence of a validation subsample. We first estimate the induced relative risk function by kernel smoothing based on the validation subsample, and then improve the estimation by utilizing the information on the incomplete observations from non-validation subsample and the auxiliary observations from the primary sample. Asymptotic normality of the proposed estimator is derived. The proposed method allows one to robustly model the failure time data with an informative multivariate auxiliary covariate. Comparison of the proposed approach with several existing methods is made via simulations. Two real datasets are analyzed to illustrate the proposed method.

Mathematics Subject Classification (MSC): 62G07, 62G20

Keywords: Auxiliary covariates; Censoring; Estimated partial likelihood; Local linear smoothing; Validation

1 Introduction

In epidemiologic studies, the exposure variable vector X is often too difficult or too expensive to measure on the full cohort, whereas an auxiliary variable vector W for X can be easily measured for all subjects in the study cohort. For example, in a large scale nutritional study, the PIN Study (Savitz et al. 2001), it would be prohibitively expensive to obtain the exact dietary iron intake on each individual recruited. Instead, a self administered quantitative food questionnaire is conducted on all subjects where a crude assessment of iron intake is obtained. The true exposure, the blood serrum ferritin concentration, is only assayed for a validation set consisting of a small subset of the full study cohort. Although the true covariates are missing for most individuals, the existence of some surrogates or auxiliary measurements conveys information about X and serves as common proxy measure. Utilizing the available auxiliary information to improve the efficiency of the effects estimation and in turns to increase the power of the study is critical for the success of the studies. In this paper, we study censored failure time regression with a continuous auxiliary covariate vector.

A variety of authors have contributed their work to this field. Related works include Prentice (1982), Pepe et al. (1989), Lin and Ying (1993), Hughes (1993), Lipsitz and Ibrahim (1996), Zhou and Wang (2000), Fan and Wang (2009), Liu et al. (2010), etc. In particular, Prentice (1982) introduced a partial likelihood estimator based on the induced relative risk function. This method was further developed by Pepe et al. (1989) using parametric modeling. Zhou and Pepe (1995) proposed an estimated partial likelihood method for discrete auxiliary covariates to relax the parametric assumptions on the frequency of events and the underlying distributions of covariates. This method was extended by Zhou and Wang (2000) to deal with continuous auxiliary variables, based on the Nadaraya-Watson kernel smoother method (Nadaraya 1964; Watson, 1964). Fan and Wang (2009), Liu et al. (2010) used the same approach for multivariate failure time data with auxiliary covariates. While Zhou and Wang's (2000) approach is useful in certain situations, there are some restrictions on it. First, the approach is effective only when the auxiliary variable W is of low dimension so that the "curse of dimensionality" in nonparametric smoothing can be avoided. Secondly, it requires that, conditionally on X, W provides no additional information about the hazard of failure; that is, all of the effects of W on failure and censoring are mediated through X, which is somewhat restricted since W may not be a true surrogate and depends on the failure given X.

Further, this method does not fully utilize the observations in the non-validation subsample and hence cannot be efficient in certain situations.

We here propose a new method to deal with the above problems associated with the method in Zhou and Wang (2000). The proposed method allows W to be multivariate and to be informative in the sense that, conditional X, it may provide additional information on the hazard of failure. We first estimate the induced relative risk function with a kernel smoother based on the validation sample, and then improve the estimation by utilizing the information on the incomplete observations from the non-validation subsample. In addition, the local linear smoother (see for example in Fan and Gijbels 1996) is employed to enhance the performance of the kernel smoother in Zhou and Wang (2000) at the boundary regions. Our method will be expected to improve the efficiency of the estimator of Zhou and Wang (2000) in various situations, for example, when auxiliary variable W is informative or not very informative about X (see also the simulation results). Asymptotic normality of our estimator is derived.

The proposed methodology can be extended to model multivariate failure time data with auxiliary covariates by following the method in Fan and Wang (2009) or Liu et al. (2010).

The paper is organized as follows. In Section 2, we introduce the hazards models. In Section 3 we introduce our new estimation approach to predicting the induced relative risk for individuals in non-validation subsample based on the kernel smoother. In Section 4 we concentrate on the asymptotic properties of the proposed estimators. We conduct simulations in Section 5 to compare the efficiencies of different estimating methods. In Section 6 we apply the proposed methodology to two real datasets.

2 Cox's proportional hazards models

To facilitate exposition, we here employ the notations in Zhou and Wang (2000). Suppose that there are n independent individuals in a study cohort. Let $\{X_i(t), Z_i(t)\}$ denote the covariate vector for the ith subject at time t $(i = 1, \cdots, n)$. Assume that $X_i(\cdot)$ is only observed in the validation subsample which is chosen at the baseline under the ignorable missing mechanism condition (Rubin 1976). Let $Z_i(\cdot)$ be the remaining covariate vector that is always observed, and $W_i(\cdot)$ the informative auxiliary variables for $X_i(\cdot)$. Let η_i be an indicator variable with $\eta_i = 1$ if the ith individual is in the validation set and 0 if in the nonvalidation set. Put $V = \{i : \eta_i = 1\}$ and $\bar{V} = \{i : \eta_i = 0\}$. We assume that individuals in the validation subsample are randomly selected and hence representative. Then observed data for the ith subject is $\{S_i, \delta_i, Z_i(\cdot), W_i(\cdot), X_i(\cdot)\}$ if $\eta_i = 1$, and $\{S_i, \delta_i, Z_i(\cdot), W_i(\cdot)\}$ if $\eta_i = 0$, where S_i is the observed event time for the ith subject, which is the minimum of the potential failure time T_i and the censoring time C_i, and δ_i is the indicator of censoring. We consider the following conditional hazard rate function of failure (Cox 1972)

$$\begin{aligned}\lambda\{t; X_i(t), Z_i(t)\} &\equiv \lim_{\Delta t \downarrow 0} \left[\frac{1}{\Delta t} \Pr\{t \leq T_i < t + \Delta t | T_i \geq t, X_i(t), Z_i(t)\}\right] \\ &= \lambda_0(t) \exp\left\{\beta_1' X_i(t) + \beta_2' Z_i(t)\right\}, \qquad (2.1)\end{aligned}$$

where $\lambda_0(\cdot) \geq 0$ is the unspecified base-line hazard and $\beta = \left(\beta_1', \beta_2'\right)'$ is the relative risk parameter vector to be estimated.

For model (2.1), the relative risk functions are $\gamma_i(\beta, t) \equiv \exp\left\{\beta_1' X_i(t) + \beta_2' Z_i(t)\right\}$, and the partial likelihood function for the parameters β is

$$PL(\beta) = \prod_{i \in V \cup \bar{V}} \left\{\frac{\gamma_i(\beta, S_i)}{\sum_{j \in \mathcal{R}(S_i)} \gamma_j(\beta, S_i)}\right\}^{\delta_i}, \qquad (2.2)$$

where $\mathcal{R}(S_i)$ is the risk set at time S_i. However, for $i \in \bar{V}$, the true variate $X_i(t)$ is not observed, and hence the corresponding relative risk function $\gamma_i(\beta, t)$ is not available and has to be imputed.

Zhou and Wang (2000) used the conditional expectation

$$\exp\left\{\beta_2' Z_i(t)\right\} E\left[\exp\left\{\beta_1' X_i(t)\right\} | S_i \geq t, Z_i(t), W_i(t)\right] \qquad (2.3)$$

for the imputation of $\gamma_i(\beta, t)$ ($i \in \bar{V}$). Based on data in V, they obtained the Nadaraya-Watson kernel estimator (Nadaraya 1964; Watson 1964) of the above imputation and replaced $\gamma_i(\beta, t)$ for $i \in \bar{V}$ in (2.2) by the kernel estimator, which leads to the estimated partial likelihood. Under the assumption that W is not informative, that is, all of the effects of W on failure and censoring are mediated through X, so that

$$\begin{aligned}\lambda\{t; X_i(t), Z_i(t), W_i(t)\} &\equiv \lim_{\Delta t \downarrow 0} \left[\frac{1}{\Delta t} Pr\{t \leq T_i < t + \Delta t | T_i \geq t, X_i(t), Z_i(t), W_i(t)\}\right] \\ &= \lim_{\Delta t \downarrow 0} \left[\frac{1}{\Delta t} Pr\{t \leq T_i < t + \Delta t | T_i \geq t, X_i(t), Z_i(t)\}\right] \\ &= \lambda_0(t) \exp\left\{\beta_1' X_i(t) + \beta_2' Z_i(t)\right\} \\ &\equiv \lambda\{t; X_i(t), Z_i(t)\},\end{aligned}$$

they derived the consistency and asymptotic normality of the estimation. However, if W is informative, their method will generally be biased (see also Section 5). In addition, since this method directly used information in the auxiliary covariate W and estimated the conditional expectation (2.3), it may encounter the so-called "curse of dimensionality" if W is of higher dimension. For the present study, we propose a new method for imputation of the relative risk function. The information in W will be used in a new way. This leads to a new estimated partial likelihood.

3 Estimated partial likelihood with a local smoother

In this section, we introduce our method to estimate the parameters in model (2.1) based on maximizing the estimated partial likelihood.

3.1 Local smoother for the relative risk function

Instead of (2.3), we use the conditional expectation of $\gamma_i(\beta, t)$,

$$\phi_i(\beta, t) = \exp\{\beta_2' Z_i(t) E\left[\exp\left\{\beta_1' X_i(t)\right\} | S_i \geq t, Z_i(t)\right],$$

as imputation of $\gamma_i(\beta, t)$. Let d be the dimension of Z and let

$$r_i(\beta, t) = \eta_i \gamma_i(\beta, t) + (1 - \eta_i)\phi_i(\beta, t) \tag{3.4}$$

be the induced risk function. Put $\zeta_i(\beta_1, t) = \exp\left(\beta_1' X_i(t)\right)$, and $\nu_i(\beta_1, t) = E[\zeta_i(\beta_1, t)|S_i \geq t, Z_i(t)]$. To use the partial likelihood (2.2), we need to estimate $\phi_i(\beta, t)$ or equivalently $\nu_i(\beta_1, t)$ for $i \in \bar{V}$. Using the local linear smoother (see for example, Fan and Gijbels 1996) leads to the following (functional) estimators of $\nu_j(\beta_1, t)$ for $j \in \bar{V}$

$$\hat{\nu}_j(\beta_1, t) = \sum_{i \in V} \omega_i(t, Z_j(t); h)\zeta_i(\beta_1, t), \tag{3.5}$$

where h is the bandwidth,

$$\omega_i(t, Z_j(t); h) = \frac{\{s_2 - (Z_i(t) - Z_j(t))s_1\} Y_i(t) K_h(Z_i(t) - Z_j(t))}{\sum_{i \in V}\{s_2 - (Z_i(t) - Z_j(t))s_1\} Y_i(t) K_h(Z_i(t) - Z_j(t))},$$

$Y_i(t) = I_{[S_i \geq t]}$ is the at-risk indicator, $s_k = \sum_{i \in V}(Z_i(t) - Z_j(t))^k Y_i(t) K_h(Z_i(t) - Z_j(t))$, and $K_h(\cdot) = h^{-d} K(\cdot/h)$ for a d-variate kernel function $K(\cdot)$.

The above estimation of the relative risk function was similarly used in Zhou and Wang (2000) for a nonparametric smoothing problem on the estimation of $E[\gamma_i(\beta, t)|S_i \geq t, Z_i(t), W_i(t)]$, where the "curse of dimensionality" problem can happen if W is multivariate. Note that this estimation method uses only the complete observations in V and neglects the important information on incomplete observations in $\bar{V}$. It follows that this approach cannot be expected to be efficient in certain situations. In addition, it is required in Zhou and Wang (2000) that, conditional on X, the auxiliary variable W provides no additional information on the the hazard of failure. This requirement may not hold if W is not a genuine surrogate of X. In the following, we propose an improved estimation approach which utilizes information from W and observations in $\bar{V}$ and does not impose the requirement. Moreover, the proposed method allows one to model the failure time data with informative multivariate auxiliary variable W without "curse of dimensionality". Note that even for one dimensional Z and W, the method in Zhou and Wang (2000) requires a two-dimensional smoother while the new method needs only one-dimensional smoothing. To have a performance comparable with that of a one-dimensional nonparametric smoother using $M_1 = 50$ data points, we need about $M = M_1^{1.2} = 109$ data points for a 2-dimensional nonparameteric smoother. Hence the loss of efficiency due to highly dimensional smoothing is large and increasing exponentially fast (see page 317 of Fan and Yao 2003).

3.2 Improved estimation of the relative risk function and the estimated partial likelihood

Recall that W is a vector of auxiliary variables for X and is hence correlated with X. Let $\xi_i(\alpha, t) = \exp(\alpha' W_i(t))$, where α is a parameter vector to be chosen. Considering the conditional expectation of $\psi_i(\alpha, t) = E[\xi_i(\alpha, t)|S_i \geq t, Z_i(t)]$, then we can estimate $\psi_i(\alpha, t)$ by running local linear smoothing based on the data in V:

$$\hat{\psi}_j(\alpha, t) = \sum_{i \in V} \omega_i(t, Z_j(t); h)\xi_i(\alpha, t). \tag{3.6}$$

The following result depicts asymptotic correlation of $\hat{\nu}_j(\beta_1, t)$ and $\hat{\psi}_j(\alpha, t)$.

Proposition 3.1. *Suppose that the conditions in Appendix 1 hold. Given* $(S_j \geq t, Z_j(t))$, $\sqrt{nh^d}[(\hat{\nu}_j(\beta_1, t) - \nu_j(\beta_1, t)), (\hat{\psi}_j(\alpha, t) - \psi_j(\alpha, t))]$ *is jointly asymptotically normal with mean zero and covariance matrix*

$$\Sigma = \nu_0(K)p^{-1}(Z_j) \begin{bmatrix} \sigma_1^2(Z_j, t) & \rho_\alpha^*(Z_j, t)\sigma_1(Z_j, t)\sigma_2(Z_j, t) \\ \rho_\alpha^*(Z_j, t)\sigma_1(Z_j, t)\sigma_2(Z_j, t) & \sigma_2^2(Z_j, t) \end{bmatrix},$$

where $\nu_0(K) = \int K^2(u)du$, $\sigma_1^2(Z_j, t) = Var[\zeta_j|S_j \geq t, Z_j]$, $\sigma_2^2(Z_j, t) = Var[\xi_j|S_j \geq t, Z_j]$, *and* $p(\cdot)$ *is the density function of* Z.

By the distribution theory for multivariate normal variates, the conditional distribution of $\sqrt{nh^d}[\hat{\nu}_j(\beta_1, t) - \nu_j(\beta_1, t)]$ given $\sqrt{nh^d}[\hat{\psi}_j(\alpha, t) - \psi_j(\alpha, t)]$ is asymptotically normal with mean

$$\rho_\alpha^*(Z_j, t)\frac{\sigma_1(Z_j, t)}{\sigma_2(Z_j, t)}\sqrt{nh^d}[\hat{\psi}_j(\alpha, t) - \psi_j(\alpha, t)].$$

The conditional mean can be estimated by substituting consistent estimators based on the validation sample for $\rho_\alpha^*(Z_j, t)$, $\sigma_1(Z_j, t)$ and $\sigma_2(Z_j, t)$, and by replacing $\psi_j(\alpha, t)$ with the primary sample based estimator

$$\bar{\psi}_j(\alpha, t) = \sum_{i \in V \cup \bar{V}} \bar{\omega}_i(t, Z_j(t); h)\xi_i(\alpha, t), \tag{3.7}$$

where

$$\bar{\omega}_i(t, Z_j(t); h) = \frac{\left\{\bar{s}_2 - (Z_i(t) - Z_j(t))\bar{s}_1\right\} Y_i(t)K_h(Z_i(t) - Z_j(t))}{\sum_{i \in V \cup \bar{V}} \left\{\bar{s}_2 - (Z_i(t) - Z_j(t))\bar{s}_1\right\} Y_i(t)K_h(Z_i(t) - Z_j(t))}$$

and $\bar{s}_k = \sum_{i \in V \cup \bar{V}} (Z_i(t) - Z_j(t))^k Y_i(t)K_h(Z_i(t) - Z_j(t))$.

By equating $\sqrt{n_v h^d}[\hat{\nu}_j(\beta_1, t) - \nu_j(\beta_1, t)]$ with its estimated conditional mean and solving for $\nu_j(\beta_1, t)$, we obtain an improved (functional) estimate

$$\bar{\nu}_j(\beta_1, t) = \hat{\nu}_j(\beta_1, t) - \hat{\rho}_\alpha^*(Z_j, t)\frac{\hat{\sigma}_1(Z_j, t)}{\hat{\sigma}_2(Z_j, t)}\left[\hat{\psi}_j(\alpha, t) - \bar{\psi}_j(\alpha, t)\right]. \tag{3.8}$$

The updated estimator $\bar{\nu}_j(\beta_1, t)$ is doomed to be more accurate than $\hat{\nu}_j(\beta_1, t)$ in (3.5), since it has used the information from W and observations in $\bar{V}$. Even though the information about W may not be utilized in a very efficient way as in Zhou and Wang's (2000) estimator when W is not informative, it is the price we have to pay for achieving robustness against informative W. Note that $\bar{\nu}_j$ depends on α which is related to efficiency of the estimator. Intuitively, one should choose α to maximize the conditional correlation coefficient between ζ_j and ξ_j, given $(S_j \geq t, Z_j)$, which is evident from the following result.

Proposition 3.2. *Assume that the conditions in Appendix 1 hold. Given* $(S_j \geq t, Z_j(t))$, *then we have*

$$\sqrt{nh^d}\left[\bar{v}_j(\beta_1, t) - v_j(\beta_1, t)\right] \xrightarrow{\mathcal{D}} \mathcal{N}(0, \Omega),$$

where $\Omega(Z_j, t) = \sigma_1^2(Z_j, t)\left[1 - (1-\rho)\rho_\alpha^{*2}(Z_j, t)\right] v_0(K) p^{-1}(Z_j)$.

When $\rho_\alpha^* = 0$, i.e. the relative risk contributed by W is not correlated to that contributed by X, given $(S \geq t, Z)$, the estimator $\bar{v}_j$ is asymptotically equivalent to $\hat{v}_j$ in (3.5).

In general, $\rho_\alpha^* > 0$. Hence, by Propositions 3.1 and 3.2, $\bar{v}_j$ is more efficient than $\hat{v}_j$. Note that the proposed estimator is consistent for any α.

The above estimation method for $v_j(\beta, t)$ was similarly used in Chen and Chen (2000) for estimating parameters in a parametric regression model. Our estimation can be regarded as an extension of their estimation approach in nonparametric regression. In addition, we do not need a working model to specify the regression relationship between the surrogate and the covariate, and hence there is no risk of misspecification of the working model.

For each given value of β, with the estimator $\bar{v}_j(\beta, t)$, we can estimate the induced relative risk $r_i(\beta, t)$ in (3.4) by

$$\hat{r}_i(\beta, t) = \eta_i \gamma_i(\beta, t) + (1 - \eta_i)\bar{\phi}_i(\beta, t), \tag{3.9}$$

where $\bar{\phi}_i(\beta, t) = \bar{v}_i(\beta_1, t) \exp\left\{\beta_2' Z_i(t)\right\}$. Then the parameters β can be estimated by maximizing the following estimated partial likelihood function (EPL):

$$EPL(\beta) = \prod_{i=1}^{n} \left\{ \frac{\hat{r}_i(\beta, S_i)}{\sum_{j \in \mathcal{R}(S_i)} \hat{r}_j(\beta, S_i)} \right\}^{\delta_i}. \tag{3.10}$$

We denote $\hat{\beta}_{EPL} = \arg\max_\beta EPL(\beta)$.

For an extreme case with $W \approx Z$, Zhou and Wang's imputation for (2.3) approximately becomes $\hat{\phi}_i(\beta, t) = \hat{v}_i(\beta, t) \exp(\beta' Z_i(t))$ and uses a two dimensional smoother, which is inferior to the improved estimator $\bar{\phi}_i(\beta, t)$, and hence by the definition of $\hat{\beta}_{EPL}$, our estimator is superior to Zhou and Wang's. However, it is generally difficult to compare these two estimators. In our estimation of the induced relative risk, we used an improved estimator $\bar{\phi}_j(\beta, t)$ for $j \in \bar{V}$. The "curse of dimensionality" problem in Zhou and Wang (2000) can be avoided for a multivariate W. Our approach would at least be useful in cases where the number of variables in Z which are correlated with the missing covariate X is low, whereas the exposure variables of interest and their auxiliary variables may be multivariate.

An alternative to $\hat{\beta}_{EPL}$ is to maximize (3.10) but with $\hat{r}_i(\beta, t)$ replaced by $\tilde{r}_i(\beta, t) = \eta_i \gamma_i(\beta, t) + (1 - \eta_i)\hat{\phi}_i(\beta, t)$, where $\hat{\phi}_i(\beta, t) = \hat{v}_i(\beta, t) \exp(\beta' Z_i(t))$. We denote the resulting estimator by $\hat{\beta}_V$, which does not use the information on W in $\bar{V}$. Intuitively, $\hat{\beta}_{EPL}$ should be better than $\hat{\beta}_V$, but this is not true in general, since comparison of the asymptotic results in Theorems 4.1 and 4.2 below could not lead to a dominated estimator. However, in small validation ratio settings, $\hat{\beta}_V$ is not expected to perform well, since it uses only the observations in the validation set for smoothing.

4 Asymptotic behaviors

Let n_v be the subsample size of the validation set and let ρ be the limit of ratio of validation observations, $\lim_{n\to\infty} n_v/n$. Assume that $\rho \in (0,1]$. Define $s^{(0)}(\beta,t) = E[Y_i(t)r_i(\beta,t)]$, $s^{(1)}(\beta,t) = (\partial/\partial\beta)s^{(0)}(\beta,t)$, $s^{(2)}(\beta,t) = (\partial/\partial\beta^\tau)s^{(1)}(\beta,t)$.

For any matrix A, we use $A^{\otimes 2}$ to denote matrix AA^τ.

Let $N_i(t) = I_{[S_i<t,\delta_i=1]}$ and

$$M_i(t) = N_i(t) - \int_0^t Y_i(u)r_i(\beta_0,u)\lambda_0(u)du.$$

Define the filters $\mathcal{F}_i(t) = \sigma\{N_i(u), Y_i(u+), X_i(u), Z_i(u) : 0 \le u \le t\}$.

The censoring time is assumed to be independent of the failure time conditioning on the true covariates in model (2.1), that is,

$$P\{t \le T < t+\Delta t | T \ge t, C \ge t, X(t), Z(t)\} = P\{t \le T < t+\Delta t | T \ge t, X(t), Z(t)\}, \tag{4.11}$$

which is different from that in Zhou and Wang (2000) where it is assumed

$$P\{t \le T < t+\Delta t | T \ge t, C \ge t, W(t), Z(t)\} = P\{t \le T < t+\Delta t | T \ge t, W(t), Z(t)\}.$$

Then, under the independent censoring assumption (4.11),

$M_i(t)$ is a mean zero martingale with repsect to $\mathcal{F}_i(t)$ (Kalbfleisch and Prentice 1980; Fleming and Harrington 1991).

In addition, the cumulative hazard $\Lambda_0(t) = \int_0^t \lambda_0(w)\,dw$ can be consistently estimated as

$$\hat\Lambda_0(t) = \int_0^t \left[\sum_{i=1}^{n_v} Y_i(u)r_i(\hat\beta_{EPL},u)\right]^{-1} \sum_{i=1}^{n_v} dN_i(u).$$

Without loss of generality, we assume that $t \in [0,1]$. Put $\Delta(\phi_i)(u) = \phi_i^{(1)}(u)/\phi_i(u) - s^{(1)}/s^{(0)}$, $\Delta(\gamma_i)(u) = \gamma_i^{(1)}(u)/\gamma_i(u) - s^{(1)}/s^{(0)}$, $Q_i = \int_0^1 \Delta(\phi_i)(u)Y_i(u)[\gamma_i(\beta_0,u) - \phi_i(\beta_0,u)]\,\lambda_0(u)du$, $Q_i^* = \int_0^1 \Delta(\phi_i)(u)Y_i(u)\theta_i(u;\alpha)\lambda_0(u)du$, where $\phi_1^{(1)}(\beta,u) = (\partial/\partial\beta)r_i(\beta,u)$, and

$$\theta_i(u;\alpha) = [\xi_i(\alpha,u) - \psi_i(\alpha,u)]\exp(\beta_2' Z_i(u))\rho_\alpha^*(Z_i,u)\sigma_1(Z_i,u)/\sigma_2(Z_i,u).$$

The following theorem shows that $\hat\beta_{EPL}$ is asymptotically normal.

Theorem 4.1. *Suppose that Condition (A) in Appendix 1 holds. Then $\hat\beta_{EPL}$ is consistent estimator for β and satisfies*

$$\sqrt{n}(\hat\beta_{EPL} - \beta_0) \xrightarrow{\mathcal{L}} \mathcal{N}(0,\Omega),$$

where $\Omega = I^{-1}(\beta_0)\Sigma(\beta_0)I^{-1}(\beta_0)$ *with* $\Sigma(\beta_0) = (1-\rho)\Sigma_1(\beta_0) + \rho\Sigma_2(\beta_0)$,

$$I(\beta_0) = -E\left[\int_0^1 \left(\frac{r_i^{(2)}(\beta_0,u)}{r_i^{(0)}(\beta_0,u)} - \left\{\frac{r_i^{(1)}(\beta_0,u)}{r_i^{(0)}(\beta_0,u)}\right\}^{\otimes 2} - \frac{s^{(2)}(\beta_0,u)}{s^{(0)}(\beta_0,u)} - \left\{\frac{s^{(1)}(\beta_0,u)}{s^{(0)}(\beta_0,u)}\right\}^{\otimes 2}\right) dN_i(t)\right],$$

$$\Sigma_1(\beta_0) = E\left[\int_0^1 \Delta(\phi_i)(u)dM_i(u) - (1-\rho)Q_i^*\right]^{\otimes 2},$$

$$\Sigma_2(\beta_0) = E\left[\int_0^1 \Delta(\gamma_i)(u)dM_i(u) - \frac{1-\rho}{\rho}\{Q_i - (1-\rho)Q_i^*\}\right]^{\otimes 2}.$$

Remark 4.1. It is interesting to note that $Q_i^* \approx Q_i$ when the auxiliary W approximates X, and hence the second term in the expectation of $\Sigma_2(\beta)$ approximates to $(1-\rho)Q_i$. Therefore, a small ρ will not result in a big $\Sigma_2(\beta_0)$. Theoretically, when $W_i = Z_i$, $Q_i^* = 0$ and the above asymptotic variance formula shares the same formula as that for the estimator in Zhou and Wang (2000) as exactly expected. However, in practice where $W_i \approx Z_i$, since Zhou and Wang (2000) used a higher dimensional smoother than us, our estimator would have better efficiency for finite samples.

When $\rho_\alpha^* = 0$, $\hat{\beta}_{EPL}$ is asymptotically equivalent to the complete-case estimator based on only the validation set V. This is also expected, since the auxiliary variable W_i contains no information on X_i at this setting. From Theorem 4.1, the asymptotic covariance matrix of $\hat{\beta}_{EPL}$ is of sandwich form, which can consistently be estimated by $\hat{\Omega}_0 = \hat{I}^{-1}(\beta_0)\hat{\Sigma}(\beta_0)\hat{I}^{-1}(\beta_0)$, where $\hat{I}(\beta)$ and $\hat{\Sigma}(\beta)$ are the corresponding empirical estimates. Specifically,

$$\hat{I}(\beta) = -n^{-1}\sum_{i=1}^{n}\int_0^1 \left(\frac{\hat{r}_i^{(2)}(\beta,u)}{\hat{r}_i^{(0)}(\beta,u)} - \left\{\frac{\hat{r}_i^{(1)}(\beta,u)}{r_i^{(0)}(\beta,u)}\right\}^{\otimes 2} - \frac{\hat{S}^{(2)}(\beta,u)}{\hat{S}^{(0)}(\beta,u)} + \left\{\frac{\hat{S}^{(1)}(\beta,u)}{\hat{S}^{(0)}(\beta,u)}\right\}^{\otimes 2}\right) dN_i(t),$$

$$\hat{\Sigma}_1(\beta) = (n-n_v)^{-1}\sum_{i\in\bar{V}}\left\{\int_0^1 \Delta(\hat{\phi}_i)(t)\left[dN_i(t) - Y_i(t)\bar{\phi}_i(\hat{\beta}_{EPL},t)\,d\hat{\Lambda}_0(t)\right] - (1-\hat{\rho})\hat{Q}_i^*\right\}^{\otimes 2},$$

$$\hat{\Sigma}_2(\beta) = n_v^{-1}\sum_{i\in V}\left\{\int_0^1 \Delta(\hat{\gamma}_i)(t)\left[dN_i(t) - Y_i(t)r_i(\hat{\beta}_{EPL},t)\,d\hat{\Lambda}_0(t)\right] - \frac{1-\hat{\rho}}{\hat{\rho}}\left[\hat{Q}_i - (1-\hat{\rho})\hat{Q}_i^*\right]\right\}^{\otimes 2},$$

where $\hat{\rho} = n_v/n$,

$$\hat{Q}_i = \int_0^1 \Delta(\hat{\phi}_i)(t)Y_i(t)\left[\hat{r}_i(\beta,t) - \hat{\phi}_i(\beta,t)\right] d\hat{\Lambda}_0(t),$$

$$\hat{Q}_i^* = \int_0^1 \Delta(\hat{\phi}_i)(t)Y_i(t)\hat{\theta}_i(t;\alpha)\,d\hat{\Lambda}_0(t),$$

$$\Delta(\hat{\phi}_i)(t) = \hat{\phi}_i^{(1)}(\beta,t)/\hat{\phi}_i(\beta,t) - \hat{S}^{(1)}(\beta,t)/\hat{S}^{(0)}(\beta,t),$$

$$\Delta(\hat{\gamma}_i)(t) = \hat{\gamma}_i^{(1)}(\beta,t)/\hat{\gamma}_i(\beta,t) - \hat{S}^{(1)}(\beta,t)/\hat{S}^{(0)}(\beta,t),$$

$$\hat{\theta}_i(t;\alpha) = \left[\xi_i(\alpha,t) - \bar{\psi}_i(\alpha,t)\right]\exp\left(\hat{\beta}_2^T Z_i(t)\right)\hat{\rho}_\alpha^*(Z_i,t)\hat{\sigma}_1(Z_i,u)/\hat{\sigma}_2(Z_i,t).$$

In summary, a constant variance estimator for $\hat{\beta}_{EPL}$ can be obtained by replacing the population quantities in the asymptotic covariance matrix $\Sigma(\beta_0)$ with their corresponding sample averages as in Zhou and Wang (2000). Hence, the asymptotic confidence intervals for β can also be constructed.

The following theorem demonstrates the asymptotic normality of $\hat{\beta}_V$.

Theorem 4.2. *Under the same conditions as in Theorem 4.1, the estimator* $\hat{\beta}_V$ *shares the same asymptotic distribution as* $\hat{\beta}_{EPL}$ *but with* $\Sigma_1(\beta_0)$ *and* $\Sigma_2(\beta_0)$ *replaced by* $\Sigma_{1V}(\beta_0)$ *and* $\Sigma_{2V}(\beta_0)$*, respectively, where* $\Sigma_{1V}(\beta_0) = E\left[\int_0^1 \Delta(\phi_i)(u)dM_i(u)\right]^{\otimes 2}$*, and* $\Sigma_{2V}(\beta_0) = E\left[\int_0^1 \Delta(\gamma_i)(u)dM_i(u) - \frac{1-\rho}{\rho}Q_i\right]^{\otimes 2}$.

4.1 Choice of the parameter vector α

The choice of α affects efficiency of $\hat{\beta}_{EPL}$, although the estimator is $\sqrt{n}$-consistent for any α. In this paper, we choose α by minimizing the variance of the estimator $\hat{\beta}_{EPL}$. Given initial value of β and α, one can estimate α by minimizing the trace of $\hat{\Omega}(\alpha)$.

Once the value of α is known, maximization of $EPL(\beta)$ can be solved via Newton-Raphson iterations. Repeating this procedure, one can find a solution to the optimization problem (3.10). To reduce the burden of computation in practice, one can employ a consistent naive estimator of β as initial value, for example the estimator of β based on only the validation sample which is easy to implement because it involves only a simple fit for the usual Cox's model. In our experience, using the naive estimator as the initial value the iterations converge in a few steps.

4.2 Choice of the bandwidth parameter

As for the bandwidths, they affect the estimator $\hat{\beta}_{EPL}$, which is true in any nonparametric smoothing problems. Fortunately, the proposed estimator $\hat{\beta}_{EPL}$ is effective for a large range of bandwidths (see Condition (6) in Appendix 1). Similar to that in Zhou and Wang (2000), we employed here the empirical bandwidth $h = (h_1, h_2)'$ with $h_1 = 2\hat{\sigma}_Z n^{-1/3}$ and $h_2 = 2\hat{\sigma}_W n^{-1/3}$, where $\hat{\sigma}_Z$ and $\hat{\sigma}_W$ are respectively the sample standard deviations of Z and W, which satisfy the bandwidth conditions required in this paper.

5 Simulations

In this section, we conduct finite-sample simulations[a] The aims of the simulations are three-fold: one is to examine the small sample behavior of $\hat{\beta}_{EPL}$, another is to compare the performance of our estimator with some existing estimators under various situations, and the third and the most important is to illustrate that the proposed estimation allows for an informative auxiliary vector W. The covariates (X, Z) are generated from the following transformation to create correlation:

$$\begin{pmatrix} X \\ Z \end{pmatrix} = \begin{pmatrix} 1 & 0.0 \\ 0.5 & 1 \end{pmatrix} \begin{pmatrix} U_1 \\ U_2 \end{pmatrix}, \tag{5.12}$$

where U_i's are independent and identically distributed as $U(0, 2)$. The failure time T conditional on covariate X is from an exponential distribution with hazard function

$$\lambda(t; X) = \lambda \exp(\beta_1 X + \beta_2 Z),$$

where λ is the baseline constant hazard. We only consider the case $\lambda = 1$. Then

$$f(t; X, Z) = \exp(\beta_1 X + \beta_2 Z) \exp(-t \exp(\beta_1 X + \beta_2 Z)).$$

The auxiliary variable W is generated from

$$W = X + \gamma \log(T) + e, \tag{5.13}$$

where $e \sim \mathcal{N}(0, \sigma^2)$ and σ^2 is the parameter controlling the strength of the association between X and W. We consider the settings with $\gamma = 0$ and 2. Model (5.13) with $\gamma = 2$ allows one to explore the effectiveness of the proposed method with an informative surrogate W. For $\gamma = 0$, it also allows us to compare the performance of the newly proposed method and that in Zhou and Wang (2000). We do simulations for $\sigma = 0.2$ and 0.8. The censoring variable is uniformly distributed and independent of the failure time. The validation set is randomly selected with $P(\eta_i = 1) = 0.5$.

We choose the Gaussian kernel function with the bandwidths $\left(h_1 = 2\hat{\sigma}_Z n^{-1/3}, h_2 = 2\hat{\sigma}_W n^{-1/3}\right)$ which satisfy the bandwidth conditions in Theorem 4.1, where $\hat{\sigma}_Z$ and $\hat{\sigma}_W$ are the sample standard deviations of Z and W respectively. In the following tables, $\beta_0 = [\log(2), 0.5]'$ denotes the true value of the parameter to be estimated, *se* is the standard error of $\hat{\beta}_{EPL}$ from simulation, $mean(\hat{se})$ denotes the mean of the estimated standard errors and *cp* denotes the 95% coverage probability.

The methods we considered are the newly proposed estimated partial likehood estimation ($\hat{\beta}_{EPL}$)and its conterpart ($\hat{\beta}_{ZW}$) in Zhou and Wang (2000), the estimator ($\hat{\beta}_V$) which does not use the information on W in $\bar{V}$, the complete-case Cox regression analysis ($\hat{\beta}_{CC}$) which uses only the validation subsample, the Cox regression with W substituted for the missing X ($\hat{\beta}_N$), and the full data Cox regression ($\hat{\beta}_F$) which assumes that X is available for all n subjects in the study.

Tables 1 and 2 summarizes the results obtained from the simulation. Note that $\hat{\beta}_F$, $\hat{\beta}_{CC}$ and $\hat{\beta}_V$ in Table 2 are the same in Table 1, so we do not report them in Table 2. For finite sample sizes, the mean, median, standard error (*se*) and 95% confidence intervals of the estimator are calculated based on 500 independent runs. We observe that $\hat{\beta}_F$ is the best estimator but it is not always obtainable for practical studies. The estimator $\hat{\beta}_N$ is biased. In all the situations $\hat{\beta}_{EPL}$ is observed to be a consistent estimator of true β_0. The estimates obtained from Zhou and Wang (2000) method are biased when $\gamma = 2$. Also,we notice that the bias in their estimates increases when σ increases. Efficiency of $\hat{\beta}_{EPL}$ relative to the complete case estimator $\hat{\beta}_{CC}$ is approximately the same for $\beta_1 = \log(2)$ but much higher for $\beta_2 = 0.5$. For $\gamma = 0$ the estimator $\hat{\beta}_{ZW}$ is more efficient than the proposed estimator for smaller values of σ but as the correlation between the exposure and auxiliary variable decreases the efficiency becomes closer (see the values of *se* for $n = 300$). Also, we notice that $\hat{\beta}_{EPL}$ has less bias than $\hat{\beta}_V$ for different values of n, but they are still comparable and even in some cases $\hat{\beta}_V$ is better in terms of the standard deviation. For $\gamma = 2$, our method stays almost equally efficient as σ increases, but $\hat{\beta}_{ZW}$ fails because of its large bias and low coverage probability (*cp*). Note that, when $\gamma = 2$, W is an informative auxiliary variable about the failure time and is not very informative about X.

We also performed simulations to see the effect of validation ratio and different bandwidths on the estimation. The proposed estimator $\hat{\beta}_{EPL}$ works well for smaller validation percentages and is not very sensitive to the bandwidth selection. In particular, $\hat{\beta}_{EPL}$ is better than $\hat{\beta}_V$ when the validation ratio is 0.25, which is evidenced in Table 3. We also conducted simulations with smaller validation ratios. Our experience indicates that, as the validation ratio gets as small as 0.2, $\hat{\beta}_V$ is very bad but $\hat{\beta}_{EPL}$ still works.

We conclude that, the proposed partial likelihood estimator can be used to make inference for β under various situations. In particular, the estimator is consistent and efficient when the auxiliary variable is informative about the hazard rate of failure time while Zhou and Wang's estimator fails.

Table 1 Comparison of simulation results with $\sigma = 0.2$ and validation fraction 0.5

n		$\gamma = 0$						$\gamma = 2$		
		$\hat{\beta}_F$	$\hat{\beta}_{CC}$	$\hat{\beta}_N$	$\hat{\beta}_V$	$\hat{\beta}_{ZW}$	$\hat{\beta}_{EPL}$	$\hat{\beta}_N$	$\hat{\beta}_{ZW}$	$\hat{\beta}_{EPL}$
50% censoring										
100	$mean - \beta_0$	0.018	0.021	-0.034	-0.036	-0.045	0.006	-0.988	-0.434	0.031
		0.004	0.030	0.022	0.035	0.013	0.006	0.169	0.174	0.010
	$median - \beta_0$	0.013	-0.001	-0.045	-0.047	0.033	-0.033	-0.983	-0.484	0.021
		-0.015	0.023	0.002	0.019	-0.001	0.011	0.160	0.172	0.008
	se	0.323	0.439	0.302	0.410	0.346	0.457	0.084	0.574	0.459
		0.281	0.391	0.276	0.307	0.283	0.312	0.237	0.324	0.338
	$mean(\hat{se})$	0.292	0.429	0.280	0.402	0.329	0.414	0.068	0.373	0.411
		0.258	0.382	0.257	0.278	0.257	0.296	0.244	0.274	0.288
	cp	0.916	0.956	0.924	0.938	0.948	0.946	0.0	0.670	0.946
		0.930	0.960	0.934	0.922	0.914	0.924	0.920	0.892	0.916
300	$mean - \beta_0$	-0.019	0.026	-0.068	-0.039	-0.007	-0.021	-0.994	-0.630	0.005
		0.006	0.017	0.024	0.013	0.010	0.002	0.166	0.202	-0.007
	$median - \beta_0$	-0.028	0.024	0.077	-0.042	-0.012	-0.018	-0.991	-0.635	-0.002
		0.007	0.001	0.028	0.021	0.018	0.002	0.169	0.199	0.003
	se	0.161	0.234	0.158	0.225	0.176	0.238	0.048	0.246	0.243
		0.146	0.217	0.146	0.162	0.150	0.163	0.127	0.137	0.166
	$mean(\hat{se})$	0.164	0.233	0.158	0.227	0.177	0.222	0.039	0.170	0.231
		0.146	0.209	0.145	0.159	0.147	0.155	0.137	0.125	0.159
	cp	0.944	0.942	0.928	0.948	0.950	0.936	0.0	0.108	0.940
		0.956	0.942	0.956	0.948	0.944	0.938	0.796	0.630	0.940
20% censoring										
100	$mean - \beta_0$	0.021	0.020	-0.031	-0.041	0.044	0.002	-1.003	-0.455	0.023
		0.001	0.014	0.018	0.036	0.013	0.005	0.155	0.163	0.011
	$median - \beta_0$	0.016	0.014	-0.029	-0.048	0.038	-0.013	-1.000	-0.466	0.003
		-0.008	-0.001	0.011	0.029	0.005	0.003	0.151	0.159	0.005
	se	0.248	0.339	0.234	0.322	0.272	0.364	0.071	0.467	0.360
		0.211	0.305	0.210	0.224	0.212	0.229	0.180	0.241	0.235
	$mean(\hat{se})$	0.232	0.340	0.223	0.306	0.263	0.313	0.062	0.318	0.315
		0.205	0.302	0.204	0.217	0.204	0.213	0.195	0.214	0.215
	cp	0.936	0.956	0.934	0.912	0.966	0.894	0.0	0.550	0.904
		0.938	0.956	0.938	0.934	0.952	0.936	0.924	0.862	0.928
300	$mean - \beta_0$	-0.007	-0.009	-0.056	-0.032	-0.006	-0.011	-1.001	-0.617	0.015
		-0.001	0.008	0.016	0.008	0.004	-0.004	0.152	0.023	-0.012
	$median - \beta_0$	-0.019	-0.016	-0.064	-0.040	-0.006	-0.022	-1.001	-0.613	0.001
		-0.002	0.002	0.011	0.007	0.003	-0.005	0.154	0.024	-0.016
	se	0.131	0.190	0.127	0.177	0.141	0.194	0.044	0.304	0.195
		0.116	0.164	0.116	0.126	0.119	0.127	0.100	0.178	0.135
	$mean(\hat{se})$	0.131	0.187	0.260	0.179	0.142	0.179	0.034	0.219	0.179
		0.116	0.166	0.150	0.125	0.117	0.122	0.110	0.161	0.124
	cp	0.948	0.960	0.928	0.952	0.966	0.926	0.0	0.244	0.926
		0.952	0.954	0.952	0.954	0.952	0.944	0.748	0.070	0.934

6 Real data analysis

6.1 Primary Biliary Cirrhosis data

We apply the proposed approach to the data from the Mayo Clinic trial in primary biliary cirrhosis (PBC) of the liver conducted between 1974 and 1984. A total of 424 PBC patients, referred to Mayo Clinic during that ten-year interval, met eligibility criteria for

Table 2 Comparison of simulation results with $\sigma = 0.8$ and validation fraction 0.5

n		$\gamma = 0$			$\gamma = 2$		
		$\hat{\beta}_N$	$\hat{\beta}_{ZW}$	$\hat{\beta}_{EPL}$	$\hat{\beta}_{ZW}$	$\hat{\beta}_{ZW}$	$\hat{\beta}_{EPL}$
50% censoring							
100	*mean* $- \beta_0$	-0.369	-0.045	0.034	-0.961	-0.325	0.027
		0.139	0.055	0.008	0.177	0.141	0.009
	median $- \beta_0$	-0.381	-0.052	0.021	-0.955	-0.368	0.010
		0.028	0.051	0.008	0.164	0.140	0.002
	se	0.206	0.374	0.457	0.076	0.549	0.450
		0.260	0.287	0.338	0.243	0.325	0.332
	mean$(\hat{se})$	0.196	0.256	0.414	0.064	0.374	0.414
		0.249	0.262	0.288	0.244	0.271	.293
	cp	0.504	0.940	0.934	0.0	0.721	0.940
		0.914	0.932	0.916	0.904	0.888	0.920
300	*mean* $- \beta_0$	-0.392	-0.056	0.012	-0.965	-0.399	0.012
		0.139	0.033	-0.011	0.175	0.147	-0.009
	median $- \beta_0$	-0.392	-0.055	0.004	-0.963	-0.395	0.004
		0.139	0.044	-0.004	0.176	0.145	-0.002
	se	0.114	0.198	0.255	0.044	0.325	0.254
		0.142	0.156	0.170	0.129	0.180	0.171
	mean$(\hat{se})$	0.108	0.223	0.227	0.036	0.213	0.228
		0.140	0.157	0.159	0.137	0.157	0.158
	cp	0.068	0.932	0.932	0.0	0.520	0.932
		0.830	0.946	0.934	0.770	0.808	0.936
20% censoring							
100	*mean* $- \beta_0$	-0.368	-0.046	0.024	-0.969	-0.328	0.022
		0.126	0.052	0.019	0.163	0.128	0.021
	median $- \beta_0$	-0.372	-0.052	0.020	-0.966	-0.354	0.016
		0.122	0.053	0.020	0.168	0.124	0.015
	se	0.165	0.272	0.360	0.064	0.450	0.348
		0.202	0.213	0.240	0.186	0.238	0.238
	mean$(\hat{se})$	0.156	0.263	0.306	0.057	0.291	0.321
		0.198	0.207	0.211	0.195	0.212	0.222
	cp	0.352	0.966	0.912	0.0	0.658	0.916
		0.918	0.942	0.928	0.910	0.878	0.924
300	*mean* $- \beta_0$	-0.390	-0.048	0.013	-0.972	-0.388	0.021
		0.123	0.026	-0.011	0.161	0.127	-0.015
	median $- \beta_0$	-0.393	-0.058	-0.004	-0.972	-0.399	0.013
		0.123	0.028	-0.017	0.164	0.128	-0.016
	se	0.092	0.159	0.195	0.039	0.262	0.200
		0.115	0.123	0.131	0.102	0.138	0.133
	mean$(\hat{se})$	0.086	0.172	0.186	0.033	0.168	0.179
		0.112	0.123	0.126	0.110	0.122	0.124
	cp	0.018	0.954	0.924	0.0	0.412	0.924
		0.792	0.944	0.934	0.716	0.784	0.942

the randomized placebo controlled trial of the drug D-penicillamine. The first 312 cases in the data set participated in the randomized trial and contain largely complete data. The additional 112 cases did not participate in the clinical trial, but agreed to have basic measurements recorded and to be followed for survival. Six of those cases were lost to

Table 3 Comparison of simulation results with $\beta = [ln(2)\ \ 0.5]'$, 50% censoring, $\sigma = 0.2$, and validation fraction 0.25

	$n = 100$		$n = 200$	
	$\hat{\beta}_V$	$\hat{\beta}_{EPL}$	$\hat{\beta}_V$	$\hat{\beta}_{EPL}$
mean $-\ \beta_0$	-0.142	0.056	-0.087	0.049
	0.107	0.018	0.056	0.001
median $-\ \beta_0$	-0.152	0.035	-0.125	0.011
	0.091	-0.002	0.065	0.002
se	0.506	0.565	0.405	0.417
	0.329	0.320	0.234	0.232
mean($\hat{se}$)	0.513	0.618	0.380	0.410
	0.306	0.333	0.220	0.224
cp	0.944	0.936	0.928	0.924
	0.934	0.954	0.934	0.942

follow-up shortly after diagnosis, so the data here are on an additional 106 cases as well as the 312 randomized participants.

A clinical background description and a more extended discussion for the trial and the covariates recorded can be found in Dickson et al. (1989) and Markus et al. (1989). The variables involved in our specify analysis include id: case number; days: number of days between registration and the earlier of death, transplantation, or study analysis time; status: status of censoring; bili: serum bilirubin (in mg/dl); chol: serum cholesterol (inmg/dl) and Age: age in days.

In this analysis, we are particularly interested in the effect of patients' serum cholesterol and age on the survival of the patients. This type of failure time data can be modeled by the Cox proportional hazards models with an unknown baseline hazard function. However, about 31% outcomes of cholesterol were missing in this data set. Removing those observations may lead to biased estimates and standard errors. We noted that the outcomes of serum bilirubin were completely obtained with no missing values. Preliminary analysis showed that there is a significant correlation between serum cholesterol and bilirubin. Also, intuitively bilirubin has some additional effect on the hazard of failure and we would like to use that information efficiently. To illustrate this effect, we performed a complete Cox regression analysis for two different situations. We take the logarithmic transformation of bilirubin for our study.

In Table 4, we observe that the coefficient and standard error estimates are quite different for both the situations and the 95% confidence intervals for the coefficient of age are nonoverlapping. We can conclude that serum bilirubin has some additional effect on the hazard of failure. Hence, our proposed method can be applied to this dataset considering serum bilirubin as the informative auxiliary covariate.

Table 4 Regression analysis of primary biliary cirrhosis (PBC) data study

	Method	Variable	Parameter	Standard error	95% Confidence interval
logbili < 1.6	CC	logchol	0.271	0.393	(-0.499, 1.040)
		age	0.055	0.012	(0.031, 0.079)
logbili ≥ 1.6	ZW	logchol	-0.635	0.345	(-1.312, 0.042)
		age	-0.005	0.016	(-0.037, 0.027)

Table 5 displays the analysis results based on the Cox's regression for the complete data (CC), the method proposed by Zhou and Wang (2000) (ZW) and the proposed method (EPL). The CC method uses only 284 complete-case observations and the other two methods use all 418 observations. Variable "logchol" denotes the logarithm of cholesterol. The estimates of the coefficients and their standard errors are given in the table.

The regression analysis confirms that both serum cholesterol and age are significantly related to the time to event. For estimating the effect of serum cholesterol and age, there is a reasonable efficiency gain by using the two methods based on partial likelihood approach over the complete case Cox regression analysis. But there is a discrepancy between the estimates from complete data and Zhou and Wangs estimate which could be due to the fact that the latter method does not consider the additional effect contributed by the auxiliary covariate. In our simulation we observed that the standard error of the estimates were underestimated in Zhou and Wangs method when auxiliary variable was informative. In the real data analysis also the standard error estimate for serum cholesterol is underestimated. Moreover, the standard error estimates in our method is comparable to Zhou and Wangs method whereas the calculation time is much less compared to their method.

6.2 Serrum Ferritin Concentration in relation to preterm delivery study

We apply the proposed approach to the data on iron intake in relation to preterm delivery study from the University of North Carolina Hospitals at Chapel Hill. A total of 1520 women were included in the study. 17 of these women were lost to follow up. So the data consist of 1503 individuals among which 270 individuals had their serrum ferritin concentration (*FERRITIN*) measured with an immunometric assay. However a crude score for dietary iron intake (*DTFE*) was collected using a dietary food frequency questionnaire for all the individuals.

A clinical background description and a more extended discussion for the trial and the covariates recorded can be found in Savitz et al. (2001). The variables involved in our specfic analysis include (i) id: case number; (ii) Gestation Time: The number of weeks from pregnancy to delivery; (iii) *DTFE*: Dietary iron intake(in 100mg/dl); (iv) *Ferritin*: Serum Ferritin (in 100 mg/dl); and (v) *Age*: age in years. By using the notations in the proposed method, X is *Ferritin*, W is *DTFE*, and Z is *Age*.

In this analysis, we are particularly interested in the effect of patients' serum ferritin and age on the delivery of the patients. This type of failure time data can be modeled by the Cox proportional hazards models.

Table 5 Regression analysis of primary biliary cirrhosis (PBC) data

Method	Variable	Estimates of parameters	Standard error	95% Confidence interval
CC	logchol	0.853	0.214	(0.432, 1.273)
	age	0.048	0.010	(0.029, 0.067)
ZW	logchol	1.142	0.154	(0.840, 1.444)
	age	0.047	0.007	(0.033, 0.061)
EPL	logchol	0.851	0.215	(0.429, 1.273)
	age	0.044	0.007	(0.029, 0.058)

However, outcomes for serum ferritin were missing in this data set. Removing those observations can lead to biased or inefficient estimates. We noted that the outcomes of dietary iron intake were completely obtained with no missing values.

Table 6 displays the analysis results based on the the CC method, the ZW method, and the proposed EPL method. The CC method used only 270 complete-case observations and the other two methods used all 1503 observations.

The regression analysis using the new method confirms that both serum ferritin and age are significantly related to the time to event. For estimating the effect of serum ferritin and age, there is also a reasonable efficiency gain by using the two methods based on partial likelihood approach over the complete case cox regression analysis. The estimate of serrum ferritin is lower by the EPL method. The estimate is significantly different from zero with p-value 0.020. In contrast, the p-value from CC method in estimation of serrum ferritin is 0.06.

7 Conclusion

We have introduced an EPL estimation method for Cox's models with informative auxiliary covariates and established asymptotic normality of our estimator. The proposed proposed methodology allows for multivariate auxiliary covariates W without suffering the curse of dimensionality.

We used the same bandwidth as suggested by Zhou and Wang (2000) in our estimation. Though it performs reasonably well, one can develop a bandwidth selection criteria like generalized cross-validation for an improved estimation. It is desirable to increase the efficiency of the estimation. In future, we can consider the optimization of α or introduce some weight structure in the score equation to achieve robustness. Further, it is worthy extending our approach to model multivariate failure time.

Endnote

[a] All numerical results in this paper are obtained using the software MATLAB and the codes are available (Additional file 1).

Appendix 1: Condition (A)

For the risk function $r_i(\beta,t)$ (as well as for $\hat{r}_i(\beta,t)$, $\gamma_i(\beta,t)$, $\hat{\phi}_i(\beta,t)$ and $\phi_i(\beta,t)$), we denote by $r_i^{(j)}(\beta,t)$ the jth derivative of $r_i(\beta,t)$ with respect to β, $j = 0,1,2$, where $r_i^{(0)}$ means the function itself. Define $S^{(0)}(\beta,t) = n^{-1}\sum_{i=1}^{n} Y_i(t)r_i(\beta,t)$, $S^{(1)}(\beta,t) = (\partial/\partial\beta)S^{(0)}(\beta,t)$, $S^{(2)}(\beta,t) = (\partial/\partial\beta)S^{(1)}(\beta,t)$, $s^{(0)}(\beta,t) = E[Y_i(t)r_i(\beta,t)]$, $s^{(1)}(\beta,t) = (\partial/\partial\beta)s^{(0)}(\beta,t)$, $s^{(2)}(\beta,t) = (\partial/\partial\beta)s^{(1)}(\beta,t)$. Let $\Delta(\phi)(u) = \frac{\phi^{(1)}(\beta,u)}{\phi(\beta,u)} - \frac{s^{(1)}(\beta,u)}{s^{(0)}(\beta,u)}$,

Table 6 Regression analysis of Iron intake in relation to preterm delivery study

Method	Variables	Estimates of parameters	Standard error	*p*-value	Hazard ratio
CC	ferritin	0.2451	0.1306	0.060	1.278
	age	0.009	0.0108	0.402	1.009
ZW	ferritin	0.2236	0.076	0.004	1.251
	age	0.0102	0.0043	0.018	1.010
EPL	ferritin	0.1797	0.0771	0.020	1.197
	age	0.0159	0.0036	0.000	1.016

$$I(\beta)=-\int_0^1 \left[\frac{r^{(2)}(\beta,u)}{r^{(0)}(\beta,u)}-\left\{\frac{r^{(1)}(\beta,u)}{r^{(0)}(\beta,u)}\right\}^{\otimes 2}-\frac{s^{(2)}(\beta,u)}{s^{(0)}(\beta,u)}+\left\{\frac{s^{(1)}(\beta,u)}{s^{(0)}(\beta,u)}\right\}^{\otimes 2}\right] s^{(0)}(\beta,u)\, dN(u),$$

$$Q=\int_0^1 \Delta(\phi)(u)Y(u)[\gamma(\beta,u)-\phi(\beta,u)]\,\lambda_0(u)du,$$

$$Q^*=\int_0^1 \Delta(\phi)(u)Y(u)\theta(Z,u;\alpha)\lambda_0(u)du.$$

The following conditions are needed for the theoretical results in the paper:

(1) $\int_0^1 \lambda_0(s)ds < \infty$.

(2) $Pr(Y(1)=1|V) > 0$ for any validation set V.

(3) There exists an open subset $\mathcal{B}$, containing the true value β_0 of β, of the Euclidean space $\mathcal{R}^p$. In addition, $r_i^{(2)}(\beta,t)$ with elements $(\partial^2/\partial\beta_i\partial\beta_j)r(\beta,t)$ exists and is continuous on $\mathcal{B}$ for each $t \in [0,1]$, uniform in t, and $\phi(\beta,t)$ is bounded away from 0 on $\mathcal{B}\times[0,1]$. Furthermore, $I(\beta)$ is positive definite.

(4)

$$E\left\{\sup_{\mathcal{B}\times[0,1]} |Y(t)r^{(j)}(\beta,t)|\right\} < \infty, \quad j=0,1,2,$$

$$E\left\{\sup_{\mathcal{B}\times[0,1]} |Y(t)\left(\frac{r^{(1)}(\beta,t)}{r(\beta,t)}\right)^{\otimes 2j} r(\beta,t)|\right\} < \infty, \quad j=1,2,$$

$$E\left\{\sup_{\mathcal{B}\times[0,1]} |Y(t)\left(\frac{r^{(2)}(\beta,t)}{r(\beta,t)}\right)^{\otimes j} r(\beta,t)|\right\} < \infty, \quad j=1,2.$$

Also observe that, $s^{(0)}(\beta,t)=E[Y(t)r(\beta,t)]=E[Y(t)r^*(\beta,t)]$.

(5) Let $F_{Y(t),Z}$ be the joint distribution of $(Y(t),Z)$, and $f(t,z) = (\partial/\partial z)F_{Y(t),z}(1,z)$. For each $t \in [0,1]$, both $f(t,z)$ and $\phi(\beta,t)$ have the 2nd continuous derivative almost everywhere.

(6) $h\to 0$, $nh^{d+4}\to 0$ and $nh^d(\log n)^{-2}\to\infty$, as $n\to\infty$.

Appendix 2: Technical Proofs

Proof of Proposition 3.1. The argument employed here is similar to that for Theorem 1 of Jiang et al. (2011). Note that $\hat{\nu}_j-\nu_j=\sum_{i\in V}\omega_i(\nu_i-\nu_j)+\sum_{i\in V}\omega_i(\zeta_i-\nu_i)$. By standard nonparametric regression techniques (see for example Härdle 1990; Fan and Gijbels 1996), it can be shown that the first term above contributes to bias and is $O_p(h^2)$, which is of order $o_p(1/\sqrt{nh^d})$, if one uses an undersmoothing bandwidth such that $nh^{d+4}\to 0$, so that $\hat{\nu}_j-\nu_j=\sum_{i\in V}\omega_i(\zeta_i-\nu_i)+o_p(1/\sqrt{nh^d})$. Similarly, $\hat{\psi}_j-\psi_j=\sum_{i\in V}\omega_i(\xi_i-\psi_i)+o_p(1/\sqrt{nh^d})$. Then the asymptotic normality can be obtained by using the Cramé-Wald device and directly computing the asymptotic mean and variance (see, for example the Lemma 6.3 in Jiang and Mack 2001).

Proof of Proposition 3.2. Note that from (3.8)

$$\sqrt{nh^d}[\bar{\nu}_j-\nu_j] = \sqrt{nh^d}[\hat{\nu}_j-\nu_j] -\rho^*(Z_j,t)\frac{\sigma_1(Z_j,t)}{\sigma_2(Z_j,t)}\sqrt{nh^d}[(\hat{\psi}_j-\psi_j)+(\bar{\psi}_j-\psi_j)]\,(1+o_p(1)).$$

The asymptotic normality of $\sqrt{nh^d}(\bar{\upsilon}_j - \nu_j)$ is obtained by the Slutsky's theorem and the asymptotic normality of $\sqrt{nh^d}(\hat{\upsilon}_j - \nu_j)$, $\sqrt{nh^d}(\hat{\psi}_j - \psi_j)$ and $\sqrt{nh^d}(\bar{\psi}_j - \psi_j)$.

Lemma 7.1. *Under Condition (A),*

$$\sup_{\beta \in \mathcal{B}} ||n^{-1}\partial \hat{U}(\beta, 1)/\partial\beta - (-I(\beta))|| \xrightarrow{P} 0.$$

Proof. By simple algebra, we have

$$n^{-1}\partial\hat{U}(\beta,1)/\partial\beta = \int_0^1 n^{-1}\sum_{i=1}^n \left[\frac{\hat{r}_i^{(2)}(\beta,t)}{\hat{r}_i^{(0)}(\beta,t)} - \left(\frac{\hat{r}_i^{(1)}(\beta,t)}{\hat{r}_i^{(0)}(\beta,t)}\right)^{\otimes 2} - \frac{\hat{S}^{(2)}(\beta,t)}{\hat{S}^{(0)}(\beta,t)} + \left(\frac{\hat{S}^{(1)}(\beta,t)}{\hat{S}^{(0)}(\beta,t)}\right)^{\otimes 2}\right] dN_i(t).$$

Note that for $j = 0, \ldots, 2$

$$\sup_{\mathcal{B}\times[0,1]} ||\hat{r}_i^{(j)}(\beta,t) - r_i^{(j)}(\beta,t)|| \xrightarrow{P} 0, \tag{7.14}$$

and

$$\sup_{\mathcal{B}\times[0,1]} ||\hat{S}_i^{(j)}(\beta,t) - s_i^{(j)}(\beta,t)|| \xrightarrow{P} 0, \tag{7.15}$$

uniformly for $i = 1, \ldots, n$ if $h \to 0$ and $nh^d \to \infty$. It follows that

$$\begin{aligned} n^{-1}\partial\hat{U}(\beta,1)/\partial\beta &= \int_0^1 n^{-1}\sum_{i=1}^n \left[\frac{r_i^{(2)}(\beta,t)}{r_i^{(0)}(\beta,t)} - \left(\frac{r_i^{(1)}(\beta,t)}{r_i^{(0)}(\beta,t)}\right)^{\otimes 2} - \frac{s^{(2)}(\beta,t)}{s^{(0)}(\beta,t)} + \left(\frac{s^{(1)}(\beta,t)}{s^{(0)}(\beta,t)}\right)^{\otimes 2}\right] dN_i(t) + o_p(1) \\ &= -I(\beta) + o_p(1), \end{aligned}$$

uniformly in $\beta \in \mathcal{B}$. □

Proof of Theorem 4.1. The proof is argued in the framework of the multivariate counting processes, the martingale theory, and the techniques commonly used in nonparametric regression. Following the same routine as in Zhou and Wang (2000), the consistency of $\hat{\beta}_{EPL}$ can be derived by using the Inverse Function Theorem (Rudin 1964; Andersen and Gill, 1982) and the argument by Foutz (1977). In the following, we give only the asymptotic normality in Theorem 4.1. The main techniques we employed are Taylor's expansion of the score function corresponding to the estimated likelihood function (3.10), Lenglart inequality, the martingale central limit theorem (see e.g. Fleming and Harrington 1991), and nonparametric regression techniques.

By using counting process notation, the score function corresponding to the estimated partial likelihood function (3.10) at time point t can be written as

$$\hat{U}(\beta,t) = \sum_{i=1}^n \int_0^t \Delta(\hat{r}_i)(\beta,u)dM_i(u) + \sum_{i=1}^n \int_0^t \Delta(\hat{r}_i)(\beta,u)r_i(\beta_0,u)Y_i(u)\lambda_0(u)du, \tag{7.16}$$

where

$$\Delta(\hat{r}_i)(u) = \frac{\hat{r}_i^{(1)}(\beta,u)}{\hat{r}_i(\beta,u)} - \frac{\sum_{i=1}^n Y_i(u)\hat{r}_i^{(1)}(\beta,u)}{\sum_{i=1}^n Y_i(u)\hat{r}_i(\beta,u)}.$$

By (7.16), $\hat{\beta}_{EPL}$ solves the equation $\hat{U}(\beta,1) = 0$. By Taylor's expansion, one gets

$$n^{-1/2}\hat{U}(\beta,1) = -n^{-1}\frac{\partial \hat{U}(\beta_*,1)}{\partial \beta}\sqrt{n}(\hat{\beta}_{EPL} - \beta_0), \tag{7.17}$$

where β_* is between $\hat{\beta}_{EPL}$ and β_0. By Lemma 7.1 and consistency of $\hat{\beta}_{EPL}$,

$$-n^{-1}\frac{\partial \hat{U}(\beta_*,1)}{\partial \beta} \xrightarrow{P} I(\beta_0).$$

Therefore, to prove the asymptotic normality in the theorem it suffices to show that $n^{-1/2}\hat{U}(\beta,1)$ is asymptotically normal with mean 0 and variance $\Sigma(\beta_0) = (1-\rho)\Sigma_1(\beta_0)+\rho\Sigma_2(\beta_0)$, which is evidenced in Lemma 7.4 below.

Proof of Theorem 4.2. Using similar arguments to Theorem 4.1, we establish the result.

Lemma 7.2. *Under Condition (A),*

$$n^{-1/2}\sum_{i=1}^n \int_0^1 \left(\hat{r}_i^{(k)}(\beta,w) - r_i^{(k)}(\beta,w)\right)^2 Y_i(w)r_i(\beta,w)\lambda_0(w)\,dw \xrightarrow{p} 0,\ \ k = 0,1$$

$$n^{-1/2}\sum_{i=1}^n \int_0^1 \left(\hat{S}^{(k)}(\beta,w) - S^{(k)}(\beta,w)\right)^2 Y_i(w)r_i(\beta,w)\lambda_0(w)\,dw \xrightarrow{p} 0,\ \ k = 0,1.$$

Proof. The result can be obtained by following the same argument as that for Lemma 2.4 of Zhou and Wang (1999).

□

Lemma 7.3. *Under Condition (A), the second term of $\hat{U}(\beta,t)$ in (7.16) admits the following decomposition*

$$\begin{aligned} &n^{-1/2}\sum_{i=1}^n \int_0^1 \Delta(\hat{r}_i)(\beta,w)Y_i(w)r_i(\beta_0,w)\lambda_0(w)\,dw \\ &= -n^{-1/2}\sum_{i=1}^n \int_0^1 \Delta(r_i)(\beta,w)Y_i(w)[\hat{r}_i(\beta,w) - r_i(\beta,w)]\,\lambda_0(w)dw + o_p(1). \end{aligned}$$

Proof. The proof uses the same argument as that for Lemma 2.5 of Zhou and Wang (1999). By the Taylor expansion

$$\begin{aligned} f(x,y) &= f(x_0,y_0) + \frac{\partial f(x,y)}{\partial x}\Big|_{x_0,y_0}(x-x_0) \\ &\quad + \frac{\partial f(x,y)}{\partial y}\Big|_{x_0,y_0}(y-y_0) + O\left((x-x_0)^2 + (y-y_0)^2\right), \end{aligned}$$

if $\frac{\partial^2 f}{\partial x^2}$, $\frac{\partial^2 f}{\partial y^2}$, and $\frac{\partial^2 f}{\partial x \partial y}$ are finite. Then

$$\frac{\hat{r}^{(1)}}{\hat{r}} = \frac{\hat{r}^{(1)}}{r} - \frac{r^{(1)}(\hat{r}-r)}{r^2} + O\left[(\hat{r}-r)^2 + \left(\hat{r}^{(1)} - r^{(1)}\right)^2\right]$$

$$\frac{\hat{S}^{(1)}}{\hat{S}^{(0)}} = \frac{\hat{S}^{(1)}}{S^{(0)}} - \frac{S^{(1)}\left(\hat{S} - S^{(0)}\right)}{S^{(0)2}} + O\left[\left(\hat{S} - S^{(0)}\right)^2 + \left(\hat{S}^{(1)} - S^{(1)}\right)^2\right].$$

Note that $\sum_i \Delta\hat{r}_i(u)\hat{r}_i(u)Y_i(u) = 0$. It follows that the left side of the result in the lemma can be expressed as

$$\begin{aligned}
&n^{-1/2}\sum_{i=1}^{n}\int_0^1 \Delta(\hat{r}_i)(\beta, w)Y_i(w)r_i(\beta, w)\lambda_0(w)\,dw \\
&= -n^{-1/2}\sum_{i=1}^{n}\int_0^1 \Delta(\hat{r}_i)(\beta, w)Y_i(w)[\,\hat{r}_i(\beta, w) - r_i(\beta, w)]\,\lambda_0(w)\,dw \\
&= -n^{-1/2}\sum_{i=1}^{n}\int_0^1 \Delta(r_i)(\beta, w)Y_i(w)[\hat{r}_i(\beta, w) - r_i(\beta, w)]\,\lambda_0(w)\,dw + o_p(1),
\end{aligned}$$

where the last equality is from Lemma 7.2. Therefore the result holds. □

Lemma 7.4. *Under Condition (A),*

$$n^{-1/2}\hat{U}(\beta, 1) \xrightarrow{\mathcal{L}} N\big(0, (1-\rho)\Sigma_1(\beta) + \rho\Sigma_2(\beta)\big).$$

Proof. Note that $\hat{r}_i - r_i = (1-\eta_i)\left(\bar{\phi}_i - \phi_i\right)$. Applying the first order approximation $x/y = x_0/y_0 + (x-x_0)/y_0 - (y-y_0)x_0/y_0^2 + O\left((x-x_0)^2 + (y-y_0)^2\right)$ to $\hat{r}^{(1)}/\hat{r}$ and $\hat{S}^{(1)}/\hat{S}^{(0)}$ around $(r^{(1)}, r)$ and $(s^{(1)}, s^{(0)})$, respectively, and by Lemma 7.3 the second term of $n^{-1/2}\hat{U}(\beta, 1)$ in (7.16) becomes

$$\begin{aligned}
&\Delta(r_i)(\beta, w)Y_i(w)[\hat{r}_i(\beta, w) - r_i(\beta, w)]\,\lambda_0(w)\,dw + o_p(1) \\
&= -n^{-1/2}\sum_{j\in\bar{V}}\int_0^1 \left(\bar{\phi}_j - \phi_j\right)\Delta(\phi_j)(u)Y_j(u)\lambda_0(u)du + o_p(1) \\
&= I_{n1} + o_p(1). \qquad (7.18)
\end{aligned}$$

Note that $\hat{\phi}_j(\beta, t) = \hat{v}_j(\beta_1, t)\exp\left\{\beta_2' Z_j(t)\right\}$. Since

$$\begin{aligned}
\bar{\phi}_j - \phi_j &= (\hat{\phi}_j - \phi_j) - \exp\left\{\beta_2' Z_j(u)\right\}\rho_\alpha^*(Z_j, u)\frac{\sigma_1(Z_j, u)}{\sigma_2(Z_j, u)}\left(\hat{\psi}_j - \bar{\psi}_j\right)(1 + o_p(1)) \\
&= \sum_{i\in V}\omega_i(\gamma_i - \phi_j) - \exp\{\beta_2' Z_j(u)\}\left[\sum_{i\in V}\omega_i(\xi_i - \psi_j)\rho_\alpha^*(Z_j, u)\frac{\sigma_1(Z_j, u)}{\sigma_2(Z_j, u)}\right. \\
&\quad \left. - \sum_{i\in V\cup\bar{V}}\bar{\omega}_i(\xi_i - \psi_j)\rho_\alpha^*(Z_j, u)\frac{\sigma_1(Z_j, u)}{\sigma_2(Z_j, u)}\right](1 + o_p(1)) + o_p\left(\frac{1}{\sqrt{n}}\right) \\
&= \sum_{i\in V}\omega_i\left[(\gamma_i - \phi_j) - \exp\left\{\beta_2' Z_j(u)\right\}\rho_\alpha^*(Z_j, u)\frac{\sigma_1(Z_j, u)}{\sigma_2(Z_j, u)}(\xi_i - \psi_j)\right](1 + o_p(1)) \\
&\quad + \sum_{i\in V\cup\bar{V}}\bar{\omega}_i(\xi_i - \psi_j)\exp\left\{\beta_2' Z_j(u)\right\}\rho_\alpha^*(Z_j, u)\frac{\sigma_1(Z_j, u)}{\sigma_2(Z_j, u)}(1 + o_p(1)) + o_p\left(\frac{1}{\sqrt{n}}\right),
\end{aligned}$$

the first term in (7.18) can be rewritten as

$$
\begin{aligned}
I_{n1} &= -n^{-1/2}\sum_{j\in\bar{V}}\int_0^1 \Delta(\phi_j)(u)Y_j(u)\lambda_0(u)\\
&\times\left\{\sum_{i\in V}\omega_i\left[(\gamma_i-\phi_j)-\exp\left\{\beta_2' Z_j(u)\right\}\rho_\alpha^*(Z_j,u)\frac{\sigma_1(Z_j,u)}{\sigma_2(Z_j,u)}(\xi_i-\psi_j)\right]\right.\\
&\left.+\sum_{i\in V\cup\bar{V}}\bar{\omega}_i(\xi_i-\psi_j)\exp\left\{\beta_2' Z_j(u)\right\}\rho_\alpha^*(Z_j,u)\frac{\sigma_1(Z_j,u)}{\sigma_2(Z_j,u)}\right\}du(1+o_p(1))+o_p(1)\\
&\equiv J_{n1}+J_{n2}+o_p(1).
\end{aligned}
$$

Note that

$$n_v^{-1}\sum_{i\in V}Y_i(t)K_h(Z_i-Z_j)=f(t,Z_j)(1+o_p(1)),$$

$$n^{-1}\sum_{i\in V\cup\bar{V}}Y_i(t)K_h(Z_i-Z_j)=f(t,Z_j)(1+o_p(1)),$$

$$\omega_i(t,Z_j;h)=f^{-1}(t,Z_j)(1+o_p(1))\,n_v^{-1}Y_i(t)K_h(Z_i-Z_j),$$

$$\bar{\omega}_i(t,Z_j;h)=f^{-1}(t,Z_j)(1+o_p(1))\,n^{-1}Y_i(t)K_h(Z_i-Z_j),$$

uniformly for $j=1,\ldots,n$. Then

$$
\begin{aligned}
J_{n1} &= -\frac{1}{\sqrt{n}}\sum_{j\in\bar{V}}\int_0^1\Delta(\phi_j)(u)Y_j(u)\lambda_0(u)f^{-1}(u,Z_j)\times\\
&\quad\frac{1}{n_v}\sum_{i\in V}Y_i(u)K_h(Z_i-Z_j)\left[(\gamma_i-\phi_j)-\exp\{\beta_2' Z_j(u)\}\right.\\
&\quad\left.\times\rho_\alpha^*(Z_j,u)\frac{\sigma_1(Z_j,u)}{\sigma_2(Z_j,u)}(\xi_i-\psi_j)\right]du+o_p(1)\\
&= -\frac{1}{\sqrt{n}}\frac{n-n_v}{n_v}\sum_{i\in V}[Q_i-Q_i^*]+o_p(1),
\end{aligned}
$$

$$
\begin{aligned}
J_{n2} &= -\frac{1}{\sqrt{n}}\sum_{j\in\bar{V}}\int_0^1\Delta(\phi_j)(u)Y_j(u)\lambda_0(u)\exp\{\beta_2' Z_j(u)\}\rho_\alpha^*(Z_j,u)\frac{\sigma_1(Z_j,u)}{\sigma_2(Z_j,u)}\\
&\quad\times\frac{1}{n}\sum_{i\in V\cup\bar{V}}Y_i(u)K_h(Z_i-Z_j)(\xi_i-\psi_j)f^{-1}(u,Z_j)du+o_p(1)\\
&= -\frac{1}{\sqrt{n}}\frac{n-n_v}{n}\sum_{i\in V\cup\bar{V}}Q_i^*+o_p(1).
\end{aligned}
$$

Therefore, the second term of $n^{-1/2}\hat{U}(\beta,1)$ in (7.16) equals

$$-\frac{1}{\sqrt{n}}\frac{n-n_v}{n_v}\sum_{i\in V}[Q_i-Q_i^*]-\frac{1}{\sqrt{n}}\frac{n-n_v}{n}\sum_{i\in V\cup\bar{V}}Q_i^*+o_p(1).$$

Hence, $n^{-1/2}\hat{U}(\beta,1)$ can be expressed as

$$n^{-1/2}\sum_{i\in\bar{V}}\left[\int_0^1\left\{\frac{\phi_i^{(1)}(\beta,u)}{\phi_i(\beta,u)}-\frac{s^{(1)}(\beta,u)}{s^{(0)}(\beta,u)}\right\}dM_i(s)-\frac{n-n_v}{n}Q_i^*\right]+o_p(1)$$

$$+\,n^{-1/2}\sum_{i\in V}\left[\int_0^1\left\{\frac{r_i^{(1)}(\beta,u)}{r_i(\beta,u)}-\frac{s^{(1)}(\beta,u)}{s^{(0)}(\beta,u)}\right\}dM_i(s)-\frac{n-n_v}{n_v}(Q_i-Q_i^*\frac{n-n_v}{n})\right].$$

For the 1st and 3rd terms above, each of them is a sum of independently distributed terms with mean zero from the nonvalidation and validation subsamples, respectively. The 1st term converges weakly to a gaussian process with covariance $(1-\rho)\Sigma_1(\beta_0)$. The 3rd term is asymptotically normal with mean zero and variance $\rho\Sigma_2(\beta_0)$. By independence of the two terms, $n^{-1/2}\hat{U}(\beta,1)\xrightarrow{P}N(0,\Sigma(\beta))$ with $\Sigma(\beta)=(1-\rho)\Sigma_1(\beta)+\rho\Sigma_2(\beta)$. □

Additional file

Additional file 1: The codes for numerical results in the paper.

Acknowledgements
The work was partially supported by NSF grant DMS-0906482 and NSFC grant 71361010. The research of Yanqing Sun was partially supported by the National Institutes of Health NIAID [grant number R37 AI054165] and by the National Science Foundation [grant number DMS-1208978]. The authors would also like to thank Dr. Savitz for making the data on serrum ferritin concentration in relation to preterm delivery study available for the application.

Author details
[1]Department of Mathematics and Statistics, University of North Carolina at Charlotte, Charlotte, NC 28223, USA.
[2]Department of Biostatistics, University of North Carolina at Chapel Hill, Chapel Hill, NC 27599, USA.

References
Andersen, PK: Gill: Cox's regression model for counting processes: a large sample study. Lifetime Data Anal. **10**, 1100–1120 (1982)
Chen, Y-H, Chen, R: A unified approach to regression analysis under double-sampling designs. J. R. Stat. Soc. B. **62**, 449–460 (2000)
Cox, DR: Regression models and life-tables (with discussion). J. R. Stat. Soc. B. **34**, 187–220 (1972)
Dickson, ER, Grambsch, PM, Fleming, TR, Fisher, LD, Langworthy, A: Prognosis in Primary Biliary Cirrhosis: Model for Decision Making. Hepatology. **10**, 1–7 (1989)
Fan, J, Gijbels, I: Local Polynomial Modeling and Its Applications. Chapman and Hall, London (1996)
Fan, J, Yao, Q: Nonlinear Time Series: Nonparametric and Parametric Methods. Springer-Verlag, New York (2003)
Fan, Z, Wang, X: Marginal hazards model for multivariate failure time data with auxiliary covariates. J. Nonparametric Stat. **21**, 771–786 (2009)
Fleming, TR: Harrington, DP Counting Process and Survival Analysis. Wiley, New York (1991)
Foutz, RV: On the unique consistent solution to the likelihood equations. J. Am. Stat. Assoc. **72**, 147–148 (1977)
Härdle, W: Applied Nonparametric Regression. Cambridge University Press, London (1990)
Hughes, MD: Regression dilution in the proportional hazards model. Biometrics. **49**, 1056–1066 (1993)
Jiang, X, Jiang, J, Liu, Y: Nonparametric regression under double-sampling designs. J. Syst. Sci. Complex. **24**, 167–175 (2011)
Jiang, J, Mack, YP: Robust local polynomial regression for dependent data. Stat. Sinica. **11**, 705–722 (2001)
Kalbfleisch, JD, Prentice, RL: The Statistical Analysis of Failure Time Data. Wiley, New York (1980)
Lin, DY, Ying, Z: Cox regression with incomplete covariate measurements. J. Am. Stat. Assoc. **88**, 1341–1349 (1993)
Lipsitz, S, Ibrahim, JG: Using the E-M algorithm for survival data with incomplete categorical covariates. Lifetime Data Anal. **2**, 5–14 (1996)
Liu, Y, Wu, Y, Zhou, H: Multivariate failure times regression with a continuous auxiliary covariate. J. Multivariate Anal. **101**, 679–691 (2010)
Markus, BH, Dickson, ER, Grambsch, PM, Fleming, TR, Mazzaferro, V, Klintmalm, GB, Wiesner, RH, Van Thiel, DH, Starzl, TE: Efficiency of liver transplantation in patients with primary biliary cirrhosis. N. Engl. J. Med. **320**, 1709–1713 (1989)
Nadaraya, EA: On estimating regression. Theory Probab. Appl. **10**, 186–190 (1964)
Pepe, MS, Self, SG, Prentice, RL: Further results on covariate measurement errors in cohort studies with time to response data. Statist. Med. **8**, 1167–1178 (1989)
Prentice, RL: Covariate measurement errors and parameter estimation in a failure time regression model. Biomtrika. **69**, 331–342 (1982)
Rubin, DB: Inference and missing data. Biomtrika. **63**, 581–592 (1976)
Rudin, W: Principle of Mathematical Analysis. McGraw-Hill Book Co., New York (1964)

Savitz, DA, Dole, N, Jr Terry, JW, Zhou, H, Jr Thorp, JM: Smoking and pregnancy outcome among African-American and white women in central North Carolina. Epidemiology. **12**, 636–642 (2001)

Watson, GS: Smooth regression analysis. Sankhya A. **26**, 359–372 (1964)

Zhou, H, Pepe, MS: Auxiliary covariate data in failure time regression analysis. Biomtrika. **82**, 139–149 (1995)

Zhou, H, Wang, C-Y: Some asymptotic results for using kernel smoother with covariate measurement error problem in survival analysis, Vol. 2200. University of North Carolina, Chapill Hill (1999)

Zhou, H, Wang, C-Y: Failure time regression with continuous covariates measured with error. J. R. Stat. Soc. B. **62**, 657–665 (2000)

7

Bivariate beta-generated distributions with applications to well-being data

José María Sarabia*, Faustino Prieto and Vanesa Jordá

*Correspondence:
sarabiaj@unican.es
Department of Economics,
University of Cantabria, Avda. de los Castros s/n, 39005 Santander, Spain

Abstract

The class of beta-generated distributions (Commun. Stat. Theory Methods 31:497–512, 2002; TEST 13:1–43, 2004) has received a lot of attention in the last years. In this paper, three new classes of bivariate beta-generated distributions are proposed. These classes are constructed using three different definitions of bivariate distributions with classical beta marginals and different covariance structures. We work with the bivariate beta distributions proposed in (J. Educ. Stat. 7:271–294, 1982; Metrika 54:215–231, 2001; Stat. Probability Lett. 62:407–412, 2003) for the first proposal, in (Stat. Methods Appl. 18: 465–481, 2009) for the second proposal and (J. Multivariate Anal. 102:1194–1202, 2011) for the third one. In each of these three classes, the main properties are studied. Some specific bivariate beta-generated distributions are studied. Finally, some empirical applications with well-being data are presented.

Mathematics Subject Classification (2000): 62E15; 60E05

Keywords: Classical beta distribution; Bivariate beta distribution; Covariance structure; GB1 and GB2 distributions

1 Introduction

In the recent statistical literature several methodologies of constructing bivariate and multivariate distributions based on marginal and conditional distributions have been proposed; see the works by Arnold et al. (1999; 2001), Kotz et al. (2000), Sarabia and Gómez-Déniz (2008) and Balakrishnan and Lai (2009) among others.

An important field of research focuses on the study of new classes of univariate distributions which contain the classical proposals, also allowing for more flexibility in fitting data. In this sense, the class of beta-generated (BG) distributions (Eugene et al. 2002; Jones 2004) has received an increasing amount of attention in recent years.

There are several reasons for studying classes of multivariate beta generated distributions. The two existing proposals of multivariate BG distributions present some drawbacks. The first proposal (Jones and Larsen 2004) is only valid for modeling data above the diagonal. The second proposal (Arnold et al. 2006) is defined in terms of the conditional distributions, and the corresponding marginal distributions do not follow, in general, beta generated distributions.

The three bivariate and multivariate models proposed in this paper present BG marginals, with high flexibility in the marginals and in the dependence structure. The marginal distributions of the first model share one of the shape parameters, and the structure of dependence satisfies TP2 condition (see Section 3). The marginals of the second model are free, and the different pairwise of marginals are associated (see Section 3). The third model is the more flexible, in the sense than all the marginals are free (they do not share any shape parameter) and the covariance structure admits correlations of any sign.

On the other hand, these classes of distributions present several fields of applicability. For example, bivariate beta generated distributions with classical beta marginals are natural choices to be used as prior distributions for the parameters of correlated binomial random variables (with any sign for the correlation) in Bayesian analysis (see Apostolakis and Moieni 1987; Arnold and Ng 2011).

In this work, three new classes of bivariate BG distributions are proposed. These classes are constructed using three alternative definitions of bivariate distributions with classical beta marginals and different covariance structures. We work with the bivariate beta distributions proposed by Libby and Novick (1982), Jones (2001) and Olkin and Liu (2003) for the first proposal, El-Bassiouny and Jones (2009) for the second proposal and Arnold and Ng (2011) for the third one. For each of these three classes, the main properties are obtained. Some specific bivariate BG distributions are studied. Finally, some empirical applications with well-being data are presented.

The contents of this paper are as follows. In Section 2 we present some basic properties of the class of the BG distributions and a brief review about two multivariate extensions of the BG distribution. Section 3 considers and studies the three classes of bivariate BG distributions and their main properties as well as to introduce three specific bivariate distributions. A number of applications of these distributions to fit well-being data are presented in Section 4. Finally, some conclusions and future research directions are given in Section 5.

2 Some univariate and multivariate beta-generated distributions

2.1 The univariate class of beta-generated distributions

In this section we present basic properties of the class of BG distributions. We begin with an initial baseline probability density function (PDF) $f(x)$, where the corresponding cumulative distribution function (CDF) is represented by $F(x)$. The class of BG distributions is defined in terms of the PDF by $(a, b > 0)$,

$$g_F(x; a, b) = [B(a, b)]^{-1} f(x) F(x)^{a-1} [1 - F(x)]^{b-1}, \tag{1}$$

where $B(a, b) = \Gamma(a)\Gamma(b)/\Gamma(a + b)$ denotes the classical beta function. A random variable X with PDF (1) will be denoted by $X \sim \mathcal{BG}(a, b; F)$.

The CDF associated to (1) is,

$$G_F(x; a, b) = I_{F(x)}(a, b),$$

where $I_{F(x)}(\cdot, \cdot)$ denotes the incomplete beta ratio.

If $B \sim \mathcal{B}e(a, b)$ represents the classical beta distribution, a simple stochastic representation of (1) is,

$$X = F^{-1}(B). \tag{2}$$

This representation permits a direct simulation of the values of a random variable with PDF (1), which can be also used for generating multivariate versions of the BG distribution. The raw moments of a BG distribution can be obtained by,

$$E\left[X^r\right] = E\left[\left\{F^{-1}(B)\right\}^r\right], \quad r > 0.$$

An important number of new classes of distributions have been proposed using this methodology. Some representatives examples of BG distributions include the generalized beta of the first kind (GB1) proposed by McDonald (1984), the generalized beta of the second kind (GB2) proposed and studied by Venter (1983) and McDonald (1984), the $\log F$ distribution (Barndorff-Nielsen et al. 1982), the beta-normal distribution (Eugene et al. 2002), the beta-exponential distribution (Nadarajah and Kotz 2006) and the Skew-t distribution (Jones 2004).

Some extensions of this family have been proposed by Alexander and Sarabia (2010), Alexander et al. (2012), Cordeiro and de Castro (2011) and Zografos (2011). Other alternative flexible families of distributions can be found in Alzaatreh et al. (2013, 2014) and Lee et al. (2013).

If $a = i$ and $b = n - i + 1$ in (1), we obtain the PDF of the i-th order statistic from F (Jones 2004). Below, we highlight some representative values of a and b,

- If $a = b = 1, g_F = f$.
- If $a = n$ and $b = 1$, we obtain the distribution of the maximum.
- If $a = 1$ and $b = n$, we obtain the distribution of the minimum.
- If $a \neq b$, we obtain a family of skew distributions.

Parameters a and b control the tailweight of the distribution. Specifically, the a parameter controls left-hand tailweight and the b parameter controls the right-hand tailweight of the distribution. On the other hand, $a = b$ yields a symmetric sub-family, with a controlling tailweight. If $a = b = 1$ the BG family is always symmetric if the baseline function $F(x)$ is symmetric. In this sense, the BG distribution accommodates several kind of tails. For example (see Jones 2004),

- Potential tails: If $f \sim x^{-(\alpha+1)}$ and $\alpha > 0$, when $x \to \infty$ $g_F \sim x^{-b\alpha-1}$,
- Exponential tails: If $f \sim e^{-\alpha x}$ and $\beta > 0$, then $g_F \sim e^{-b\beta x}$ if $x \to \infty$.

2.2 Two previous classes of multivariate extensions of beta-generated distributions

There are two proposals for multivariate extensions of BG distributions. The first proposal used the joint PDF of a subset of order statistics, and has been proposed by Jones and Larsen (2004). The second proposal used the so called Rosenblatt construction (Rosenblatt 1952), and has been proposed by Arnold et al. (2006). These two alternatives are not related with the multivariate BG distributions studied in this paper.

3 Three classes of bivariate beta-generated distributions

In this section we introduce three new classes of bivariate BG distributions. These three classes are constructed combining the basic stochastic representation (2) with three recent definitions of bivariate beta distributions proposed in the literature. The three definitions differ from each other in the marginal distributions and in the flexibility of the covariance structure. First, we need some previous notation. Let X be a random variable distributed as the classical gamma denoted by $X \sim G_a$, with PDF $f(x) = [\Gamma(a)]^{-1} x^{a-1} e^{-x}$, with $x \geq 0$ and $a > 0$. Then, if $X_1 \sim G_a$ and $X_2 \sim G_b$ are independent gamma random variables, the transformed random variable $X = X_1/(X_1 + X_2)$ is distributed as the classical beta distribution with parameters (a, b).

3.1 The first class of bivariate beta-generated distributions

The first class of bivariate beta-generated distribution is based on the following class of bivariate beta distribution.

Definition 1. Let G_{a_1}, G_{a_2} and G_b be three independent gamma random variables with $a_1, a_2, b > 0$. The first class of bivariate beta distribution is defined by the stochastic representation,

$$(Z_1, Z_2)^\top = \left(\frac{G_{a_1}}{G_{a_1} + G_b}, \frac{G_{a_2}}{G_{a_2} + G_b} \right)^\top . \tag{3}$$

This class was initially proposed by Libby and Novick (1982) and then studied by Jones (2001) and Olkin and Liu (2003).

Now, using (3) we define the following class of bivariate BG distributions.

Definition 2. Let G_{a_1}, G_{a_2} and G_b be three independent gamma random variables with $a_1, a_2, b > 0$. The first class of bivariate BG distribution is defined by the stochastic representation,

$$(X_1, X_2)^\top = \left(F_1^{-1} \left\{ \frac{G_{a_1}}{G_{a_1} + G_b} \right\}, F_2^{-1} \left\{ \frac{G_{a_2}}{G_{a_2} + G_b} \right\} \right)^\top , \tag{4}$$

where $F_1(\cdot), F_2(\cdot)$ are genuine CDF.

3.1.1 *Basic properties*

In this section we study some basic properties of the bivariate BG distribution defined in (4). The marginal distributions of (4) are BG distributions,

$$X_1 \sim \mathcal{BG}(a_1, b; F_1),$$
$$X_2 \sim \mathcal{BG}(a_2, b; F_2).$$

Note that both marginals share the second shape parameter b. However, this fact does not make the model less flexible, since both baseline distributions F_1 and F_2 are different.

Using the joint PDF of the bivariate beta distribution (3) (See Appendix), we obtain the joint PDF of the first class of bivariate BG distribution given by,

$$f(x_1, x_2) = \frac{F_1(x_1)^{a_1-1} F_2(x_2)^{a_2-1} [1 - F_1(x_1)]^{a_2+b-1} [1 - F_2(x_2)]^{a_1+b-1} f_1(x_1) f_2(x_2)}{B(a_1, a_2, b) [1 - F_1(x_1) F_2(x_2)]^{a_1+a_2+b}}, \tag{5}$$

where $a_1, a_2, b > 0$ and $B(a_1, a_2, b) = \Gamma(a_1)\Gamma(a_2)\Gamma(b)/\Gamma(a_1 + a_2 + b)$. An alternative expression of (5) in terms of the PDF of the BG distribution is,

$$f(x_1, x_2) = \frac{\Gamma(a_1 + b)\Gamma(a_2 + b)}{\Gamma(b)\Gamma(a_1 + a_2 + b)} \frac{g_{F_1}(x_1; a_1, a_2 + b) g_{F_2}(x_2; a_2, a_1 + b)}{[1 - F_1(x_1) F_2(x_2)]^{a_1+a_2+b}},$$

where $g_F(x; a, b)$ represents the PDF defined in (1). The joint PDF can be also written as an infinite mixture,

$$f(x_1, x_2) = \sum_{j=0}^{\infty} dA(j) g_{F_1}(x_1; a_1 + j, a_2 + b) \cdot g_{F_2}(x_2; a_2 + j, a_1 + b), \tag{6}$$

where $d = \frac{\Gamma(a_1+b)\Gamma(a_2+b)}{\Gamma(b)\Gamma(a_1+a_2+b)}$ and

$$A(j) = \frac{\Gamma(a_1 + j)}{\Gamma(a_1)} \cdot \frac{\Gamma(a_2 + j)}{\Gamma(a_2)} \cdot \frac{\Gamma(a_1 + a_2 + b)}{\Gamma(a_1 + a_2 + b + j)} \cdot \frac{1}{j!}.$$

The conditional distribution of X_1 given X_2 is,

$$f(x_1|x_2) = \frac{\Gamma(a_1 + a_2 + b)}{\Gamma(a_1)\Gamma(a_2)\Gamma(b)} \frac{F_1(x_1)^{a_1-1}[1 - F_1(x_1)]^{a_2+b-1}[1 - F_2(x_2)]^{a_1} f_1(x_1)}{[1 - F_1(x_1) F_2(x_2)]^{a_1+a_2+b}},$$

and the regression function of X_1 given X_2 is,

$$E(X_1|X_2 = x_2) = \frac{\Gamma(a_1 + a_2 + b)}{\Gamma(a_1)\Gamma(a_2)\Gamma(b)} (1 - F_2(x_2))^{a_2} \int_0^1 \frac{F_1^{-1}(t) t^{a_1-1}(1-t)^{a_2+b-1}}{[1 - tF_2(x_2)]^{a_1+a_2+b}} dt.$$

In order to study the dependence between X_1 and X_2, we consider the local dependence function defined by (Holland and Wang 1987),

$$\gamma(x_1, x_2) = \frac{\partial^2}{\partial x_1 \partial x_2} \log f(x_1, x_2).$$

We use the definitions of total positivity of order 2 (TP_2) functions and reverse rule of order 2 (RR_2) functions, which are the following.

Definition 3. A joint PDF $f(x, y)$ is said to be TP_2 (RR_2) if

$$f(x, y)f(u, v) - f(x, v)f(u, y) \geq 0 \quad (\leq 0)$$

for all $x \leq u$ and $y \leq v$.

The following result relates the local dependence function $\gamma(x_1, x_2)$ with the TP_2 and RR_2 (see Theorem 7.1 in Holland and Wang 1987).

Theorem 1. Let $f(x_1, x_2)$ be the joint PDF of (X_1, X_2) with support on a set S where the set $S = S_1 \times S_2$. Then, $f(x_1, x_2)$ is TP_2 (RR_2) if and only if $\gamma(x_1, x_2) \geq 0$ (≤ 0).

For the first class of BG distribution, it can be verified that

$$\gamma(x_1, x_2) = \frac{(a_1 + a_2 + b) f_1(x_1) f_2(x_2)}{[1 - F_1(x_1) F_1(x_1)]^2} > 0,$$

and then X_1 and X_2 are TP_2. As a consequence, the linear correlation coefficient between X_1 and X_2 is always positive.

It can be proved (see Shaked 1977) that if the joint PDF $f(x_1, x_2)$ is TP_2 (RR_2), then the conditional hazard rate of $X_1|X_2 = x_2$ is decreasing (increasing) in x_2. A similar property holds for the other conditional distribution $X_2|X_1 = x_1$. As the PDF in (5) is TP_2, this property shows the monotonicity properties of the hazard rate functions of the conditional distributions of $X_1|X_2 = x_2$ as a function of x_2 and the $X_2|X_1 = x_1$ as a function of X_1.

On the other hand, because X_1 and X_2 are increasing functions of independent random variables, X_1 and X_2 are associated random variables (Esary et al. 1967).

Expressions for the cross moments $E\left[X_1^{r_1}X_2^{r_2}\right]$ can be obtained from (5) or in terms of an infinite mixture from (6). On the other hand, if $b > r$, it can be shown that,

$$E\left\{\frac{F_1(X_1)^r F_2(X_2)^r}{[1-F_1(X_1)F_2(X_2)]^r}\right\} = \frac{\Gamma(a_1+r)\Gamma(a_2+r)\Gamma(b-r)\Gamma(a_1+a_2+b)}{\Gamma(a_1)\Gamma(a_2)\Gamma(b)\Gamma(a_1+a_2+b+r)}.$$

3.1.2 Extension to higher dimensions

The extension to higher dimensions is direct. The m-dimensional random vector is defined as,

$$(X_1,\ldots,X_m)^\top = \left(F_i^{-1}\left\{\frac{G_{a_i}}{G_{a_i}+G_b}\right\}, i=1,2,\ldots,m\right)^\top. \tag{7}$$

The marginal distributions are $X_i \sim \mathcal{BG}(a_i, b; F_i)$, $i = 1, 2, \ldots, m$. The joint PDF of (7) is given by,

$$f(x_1,\ldots,x_m) = c\prod_{i=1}^{m}\frac{F_i(x_i)^{a_i-1}f_i(x_i)}{[1-F_i(x_i)]^{a_i+1}}\left[1+\sum_{i=1}^{m}\frac{F_i(x_i)}{1-F_i(x_i)}\right]^{-b},$$

where $c^{-1} = B(a_1,\ldots,a_m,b) = \Gamma(a_1)\cdots\Gamma(a_m)\Gamma(b)/\Gamma(a_1+\ldots+a_m+b)$.

3.2 The second type of bivariate beta-generated distributions

The second type of bivariate BG distribution is motivated by the fact of having a bivariate distribution with arbitrary BG marginals. This second class is based on the following class of bivariate beta distribution, which was proposed by El-Bassiouny and Jones (2009).

Definition 4. Let G_{a_i}, $i = 1,2,3,4$ be independent gamma random variables, where $a_i > 0$, $i = 1,2,3,4$. The second class of bivariate beta distribution is defined by the stochastic representation,

$$(Z_1,Z_2)^\top = \left(\frac{G_{a_1}}{G_{a_1}+G_{a_3}}, \frac{G_{a_2}}{G_{a_2}+G_{a_3}+G_{a_4}}\right)^\top. \tag{8}$$

Now, we define the second class of BG distributions.

Definition 5. Let G_{a_i}, $i = 1,2,3,4$ be independent gamma random variables, where $a_i > 0$, $i = 1,2,3,4$. The second class of bivariate BG distribution is defined by the stochastic representation,

$$(X_1,X_2)^\top = \left(F_1^{-1}\left\{\frac{G_{a_1}}{G_{a_1}+G_{a_3}}\right\}, F_2^{-1}\left\{\frac{G_{a_2}}{G_{a_2}+G_{a_3}+G_{a_4}}\right\}\right)^\top, \tag{9}$$

where $F_1(\cdot), F_2(\cdot)$ are genuine CDF.

3.2.1 Basic properties

The marginal distributions are BG distributions with arbitrary parameters,

$$X_1 \sim \mathcal{BG}(a_1,a_3;F_1),$$
$$X_2 \sim \mathcal{BG}(a_2,a_3+a_4;F_2).$$

Using the joint PDF (8) (see Appendix), the joint PDF of this second class of bivariate BG distributions is given by,

$$f(x_1,x_2) = k\frac{g_{F_1}(x_1;a_1,A-a_1)\,g_{F_2}(x_2;a_2,A-a_2)}{[1-F_1(x_1)F_2(x_2)]^A}a_{12}(x_1,x_2),$$

where $k^{-1} = B(a_1,a_3)\,B(a_2,a_1+a_3+a_4), A = \sum_{i=1}^{4} a_i, a_{12}(x_1,x_2) = H[F_1(x_1),F_2(x_2)]$ and

$$H(z_1,z_2) = {}_2F_1\left[A,a_4;A-a_2;\frac{z_1(1-z_2)}{1-z_1z_2}\right],$$

being ${}_2F_1[..;.;]$ the Gauss confluent hypergeometric function.

The conditional density function of $X_1|X_2 = x_2$ is,

$$f(x_1|x_2) = k'\frac{f_1(x_1)\,F_1(x_1)^{a_1-1}[1-F_1(x_1)]^{A-a_1-1}[1-F_2(x_2)]^{a_1}}{[1-F_1(x_1)F_2(x_2)]^A}a_{12}(x_1,x_2),$$

where $k' = k\frac{B(a_2,a_3+a_4)}{B(a_1,A-a_1)B(a_2,A-a_2)}$ and the conditional density function of $X_2|X_1 = x_1$ is,

$$f(x_2|x_1) = k''\frac{f_2(x_2)\,F_2(x_2)^{a_2-1}[1-F_2(x_2)]^{A-a_2-1}[1-F_1(x_1)]^{A-a_1-a_3}}{[1-F_1(x_1)F_2(x_2)]^A}a_{12}(x_1,x_2),$$

where $k'' = k\frac{B(a_1,a_3)}{B(a_1,A-a_1)B(a_2,A-a_2)}$.

The regression function of X_1 given X_2 is,

$$E(X_1|X_2 = x_2) = a_1(x_2)\int_{-\infty}^{\infty}\frac{x_1f_1(x_1)\,F_1(x_1)^{a_1-1}[1-F_1(x_1)]^{A-a_1-1}}{[1-F_1(x_1)F_2(x_2)]^A}a_{12}(x_1,x_2)\,dx_1,$$

where $a_1(x_2) = k'[1-F_2(x_2)]^{a_1}$ and the regression function of X_2 given X_1 is,

$$E(X_2|X_1 = x_1) = a_2(x_1)\int_{-\infty}^{\infty}\frac{f_2(x_2)\,F_2(x_2)^{a_2-1}[1-F_2(x_2)]^{A-a_2-1}}{[1-F_1(x_1)F_2(x_2)]^A}a_{12}(x_1,x_2)\,dx_2,$$

being $a_2(x_1) = k''[1-F_1(x_1)]^{A-a_1-a_3}$.

The cross-product moments can be obtained as,

$$E\left(X_1^{r_1}X_2^{r_2}\right) = E_{(Z_1,Z_2)}\left\{\left[F_1^{-1}(Z_1)\right]^{r_1}\left[F_2^{-1}(Z_2)\right]^{r_2}\right\}, \tag{10}$$

where (Z_1,Z_2) is the bivariate random variable with joint PDF given by equation (23). Note that (10) can be computed easily by simulation from samples of the random variable (Z_1,Z_2). The local dependence function is given by,

$$\gamma(x_1,x_2) = \frac{Af_1(x_1)f_2(x_2)}{[1-F_1(x_1)F_2(x_2)]^2}+\frac{\partial^2 a_{12}(x_1,x_2)}{\partial x_1\partial x_2}. \tag{11}$$

The second term in (11) is long and will not be included here.

The random variables X_1 and X_2 are associated and then the linear correlation coefficient is always non-negative (see Definition I.11 and Proposition I.13 in Marshall and Olkin (2007)).

3.2.2 *Multivariate extensions*

A multivariate extension of (9) is also possible. We define,

$$(X_1,\ldots,X_m)^\top = \left(F_i^{-1}\left\{\frac{G_{a_i}}{G_{a_i}+\sum_{j=1}^{i}G_{b_j}}\right\}, i = 1,2,\ldots,m\right)^\top,$$

where the marginal distributions are BG distributions with parameters,

$$X_i \sim \mathcal{BG}\left(a_i, b_1 + \ldots + b_i; F_i\right), \quad i = 1, 2, \ldots, m.$$

3.3 The third type of bivariate beta-generated distributions

The next class of bivariate beta distribution is the more general class in the sense that the marginal distributions have arbitrary parameters and admits any sign for the linear correlation coefficient. The following definition was proposed by Arnold and Ng (2011).

Definition 6. The third class of bivariate beta distribution is defined by the stochastic representation,

$$(Z_1, Z_2)^\top = \left(\frac{G_{a_1} + G_{a_3}}{G_{a_1} + G_{a_3} + G_{a_4} + G_{a_5}}, \frac{G_{a_2} + G_{a_4}}{G_{a_2} + G_{a_3} + G_{a_4} + G_{a_5}}\right)^\top,$$

where G_{a_i}, $i = 1, 2, 3, 4, 5$ are independent gamma random variables, where $a_i > 0$, $i = 1, 2, 3, 4, 5$.

Now, we define the third class of BG distributions.

Definition 7. Let G_{a_i}, $i = 1, 2, 3, 4, 5$ be independent gamma random variables with $a_i > 0$, $i = 1, 2, 3, 4, 5$. The third class of bivariate BG distribution is defined by the stochastic representation,

$$(X_1, X_2)^\top = \left(F_1^{-1}\left\{\frac{G_{a_1} + G_{a_3}}{G_{a_1} + G_{a_3} + G_{a_4} + G_{a_5}}\right\}, F_2^{-1}\left\{\frac{G_{a_2} + G_{a_4}}{G_{a_2} + G_{a_3} + G_{a_4} + G_{a_5}}\right\}\right)^\top, \tag{12}$$

where $F_1(\cdot), F_2(\cdot)$ are genuine CDF.

3.3.1 Basic properties

The marginal distributions of (12) are $X_1 \sim \mathcal{BG}(a_1 + a_3, a_4 + a_5; F_1)$ and $X_2 \sim \mathcal{BG}(a_2 + a_4, a_3 + a_5; F_2)$.

The joint PDF of (12) is given by,

$$f_{X_1,X_2}\left(x_1, x_2\right) = \frac{f_1\left(x_1\right) f_2\left(x_2\right)}{\left(1 - F_1\left(x_1\right)\right)^2 \left(1 - F_2\left(x_2\right)\right)^2} f_{V,W}\left(\frac{F_1\left(x_1\right)}{1 - F_1\left(x_1\right)}, \frac{F_2\left(x_2\right)}{1 - F_2\left(x_2\right)}\right),$$

where $f_{V,W}(\cdot,\cdot)$ is defined in equation (24) in the Appendix.

The conditional density function of $X_1|X_2 = x_2$ is,

$$f\left(x_1|x_2\right) = \frac{k' f_1\left(x_1\right) f_{V,W}\left(\frac{F_1(x_1)}{1-F_1(x_1)}, \frac{F_2(x_2)}{1-F_2(x_2)}\right)}{\left[1 - F_1\left(x_1\right)\right]^2 \left[1 - F_2\left(x_2\right)\right]^{a_3+a_5+1} F_2\left(x_2\right)^{a_2+a_4-1}},$$

where $k' = B(a_2 + a_4, a_3 + a_5)$ and the conditional density function of $X_2|X_1 = x_1$ is,

$$f\left(x_2|x_1\right) = \frac{k'' f_2\left(x_2\right) f_{V,W}\left(\frac{F_1(x_1)}{1-F_1(x_1)}, \frac{F_2(x_2)}{1-F_2(x_2)}\right)}{\left[1 - F_1\left(x_1\right)\right]^{a_4+a_5+1} \left[1 - F_2\left(x_2\right)\right]^2 F_1\left(x_1\right)^{a_1+a_3-1}},$$

where $k'' = B(a_1 + a_3, a_4 + a_5)$.

The regression function of X_1 given X_2 is,

$$E\left(X_1|X_2=x_2\right)=b_1\left(x_2\right)\int_{-\infty}^{\infty}\frac{x_1f_1\left(x_1\right)f_{V,W}\left(\frac{F_1(x_1)}{1-F_1(x_1)},\frac{F_2(x_2)}{1-F_2(x_2)}\right)}{\left[1-F_1\left(x_1\right)\right]^2}dx_1,$$

where $b_1(x_2)=\frac{k'}{[1-F_2(x_2)]^{a_3+a_5+1}F_2(x_2)^{a_2+a_4-1}}$ and the regression function of X_2 given X_1 is,

$$E\left(X_2|X_1=x_1\right)=b_2\left(x_1\right)\int_{-\infty}^{\infty}\frac{x_2f_2\left(x_2\right)f_{V,W}\left(\frac{F_1(x_1)}{1-F_1(x_1)},\frac{F_2(x_2)}{1-F_2(x_2)}\right)}{\left[1-F_2\left(x_2\right)\right]^2}dx_2,$$

with $b_2(x_1)=\frac{k''}{[1-F_1(x_1)]^{a_4+a_5+1}F_1(x_1)^{a_1+a_3-1}}$.

The local dependence function can be written as,

$$\gamma\left(x_1,x_2\right)=h\left(x_1,x_2\right)\cdot\frac{\frac{\partial^2 f_{V,W}(u_1,u_2)}{\partial v\partial w}f_{V,W}\left(u_1,u_2\right)-\frac{\partial f_{V,W}(u_1,u_2)}{\partial v}\frac{\partial f_{V,W}(u_1,u_2)}{\partial w}}{f_{V,W}^2\left(u_1,u_2\right)},$$

where $h(x_1,x_2)=\frac{f_1(x_1)f_2(x_2)}{[1-F_1(x_1)][1-F_2(x_2)]}$ and $u_i=\frac{F_i(x_i)}{1-F_i(x_i)}$, $i=1,2$.

In a similar way, the cross-product moments can be obtained using formula (10), where now (Z_1,Z_2) is the bivariate random variable with joint PDF given by (24).

The covariance structure of (12) is flexible and the sign of the linear correlation coefficient can be positive or negative.

3.3.2 *Multivariate extensions*

The general multivariate version of the third class of BG distribution is based on the multivariate beta distribution proposed by Arnold and Ng (2011). Using this definition, the extension of (12) to dimensions higher than two is,

$$\left(X_1,\ldots,X_m\right)^\top=\left(F_i^{-1}\left\{\frac{G_{a_i}+G_{b_i}}{G_{a_i}+\sum_{l=1}^m G_{b_l}+G_c}\right\},i=1,2,\ldots,m\right)^\top,$$

where G_{a_i}, G_{b_i}, $i=1,2,\ldots,m$ and G_c are independent gamma random variables.

3.4 Estimation

Here we derive the maximum likelihood (ML) estimator of the parameters of the BG family of the first type defined in (4). Let (x_{1i},x_{2i}), $i=1,2,\ldots,n$ be a random sample of size n from (5), where we assume that both baseline functions are $F_1(x_1;\boldsymbol{\tau}_1)$ and $F_2(x_2;\boldsymbol{\tau}_2)$, where $\boldsymbol{\tau}_i$, $i=1,2$ is a $p_i\times 1$, $i=1,2$ vector of unknown parameters of the parent distributions. The log-likelihood function for $\boldsymbol{\theta}=(\boldsymbol{\tau}_1,\boldsymbol{\tau}_2,a_1,a_2,b)$ may be written,

$$\begin{aligned}\ell(\boldsymbol{\theta}) &= -n\log B\left(a_1,a_2,b\right)+a_1\sum_{i=1}^n\log\left[\frac{F_1\left(x_{1i};\boldsymbol{\tau}_1\right)\bar F_2\left(x_{2i};\boldsymbol{\tau}_2\right)}{1-F_1\left(x_{1i};\boldsymbol{\tau}_1\right)F_2\left(x_{2i};\boldsymbol{\tau}_2\right)}\right]\\ &+a_2\sum_{i=1}^n\log\left[\frac{F_2\left(x_{2i};\boldsymbol{\tau}_2\right)\bar F_1\left(x_{1i};\boldsymbol{\tau}_1\right)}{1-F_1\left(x_{1i};\boldsymbol{\tau}_1\right)F_2\left(x_{2i};\boldsymbol{\tau}_2\right)}\right]\\ &+b\sum_{i=1}^n\log\left[\frac{\bar F_1\left(x_{1i};\boldsymbol{\tau}_1\right)\bar F_2\left(x_{2i};\boldsymbol{\tau}_2\right)}{1-F_1\left(x_{1i};\boldsymbol{\tau}_1\right)F_2\left(x_{2i};\boldsymbol{\tau}_2\right)}\right]\\ &+\sum_{i=1}^n\log\left[\frac{f_1\left(x_{1i};\boldsymbol{\tau}_1\right)f_2\left(x_{2i};\boldsymbol{\tau}_2\right)}{\bar F_1\left(x_{1i};\boldsymbol{\tau}_1\right)F_1\left(x_{1i};\boldsymbol{\tau}_1\right)\bar F_1\left(x_{1i};\boldsymbol{\tau}_1\right)\bar F_2\left(x_{2i};\boldsymbol{\tau}_2\right)}\right],\end{aligned}\tag{13}$$

where we have used the notation $\bar F_i\left(\cdot;\boldsymbol{\tau}_i\right)=1-F_i\left(\cdot;\boldsymbol{\tau}_i\right)$, $i=1,2$.

This expression may be maximized either directly, e.g. using the Mathematica software function `FindMaximum` (see Wolfram Research, Inc. 2010), the SAS procedure `NLMIXED` (SAS Institute, Inc. 2010), the `R` software functions `nlm` or `optim` (R Development Core Team 2011), or the MATLAB function `fmincon` (The Mathworks, Inc. 2011), among others, which provides numerical algorithms for nonlinear optimization), or by solving the nonlinear equations obtained by differentiating expression (13).

Initial estimates of the parameters a_1, a_2 and b can be inferred from estimates of $\boldsymbol{\tau}_1$ and $\boldsymbol{\tau}_2$, since if $(X_1, X_2)^\top$ is distributed as (4), then $(F_1(X_1), F_2(X_2))^\top$ is distributed as (3). If we define the random variables $Y_1 = F_1(X_1)$ and $Y_2 = F_2(X_2)$, we obtain the following expressions,

$$E[Y_1] = \frac{a_1}{a_1 + b}, \tag{14}$$

$$E[Y_2] = \frac{a_2}{a_2 + b}, \tag{15}$$

$$E\left[\frac{1 - Y_1 Y_2}{(1 - Y_1)(1 - Y_2)}\right] = \frac{a_1 + a_2 + b - 1}{b - 1}, \quad b > 1 \tag{16}$$

If m_1, m_2 and m_{12} are the sample versions of previous moments, solving Equations (14) to (16) for a_1, a_2 and b we have:

$$\hat{a}_1 = \frac{m_1(1 - m_2)(1 - m_{12})}{1 - m_1 m_2 + m_{12}(m_1 + m_2 - m_1 m_2 - 1)}, \tag{17}$$

$$\hat{a}_2 = \frac{m_2(1 - m_1)(1 - m_{12})}{1 - m_1 m_2 + m_{12}(m_1 + m_2 - m_1 m_2 - 1)}, \tag{18}$$

$$\hat{b} = \frac{(1 - m_1)(1 - m_2)(1 - m_{12})}{1 - m_1 m_2 + m_{12}(m_1 + m_2 - m_1 m_2 - 1)}. \tag{19}$$

The components of the score vector $U(\boldsymbol{\theta})$ are given by,

$$
\begin{aligned}
U_{a_1}(\boldsymbol{\theta}) &= n\left[\psi(a_1 + a_2 + b) - \psi(a_1)\right] + \sum_{i=1}^{n} \log\left[\frac{F_1(x_{1i};\boldsymbol{\tau}_1)\bar{F}_2(x_{2i};\boldsymbol{\tau}_2)}{1 - F_1(x_{1i};\boldsymbol{\tau}_1)F_2(x_{2i};\boldsymbol{\tau}_2)}\right],\\
U_{a_2}(\boldsymbol{\theta}) &= n\left[\psi(a_1 + a_2 + b) - \psi(a_2)\right] + \sum_{i=1}^{n} \log\left[\frac{F_2(x_{2i};\boldsymbol{\tau}_2)\bar{F}_1(x_{1i};\boldsymbol{\tau}_1)}{1 - F_1(x_{1i};\boldsymbol{\tau}_1)F_2(x_{2i};\boldsymbol{\tau}_2)}\right],\\
U_{b}(\boldsymbol{\theta}) &= n\left[\psi(a_1 + a_2 + b) - \psi(b)\right] + \sum_{i=1}^{n} \log\left[\frac{\bar{F}_1(x_{1i};\boldsymbol{\tau}_1)\bar{F}_2(x_{2i};\boldsymbol{\tau}_2)}{1 - F_1(x_{1i};\boldsymbol{\tau}_1)F_2(x_{2i};\boldsymbol{\tau}_2)}\right],\\
U_{\boldsymbol{\tau}_1}(\boldsymbol{\theta}) &= (a_1 - 1)\sum_{i=1}^{n}\frac{\dot{F}_1(x_{1i})_{\boldsymbol{\tau}_1}}{F_1(x_{1i};\boldsymbol{\tau}_1)} - (a_2 + b - 1)\sum_{i=1}^{n}\frac{\dot{F}_1(x_{1i})_{\boldsymbol{\tau}_1}}{\bar{F}_1(x_{1i};\boldsymbol{\tau}_1)}\\
&\quad + (a_1 + a_2 + b)\sum_{i=1}^{n}\frac{\dot{F}_1(x_{1i})_{\boldsymbol{\tau}_1}F_2(x_{2i};\boldsymbol{\tau}_2)}{1 - F_1(x_{1i};\boldsymbol{\tau}_1)F_2(x_{2i};\boldsymbol{\tau}_2)} + \sum_{i=1}^{n}\frac{\dot{f}_1(x_{1i})_{\boldsymbol{\tau}_1}}{f_1(x_{1i};\boldsymbol{\tau}_1)},\\
U_{\boldsymbol{\tau}_2}(\boldsymbol{\theta}) &= (a_2 - 1)\sum_{i=1}^{n}\frac{\dot{F}_2(x_{2i})_{\boldsymbol{\tau}_2}}{F_2(x_{2i};\boldsymbol{\tau}_2)} - (a_1 + b - 1)\sum_{i=1}^{n}\frac{\dot{F}_2(x_{2i})_{\boldsymbol{\tau}_2}}{\bar{F}_2(x_{2i};\boldsymbol{\tau}_2)}\\
&\quad + (a_1 + a_2 + b)\sum_{i=1}^{n}\frac{\dot{F}_2(x_{2i})_{\boldsymbol{\tau}_2}F_1(x_{1i};\boldsymbol{\tau}_1)}{1 - F_1(x_{1i};\boldsymbol{\tau}_1)F_2(x_{2i};\boldsymbol{\tau}_2)} + \sum_{i=1}^{n}\frac{\dot{f}_2(x_{2i})_{\boldsymbol{\tau}_2}}{f_2(x_{2i};\boldsymbol{\tau}_2)},
\end{aligned}
$$

where $\dot{f}_j(x_{ji})_{\tau_j} = \partial f_j(x_{ji};\boldsymbol{\tau}_j)/\partial\boldsymbol{\tau}_j$ and $\dot{F}_j(x_{ji})_{\tau_j} = \partial F_j(x_{ji};\boldsymbol{\tau}_j)/\partial\boldsymbol{\tau}_j$ are $p_j \times 1$ vectors, with $j = 1, 2$ and $\psi(s) = d \log \Gamma(s)/ds$ is the digamma function.

For obtaining interval estimation and hypothesis testing on the model parameters, we need the observed information matrix. The $(p_1 + p_2 + 3, p_1 + p_2 + 3)$ observed matrix $J = J(\boldsymbol{\theta})$ can be obtained by taking partial second derivatives in the score vector $U(\boldsymbol{\theta})$. Assuming conditions that are fulfilled for parameters in the interior of the parameter space (but not in the boundary), the distribution of $\sqrt{n}\left(\hat{\boldsymbol{\theta}} - \boldsymbol{\theta}\right)$ is asymptotically normal $\mathcal{N}_{p_1+p_2+3}\left(\mathbf{0}, I(\boldsymbol{\theta})^{-1}\right)$, where $I(\boldsymbol{\theta})$ denotes the expected information matrix. As usual, we can substitute $I(\boldsymbol{\theta})$ by $J\left(\hat{\boldsymbol{\theta}}\right)$, that is, the observed information matrix evaluated at $\hat{\boldsymbol{\theta}}$ and then, the distribution $\mathcal{N}_{p_1+p_2+3}\left(\mathbf{0}, J(\hat{\boldsymbol{\theta}})^{-1}\right)$ can be used to construct approximate confidence intervals for the parameters.

The estimation of the other two models (9) and (12) requires a detailed study, which is beyond the scope of this paper and will be object of future research.

To finish this section, it should be mentioned that all the models proposed in this paper (4, 9 and 12) and their multivariate extensions can be enriched including location and scale parameters.

3.5 Some specific bivariate distributions

In this section we propose three specific bivariate BG models.

3.5.1 *Bivariate Beta-Normal distributions*

This model is a direct bivariate extension of the beta-normal distribution considered by Eugene et al. (2002). If $F_i(x_i) = \Phi(z_i)$, where $z_i = (x_i - \mu_i)/\sigma_i$, where $\mu_i \in \mathbb{R}$ and $\sigma_i > 0$, $i = 1, 2$ and $\Phi(z)$ is the CDF of a standard normal distribution, we obtain the bivariate joint PDF,

$$f(x_1, x_2) = \frac{\Phi(z_1)^{a_1-1}\,\Phi(z_2)^{a_2-1}\,[1 - \Phi(z_1)]^{a_2+b-1}\,[1 - \Phi(z_2)]^{a_1+b-1}\,\phi(z_1)\,\phi(z_2)}{\sigma_1\sigma_2 B(a_1, a_2, b)\,[1 - \Phi(z_1)\,\Phi(z_2)]^{a_1+a_2+b}},$$

where $a_1, a_2, b > 0$.

3.5.2 *Bivariate GB1 income distributions*

If we take $F(x) = x^a$ in (1), we obtain the generalized beta distribution of the first kind (GB1) (see McDonald 1984), which will be denoted by $X \sim GB1(p, q, a)$. Then, if $F_i(x_i) = x_i^{a_i}$, with $i = 1, 2$ and using Equation (5) we obtain,

$$f(x_1, x_2) = \frac{a_1 a_2 x_1^{a_1 p_1 - 1}\left(1 - x_1^{a_1}\right)^{p_2+q-1} x_2^{a_2 p_2 - 1}\left(1 - x_2^{a_2}\right)^{p_1+q-1}}{B(p_1, p_2, q)\left(1 - x_1^{a_1} x_2^{a_2}\right)^{p_1+p_2+q}}, \tag{20}$$

with $0 \le x_1, x_2 \le 1$. The marginal distributions are $X_1 \sim GB1(p_1, q, a_1)$ and $X_2 \sim GB1(p_2, q, a_2)$. If we set $a_1 = a_2 = 1$ in (20), we obtain the bivariate beta proposed by Olkin and Liu (2003).

3.5.3 Bivariate GB2 income distributions

Now if we take $F(x) = 1 - 1/(1 + x^a)$ in (1), we obtain the generalized beta distribution of the second kind (GB2) (see McDonald 1984), which will be denoted by $X \sim GB2(p, q, a)$. Then, if $F_i(x_i) = 1 - 1/(1 + x_i^{a_i})$, with $i = 1, 2$ and using formula (5) we have,

$$f(x_1, x_2) = \frac{a_1 a_2}{B(p_1, p_2, q)} \cdot \frac{x_1^{a_1 p_1 - 1} x_2^{a_2 p_2 - 1}}{\left(1 + x_1^{a_1} + x_2^{a_2}\right)^{p_1 + p_2 + q}}, \quad x_1, x_2 \geq 0,$$

where the marginal distributions are $X_1 \sim GB2(p_1, q, a_1)$ and $X_2 \sim GB2(p_2, q, a_2)$.

4 Applications

To illustrate the methodology developed in this paper, we have fitted the bivariate BG model of the fist type defined in (4) to estimate the international distribution of well-being for the period 1980-2010. The estimation method is based on the formulation developed in Section 3.4. It should be worth noting that we have focused on three dimensions of well-being, namely income, health and education. Since these components present a positive correlation, the first type of BG distributions given by (4) is specially suitable in this case.

4.1 The data

We have used the most recent available data from International Human Development Indicators (UNDP 2012) on the HDI and its three components for the period 1980-2010 with five years intervals.

Note that we consider well-being as a multidimensional process which, in addition to economic variables, also involves social aspects such as health and education. In this context, the Human Development Index provides an excellent theoretical benchmark to make multidimensional assessments of well-being. Then, we have focused on three dimensions of quality of life: income, educational standards and health. In particular, we focus on

Table 1 Parameter estimates for the BG models (with beta and GB1 marginals) fitted to the variables education & health by maximum likelihood (standard errors in parenthesis)

Variables: education & health								
Year	BG model with beta marginals			BG model with GB1 marginals				
	$\hat{p}_1$	$\hat{p}_2$	$\hat{q}$	$\hat{p}_1$	$\hat{p}_2$	$\hat{q}$	$\hat{a}_1$	$\hat{a}_2$
1980	2.6512	7.0544	3.4327	22.5595	10.5446	2.6651	0.1177	0.5381
	(0.2358)	(0.6334)	(0.3063)	(9.1385)	(2.4534)	(0.2890)	(0.0481)	(0.1299)
1985	2.9384	7.5581	3.2621	11.1697	14.2985	2.6168	0.2527	0.4416
	(0.2614)	(0.6780)	(0.2906)	(3.6937)	(6.2408)	(0.2810)	(0.0818)	(0.1918)
1990	2.9333	7.1301	2.8442	26.6132	6.6869	2.3141	0.1070	0.8566
	(0.2615)	(0.6410)	(0.2534)	(20.0869)	(1.7468)	(0.2473)	(0.0800)	(0.2243)
1995	2.8508	6.1734	2.3377	17.3255	4.4878	2.0254	0.1606	1.1454
	(0.2551)	(0.5572)	(0.2085)	(7.0388)	(0.9738)	(0.2125)	(0.0649)	(0.2588)
2000	2.7879	5.5716	1.9464	4.6151	5.7575	1.7644	0.5647	0.8677
	(0.2505)	(0.5051)	(0.1736)	(1.7265)	(2.8469)	(0.1811)	(0.2029)	(0.4191)
2005	2.6539	5.0138	1.5705	13.1632	2.3387	1.5422	0.2096	1.9972
	(0.2398)	(0.4577)	(0.1401)	(7.3975)	(0.5126)	(0.1578)	(0.1173)	(0.4455)
2010	2.6591	5.2296	1.4064	3.4113	3.7361	1.4084	0.7889	1.3821
	(0.2411)	(0.4794)	(0.1254)	(1.1784)	(1.4461)	(0.1418)	(0.2710)	(0.5291)

Table 2 Parameter estimates for the BG models (with beta and GB1 marginals) fitted to the variables education & income by maximum likelihood (standard errors in parenthesis)

Variables: education & income

Year	BG model with beta marginals			BG model with GB1 marginals				
	$\hat{p}_1$	$\hat{p}_2$	$\hat{q}$	$\hat{p}_1$	$\hat{p}_2$	$\hat{q}$	$\hat{a}_1$	$\hat{a}_2$
1980	2.2431	3.3973	2.8809	2.1813	16.6110	2.6533	0.9603	0.2195
	(0.2008)	(0.3063)	(0.2591)	(0.5790)	(13.5230)	(0.2743)	(0.2243)	(0.1715)
1985	2.8775	3.7222	3.1934	4.3394	10.2173	2.7715	0.6321	0.3586
	(0.2572)	(0.3338)	(0.2858)	(1.4677)	(5.6252)	(0.2881)	(0.1955)	(0.1857)
1990	3.1691	3.6855	3.0715	6.4501	7.6217	2.6157	0.4702	0.4541
	(0.2834)	(0.3302)	(0.2746)	(2.7530)	(3.5058)	(0.2725)	(0.1880)	(0.1960)
1995	3.3196	3.3074	2.7070	6.3618	6.2069	2.3322	0.4907	0.4965
	(0.2978)	(0.2967)	(0.2421)	(2.4599)	(2.2291)	(0.2418)	(0.1818)	(0.1698)
2000	3.4652	3.0623	2.3826	5.8748	5.8042	2.0680	0.5420	0.4885
	(0.3119)	(0.2751)	(0.2131)	(2.2960)	(2.0882)	(0.2139)	(0.2052)	(0.1691)
2005	3.6366	2.8670	2.0777	7.2438	3.4311	1.8787	0.4773	0.7694
	(0.3285)	(0.2580)	(0.1858)	(3.2031)	(0.8872)	(0.1918)	(0.2070)	(0.1937)
2010	3.7596	2.7789	1.8944	3.4581	4.0863	1.8134	1.0309	0.6702
	(0.3406)	(0.2505)	(0.1693)	(1.0479)	(1.1725)	(0.1835)	(0.3034)	(0.1870)

the single-dimensional indices of the HDI, which are three normalized variables placed on scale 1-0. This structure of the data is specially representative in this case since we consider Beta and GB1 marginals for the BG models.

Income is represented by Gross National Income per capita measured in PPP 2005 US dollars, to make incomes comparable across countries and over time. The health component is represented by life expectancy at birth, which is considered an indicator of the health level.The education index is made up of two indicators, expected years of schooling and mean years of schooling, which are aggregated using the geometric mean. The first educational variable informs about the number of years that a child of school entrance

Table 3 Parameter estimates for the BG models (with beta and GB1 marginals) fitted to the variables income & health by maximum likelihood (standard errors in parenthesis)

Variables: income & health

Year	BG model with beta marginals			BG model with GB1 marginals				
	$\hat{p}_1$	$\hat{p}_2$	$\hat{q}$	$\hat{p}_1$	$\hat{p}_2$	$\hat{q}$	$\hat{a}_1$	$\hat{a}_2$
1980	3.4799	5.9823	2.9491	7.2415	4.5699	2.7798	0.4964	1.1929
	(0.3103)	(0.5360)	(0.2624)	(2.1497)	(1.0817)	(0.2855)	(0.1417)	(0.2674)
1985	3.8895	7.7393	3.3336	19.9607	7.1397	2.7940	0.1918	0.9025
	(0.3458)	(0.6913)	(0.2959)	(11.3280)	(1.8739)	(0.2993)	(0.1072)	(0.2316)
1990	3.7983	8.0223	3.1630	19.2882	10.1747	2.5143	0.18067	0.63504
	(0.3379)	(0.7173)	(0.2808)	(10.8390)	(3.4854)	(0.2721)	(0.1003)	(0.2148)
1995	3.4082	7.5345	2.7867	12.4184	11.5787	2.2338	0.2457	0.5300
	(0.3039)	(0.6760)	(0.2478)	(4.6027)	(4.5656)	(0.2401)	(0.0901)	(0.2104)
2000	3.0781	7.0973	2.3944	11.9092	7.5709	1.9876	0.2346	0.7667
	(0.2752)	(0.6396)	(0.2132)	(4.7419)	(2.2840)	(0.2109)	(0.0935)	(0.2342)
2005	2.7891	6.8378	2.0250	19.9351	4.0164	1.7675	0.1323	1.4058
	(0.2501)	(0.6197)	(0.1805)	(19.3729)	(1.2603)	(0.1863)	(0.1265)	(0.4381)
2010	2.4296	6.5389	1.6754	4.4081	4.2348	1.5941	0.5388	1.4186
	(0.2188)	(0.5974)	(0.1496)	(1.2981)	(1.4324)	(0.1627)	(0.1582)	(0.4772)

Table 4 Parameter estimates for the BG model (BG truncated exponential marginals) fitted to the variables education & health by maximum likelihood (standard errors in parenthesis)

Variables: education & health					
Year	Truncated exponential model				
	$\hat{p}_1$	$\hat{p}_2$	$\hat{q}$	$\hat{a}_1$	$\hat{a}_2$
1980	3.4104	19.2541	1.3217	3.1388	4.1114
	(0.3671)	(4.0935)	(0.2372)	(0.4543)	(0.6456)
1985	3.9159	17.3862	1.4386	2.7821	3.4638
	(0.4287)	(3.7250)	(0.2519)	(0.4325)	(0.6248)
1990	4.1159	10.5428	1.6196	2.2094	2.1086
	(0.4795)	(2.0906)	(0.2652)	(0.4185)	(0.5804)
1995	4.0476	6.7782	1.6363	1.7177	1.0854
	(0.5304)	(1.3450)	(0.2499)	(0.4271)	(0.5825)
2000	3.8541	5.3537	1.5412	1.3568	0.5281
	(0.5772)	(1.1472)	(0.2193)	(0.4437)	(0.6087)
2005	4.3824	3.8719	1.2821	1.6870	0.0000
	(0.7229)	(0.4862)	(0.1420)	(0.4704)	(0.0427)
2010	3.9681	4.1826	1.1884	1.3861	0.0001
	(0.7878)	(1.0311)	(0.1607)	(0.5223)	(0.6912)

age can expect to receive if prevailing patterns of age-specific enrollment rates persist throughout the child's life (UNDP 2012). The second indicator reports the average number of years of education received by people aged 25 and older, converted from education attainment levels using official durations of each level (Barro and Lee, 2013).

Originally, our sample comprised only 105 countries, covering less than the 75 percent of global population. We had non-available data for 26 countries for one or more years before 1995. In order to offer comparable results across periods and to not restricting the sample considerably, missing values have been estimated. The estimation of these missing

Table 5 Parameter estimates for the BG model (BG truncated exponential marginals) fitted to the variables education & income by maximum likelihood (standard errors in parenthesis)

Variables: education & income					
Year	Truncated exponential model				
	$\hat{p}_1$	$\hat{p}_2$	$\hat{q}$	$\hat{a}_1$	$\hat{a}_2$
1980	2.3477	6.0323	1.5187	1.7859	2.8785
	(0.2801)	(0.9499)	(0.2891)	(0.5446)	(0.6098)
1985	3.1220	6.4260	1.6681	1.8322	2.7940
	(0.3847)	(0.9671)	(0.3094)	(0.5189)	(0.5639)
1990	3.5887	6.2829	1.5834	1.9604	2.8090
	(0.4533)	(0.9215)	(0.2838)	(0.5057)	(0.5350)
1995	3.6986	5.9955	1.3892	1.9633	3.0090
	(0.4814)	(0.8708)	(0.2429)	(0.5081)	(0.5264)
2000	3.7836	5.7931	1.2656	1.8579	3.0539
	(0.5115)	(0.8442)	(0.2131)	(0.5086)	(0.5110)
2005	3.9060	5.3306	1.1572	1.7335	2.9354
	(0.5793)	(0.8152)	(0.1925)	(0.5332)	(0.5233)
2010	3.3507	5.8944	1.1977	0.9970	2.8693
	(0.5127)	(0.9557)	(0.1937)	(0.5579)	(0.5168)

Table 6 Parameter estimates for the BG model (BG truncated exponential marginals) fitted to the variables income & health by maximum likelihood (standard errors in parenthesis)

Variables: income & health					
Year	Truncated exponential model				
	$\hat{p}_1$	$\hat{p}_2$	$\hat{q}$	$\hat{a}_1$	$\hat{a}_2$
1980	5.2910	5.6187	2.1441	1.7075	0.6332
	(0.7765)	(1.0281)	(0.3450)	(0.4689)	(0.5700)
1985	6.0810	8.9083	2.0766	2.1172	1.3880
	(0.7943)	(1.6258)	(0.3272)	(0.4063)	(0.5318)
1990	6.6056	9.9881	1.7975	2.5828	1.7600
	(0.8013)	(1.7123)	(0.2787)	(0.3801)	(0.5043)
1995	6.6121	8.9254	1.5662	2.8884	1.7273
	(0.8033)	(1.4908)	(0.2377)	(0.3774)	(0.4970)
2000	6.2010	7.4495	1.4527	2.8150	1.3495
	(0.7880)	(1.2753)	(0.2108)	(0.3768)	(0.4997)
2005	6.5460	5.5399	1.3561	2.9102	0.6186
	(0.9084)	(0.9529)	(0.1881)	(0.3852)	(0.5150)
2010	6.6966	4.1236	1.1760	3.1741	0.0000
	(1.0480)	(0.7957)	(0.1720)	(0.4275)	(0.6583)

values has been based on two complementary methodologies which jointly offered feasible and consistent results according to the sample: piecewise cubic Hermite interpolating polynomial (PCHI) and the average rate of change, which was used when PCHI offered unfeasible estimations or out of range results. The interpolated values have been obtained using the command `pchip` of the `R` package `Signal`, which uses the methodology described by Fritsch and Carlson (1980). After this procedure, our data set includes 132 countries whose indicators of income, health and education are available for eight points of time (see Appendix for details). Consequently, the sample covers over 90 percent of the world population during the whole period.

Table 7 Confidence intervals (95%) for the BG models (with beta and GB1 marginals) fitted to the variables education & health by maximum likelihood

Variables: education & health										
Year	Limit	BG model with beta marginals			BG model with GB1 marginals					
		$\hat{p}_1$	$\hat{p}_2$	$\hat{q}$	$\hat{p}_1$	$\hat{p}_2$	$\hat{q}$	$\hat{a}_1$	$\hat{a}_2$
1980	Lower	2.1890	5.8129	2.8324	4.6480	5.7359	2.0987	0.0234	0.2835
	Upper	3.1134	8.2959	4.0330	40.4710	15.3533	3.2315	0.2120	0.7927
1985	Lower	2.4261	6.2292	2.6925	3.9300	2.0665	2.0660	0.0924	0.0657
	Upper	3.4507	8.8870	3.8317	18.4094	26.5305	3.1676	0.4130	0.8175
1990	Lower	2.4208	5.8737	2.3475	-12.7571	3.2632	1.8294	-0.0498	0.4170
	Upper	3.4458	8.3865	3.3409	65.9835	10.1106	2.7988	0.2638	1.2962
1995	Lower	2.3508	5.0813	1.9290	3.5295	2.5792	1.6089	0.0334	0.6382
	Upper	3.3508	7.2655	2.7464	31.1215	6.3964	2.4419	0.2878	1.6526
2000	Lower	0.6984	2.8142	0.3379	7.9680	16.3910	0.3195	0.1146	0.3637
	Upper	3.2789	6.5616	2.2867	7.9990	11.3374	2.1194	0.9624	1.6891
2005	Lower	2.1839	4.1167	1.2959	-1.3359	1.3340	1.2329	-0.0203	1.1240
	Upper	3.1239	5.9109	1.8451	27.6623	3.3434	1.8515	0.4395	2.8704
2010	Lower	2.1865	4.2900	1.1606	1.1016	0.9017	1.1305	0.2577	0.3451
	Upper	3.1317	6.1692	1.6522	5.7210	6.5705	1.6863	1.3201	2.4191

Table 8 Confidence Intervals (95%) for the BG models (with beta and GB1 marginals) fitted to the variables education & income by maximum likelihood

Variables: education & health

Year	Limit	BG model with beta marginals			BG model with GB1 marginals				
		$\hat{p}_1$	$\hat{p}_2$	$\hat{q}$	$\hat{p}_1$	$\hat{p}_2$	$\hat{q}$	$\hat{a}_1$	$\hat{a}_2$
1980	Lower	1.8495	2.7970	2.3731	1.0465	-9.8941	2.1157	0.5207	-0.1166
	Upper	2.6367	3.9976	3.3887	3.3161	43.1161	3.1909	1.3999	0.5556
1985	Lower	2.3734	3.0680	2.6332	1.4627	-0.8081	2.2068	0.2489	-0.0054
	Upper	3.3816	4.3764	3.7536	7.2161	21.2427	3.3362	1.0153	0.7226
1990	Lower	2.6136	3.0383	2.5333	1.0542	0.7503	2.0816	0.1017	0.0699
	Upper	3.7246	4.3327	3.6097	11.8460	14.4931	3.1498	0.8387	0.8383
1995	Lower	2.7359	2.7259	2.2325	1.5404	1.8379	1.8583	0.1344	0.1637
	Upper	3.9033	3.8889	3.1815	11.1832	10.5759	2.8061	0.8470	0.8293
2000	Lower	1.0808	0.8424	0.5077	13.4885	12.1203	0.4423	0.1112	0.0826
	Upper	4.0765	3.6015	2.8003	10.3750	9.8971	2.4872	0.9442	0.8199
2005	Lower	2.9927	2.3613	1.7135	0.9657	1.6922	1.5028	0.0716	0.3897
	Upper	4.2805	3.3727	2.4419	13.5219	5.1700	2.2546	0.8830	1.1491
2010	Lower	3.0920	2.2879	1.5626	1.4042	2.5649	1.4537	0.4362	0.3037
	Upper	4.4272	3.2699	2.2262	5.5120	7.1611	2.1731	1.6256	1.0367

4.2 Fitted models and results

The bivariate data consist of three pairs of variables (income,education), (income,health) and (education, health).

We have fitted the class of models given by Equation (4) with three specifications for the baseline CDFs:

- $F_i(x_i) = x_i$, with $0 \le x_i \le 1$, $i = 1, 2$ (classical beta marginals),
- $F_i(x_i) = x_i^{a_i}$, with $0 \le x_i \le 1$, $a_i > 0$, $i = 1, 2$ (GB1 marginals) and
- $F_i(x_i) = \frac{1-\exp(-a_i x_i)}{1-\exp(-a_i)}$, with $0 \le x_i \le 1$, $a_i > 0$, $i = 1, 2$ (BG truncated exponential marginals).

Table 9 Confidence intervals (95%) for the BG models (with beta and GB1 marginals) fitted to the variables income & health by maximum likelihood

Variables: education & health

Year	Limit	BG model with beta marginals			BG model with GB1 marginals				
		$\hat{p}_1$	$\hat{p}_2$	$\hat{q}$	$\hat{p}_1$	$\hat{p}_2$	$\hat{q}$	$\hat{a}_1$	$\hat{a}_2$
1980	Lower	2.8717	4.9317	2.4348	3.0281	2.4498	2.2202	0.2187	0.6688
	Upper	4.0881	7.0329	3.4634	11.4549	6.6900	3.3394	0.7741	1.7170
1985	Lower	3.2117	6.3844	2.7536	-2.2422	3.4669	2.2074	-0.0183	0.4486
	Upper	4.5673	9.0942	3.9136	42.1636	10.8125	3.3806	0.4019	1.3564
1990	Lower	3.1360	6.6164	2.6126	-1.9562	3.3433	1.9810	-0.0159	0.2140
	Upper	4.4606	9.4282	3.7134	40.5326	17.0061	3.0476	0.3773	1.0560
1995	Lower	2.8126	6.2095	2.3010	3.3971	2.6301	1.7632	0.0691	0.1176
	Upper	4.0038	8.8595	3.2724	21.4397	20.5273	2.7044	0.4223	0.9424
2000	Lower	0.8471	4.5394	0.5105	56.4722	17.2919	0.4192	0.0219	0.1796
	Upper	3.6175	8.3509	2.8123	21.2033	12.0475	2.4010	0.4179	1.2257
2005	Lower	2.2989	5.6232	1.6712	-18.0358	1.5462	1.4024	-0.1156	0.5471
	Upper	3.2793	8.0524	2.3788	57.9060	6.4866	2.1326	0.3802	2.2645
2010	Lower	2.0008	5.3680	1.3822	1.8638	1.4273	1.2752	0.2287	0.4833
	Upper	2.8584	7.7098	1.9686	6.9524	7.0423	1.9130	0.8489	2.3539

Table 10 AIC statistics obtained by maximum likelihood for the BG models with beta (3 parameters) and GB1 marginals (5 parameters) fitted to pairs of the variables: Education, Health and Income

	Education & health		Education & income		Income & health	
Year	Model (3 par)	Model (5 par)	Model (3 par)	Model (5 par)	Model (3 par)	Model (5 par)
1980	-292.88	-352.59	-174.12	-180.00	-279.48	-288.67
1985	-309.52	-354.03	-206.89	-217.35	-332.58	-354.54
1990	-304.40	-340.63	-213.99	-228.01	-344.03	-380.03
1995	-291.33	-313.41	-206.65	-217.94	-332.39	-370.94
2000	-291.77	-296.28	-207.64	-217.52	-325.00	-354.22
2005	-296.81	-306.00	-210.98	-213.52	-323.60	-340.72
2010	-330.82	-327.52	-215.63	-213.91	-317.16	-318.24

Smaller values indicate better fitted models.

The first model (with classical beta marginals) depends on 3 parameters, and the second and third models (with GB1 and BG truncated exponential marginals) are characterized by 5 parameters. The three models have been estimated by maximum likelihood using the equations given in Section 3.4. In total, we have fitted $7 \times 3 \times 3 = 63$ different models. The initial estimates of the parameters have been obtained using Equations (17) to (19). In the case of the model with classical beta marginals, initial estimates are quite close to the ML estimators because they are based on sufficient statistics.

For the three pairs of variables, we have compared both models using the Akaike information criterion (AIC), defined by (Akaike 1974)

$$AIC = -2\log L + 2d; \tag{21}$$

where $\log L = \ell\left(\hat{\theta}\right)$ is the log-likelihood of the model evaluated at the maximum likelihood estimates and d is the number of parameters. We chose the model with the smallest value of AIC statistic.

Tables 1, 2, 3, 4, 5 and 6 show the parameter estimates and their standard errors for the three alternative models considered: BG distribution with classical beta marginals (3 parameter model p_1, p_2, q) and with GB1 and BG truncated exponential marginals (5 parameter model p_1, p_2, q, a_1, a_2), which have been fitted to pairs of the variables:

Table 11 AIC statistics obtained by maximum likelihood for the BG model (BG truncated exponential maginals) with 5 parameters fitted to pairs of the variables: education, health and income

	Education & health	Education & income	Income & health
Year	Model (5 par)	Model (5 par)	Model (5 par)
1980	-333.49	-188.88	-286.86
1985	-342.58	-223.42	-350.60
1990	-323.42	-233.30	-375.72
1995	-300.82	-230.08	-373.12
2000	-295.86	-233.28	-363.92
2005	-303.35	-233.61	-364.73
2010	-332.99	-237.16	-359.70

Smaller values indicate better fitted models.

Education, Health and Income. In particular, Tables 1 and 4 show the results obtained for Education & Health, Tables 2 and 5 show the corresponding results for Education & Income, and Tables 3 and 6 for Income & Health. The estimations have been performed by maximum likelihood, focusing on quinquennial periods from 1980 to 2010. It can be seen that, assuming the asymptotic normality of the maximum likelihood estimates, most of the estimates are statistically significant at a 0.05 level of significance.

In order to illustrate the interval estimation of the parameters in Section 3.4, we have included the asymptotic confidence intervals at 95 percent for the models with beta and GB1 marginals (Tables 7, 8 and 9).

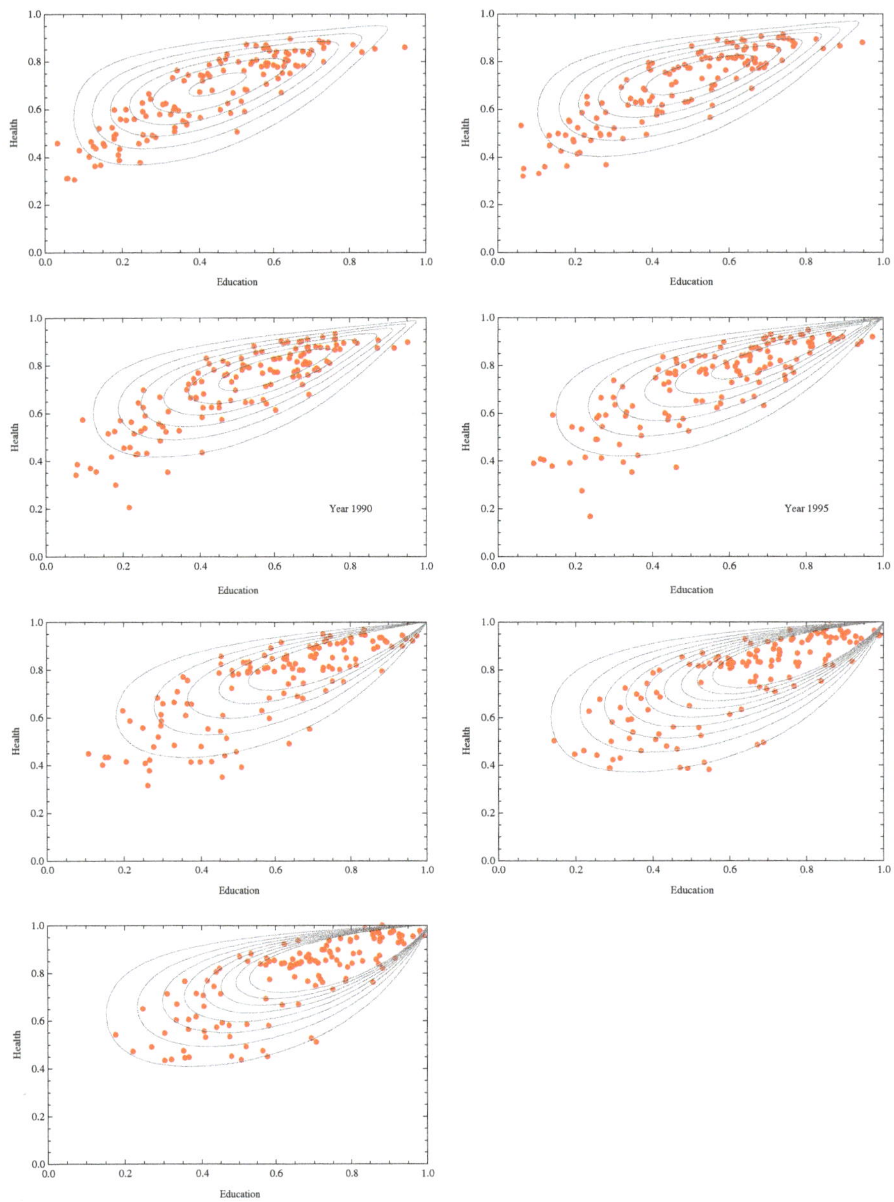

Figure 1 Contour plots for the BG model with beta marginals fitted to the variables Education & Health.

Tables 10 and 11 show the values of the *AIC* statistic (Equation (21)) obtained for both candidate models fitted to three pairs of variables: Education & Health, Education & Income and Income & Health. Our estimates point out that the values of *AIC* statistic for the BG model with GB1 marginals are lower than those observed for the BG distribution with classical beta marginals, except in the case of Education & Health and Education & Income for the year 2010. For the Education & Health data the best model is the model with GB1 marginals (except in 2000). In the case of Education & Income data the model with BG truncated exponential marginals outperforms the other two models. Finally, for Income & Health data, the model with GB1 marginals outperforms the BG truncated

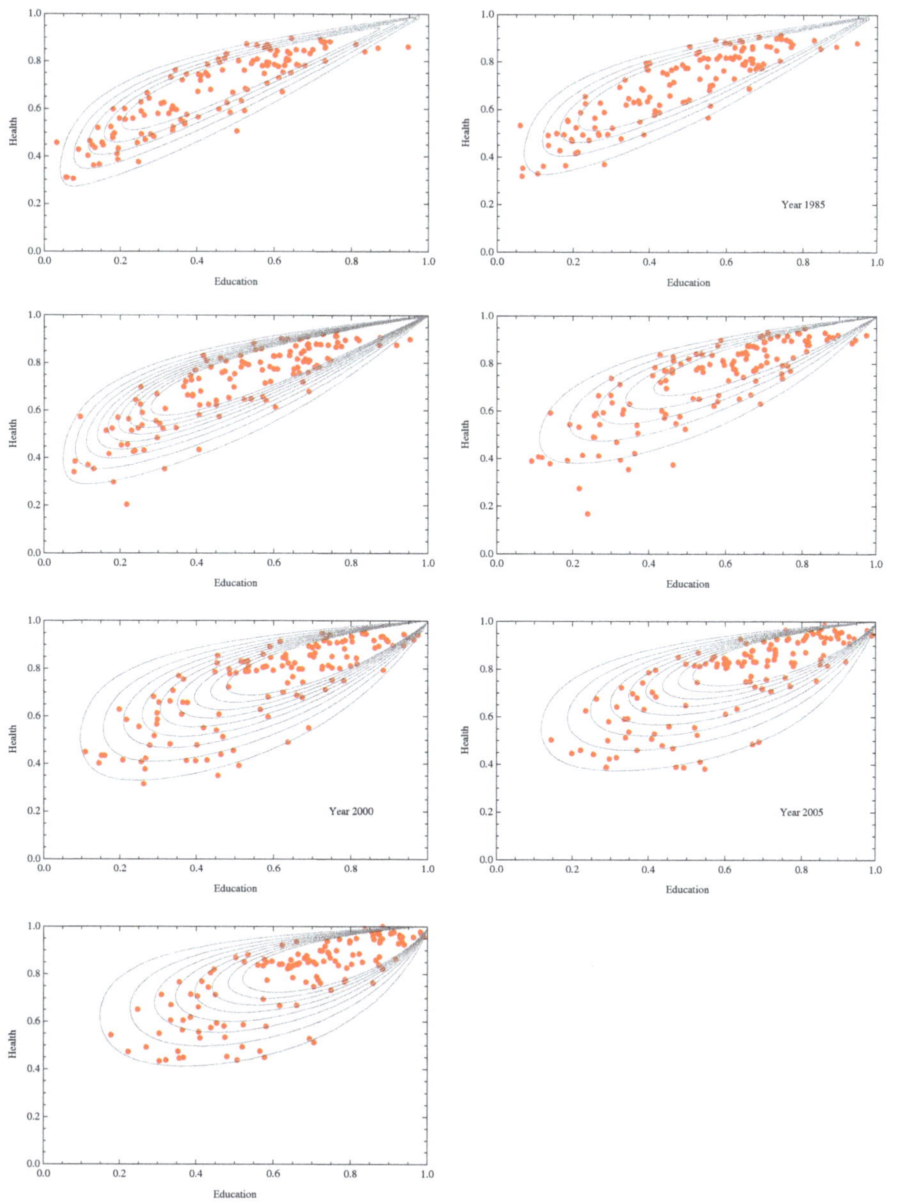

Figure 2 Contour plots for the BG model with GB1 marginals fitted to the variables Education & Health.

exponential marginals model in 1980, 1985 and 1995 while in the other four years the BG truncated exponential marginals is the model that provides the best fit. These results imply that, in general terms, the accuracy of the estimates is higher for the models with 5 parameters.

As an illustration, Figures 1, 2, 3, 4, 5 and 6 present the contour plots for the BG distributions with classical beta marginals and GB1 marginals fitted to the pairs of variables: Education & Health, Education & Income and Income & Health for every five years during the period 1980-2010. The shape of this graphics supports the existence of a positive correlation among the variables considered, thus pointing out the suitability of the first type of BG model. The contour plots also reveal that the proposed models represent the

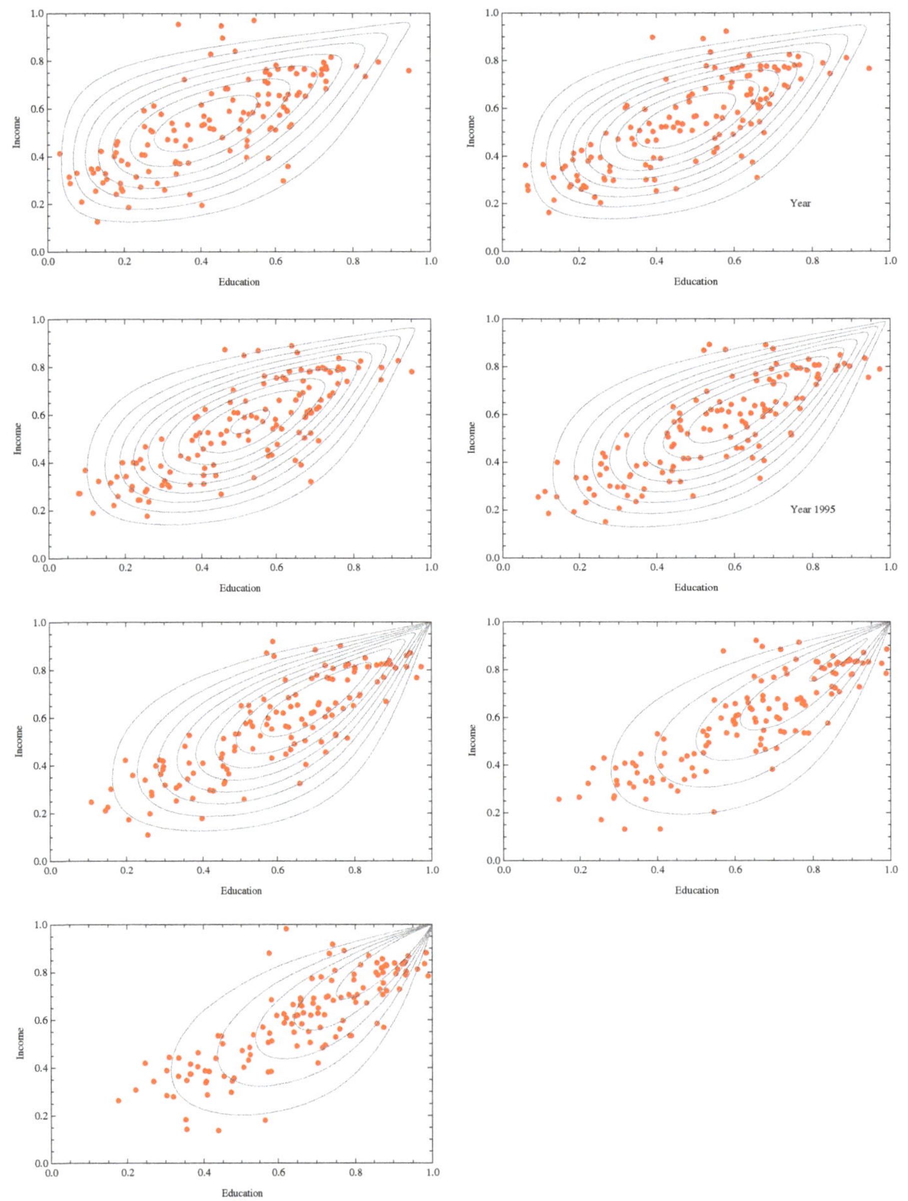

Figure 3 Contour plots for the BG model with beta marginals fitted to the variables Education & Income.

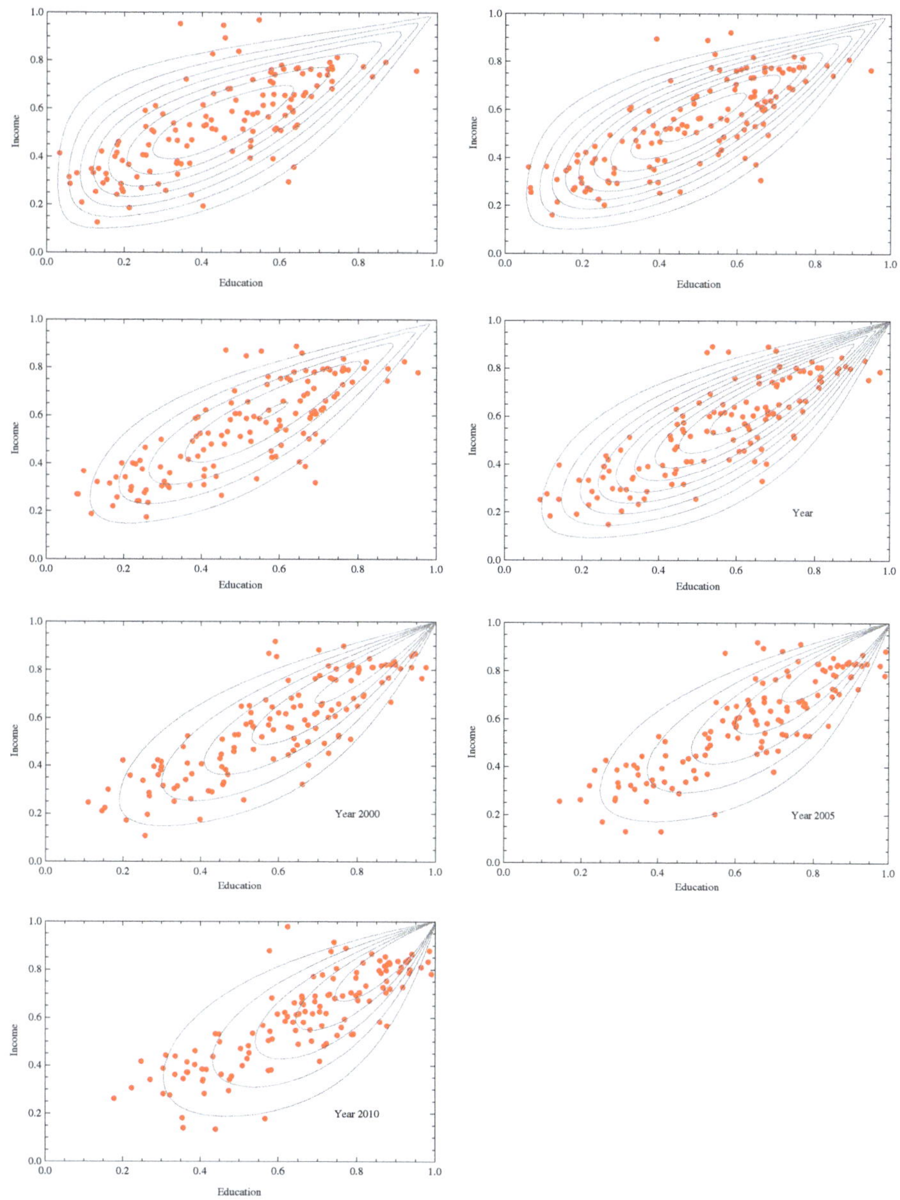

Figure 4 Contour plots for the BG model with GB1 marginals fitted to the variables Education & Income.

geography of the bivariate data adequately, being more accurate in the case of the BG distribution with GB1 marginals, as concluded from the results of the Akaike information criteria.

5 Conclusions

The main conclusions of this paper are the following. Three different classes of bivariate BG distributions have been presented. These classes have been constructed using three different definitions of bivariate beta distributions, proposed by Libby and Novick (1982), Jones (2001) and Olkin and Liu (2003) for the first proposal, El-Bassiouny and Jones (2009) for the second proposal and Arnold and Ng (2011) for the third proposal.

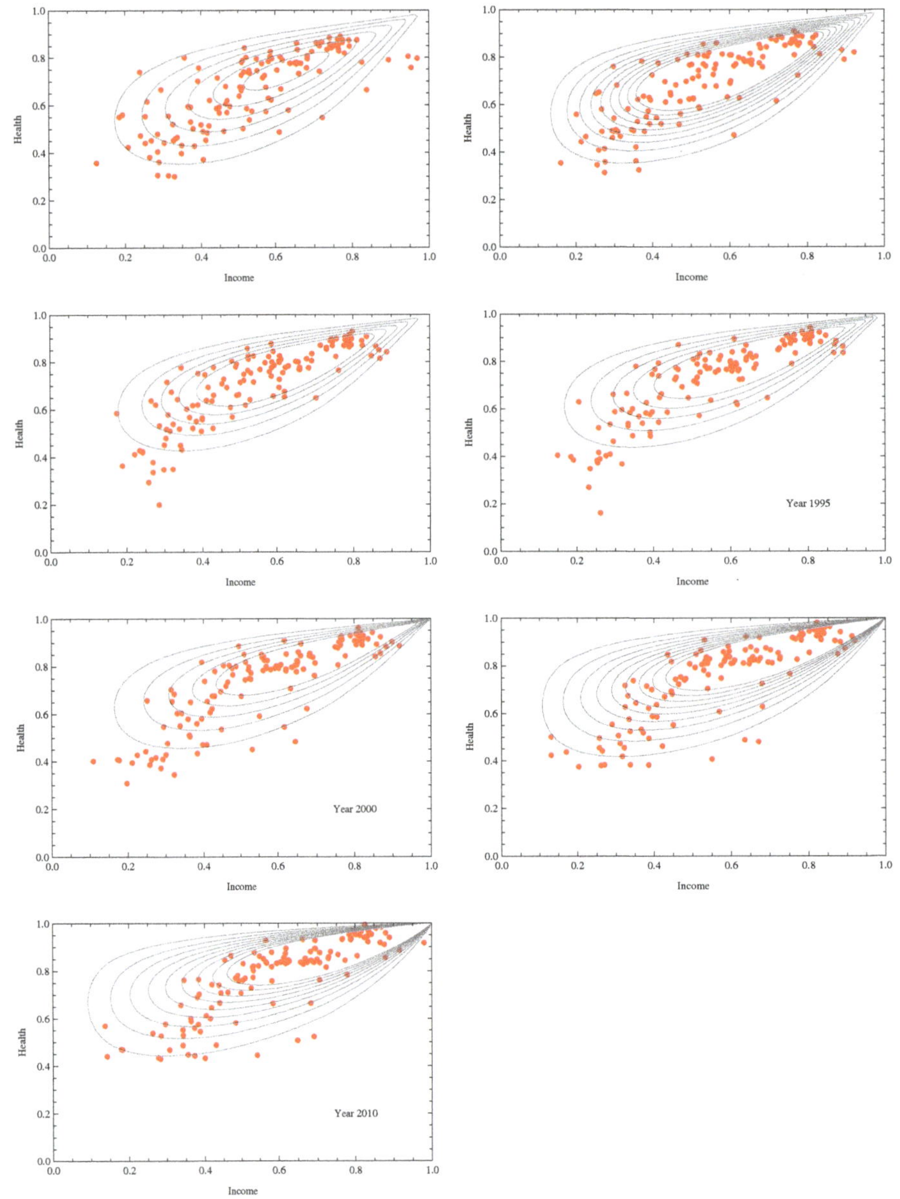

Figure 5 Contour plots for the BG model with beta marginals fitted to the variables Income & Health.

The main properties of these three classes have been studied. Three specific bivariate BG distributions have been obtained. Finally, an empirical application with well-being data has been presented.

The future research about bivariate BG distributions moves in three directions. The first line research is to propose specific models for their practical use in statistical modeling. The study of these possible models in any dimension could be an interesting field of research. Secondly, we propose to study statistical inference methodologies for bivariate (and, more generally, multivariate) BG distributions in (9) and (12). Finally, we propose to establish a model competition between BG distributions in (4), (9) and (12) for different choices of F_1 and F_2.

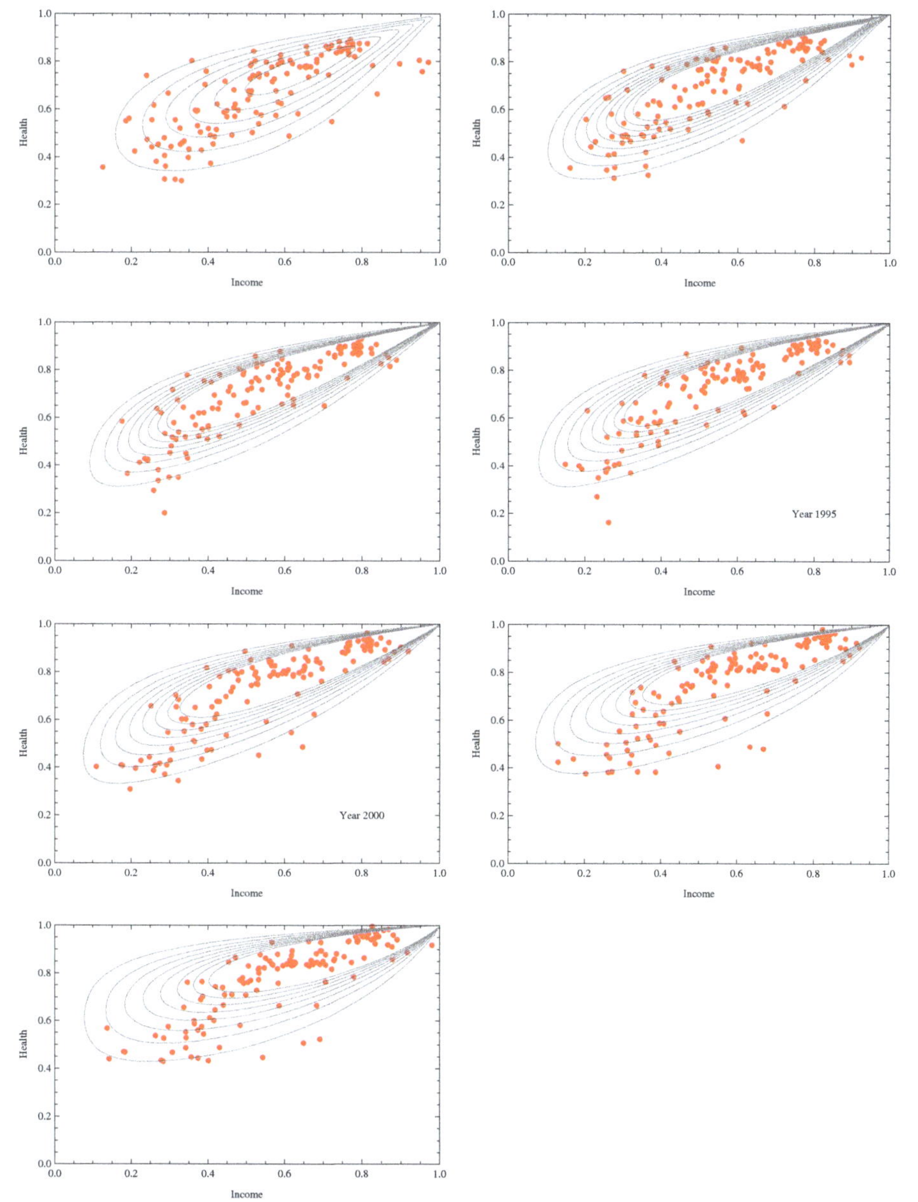

Figure 6 Contour plots for the BG model with GB1 marginals fitted to the variables Income & Health.

Appendix

The joint PDF of the different classes of bivariate beta distribution

The joint PDF of the bivariate random variable in (3) is (see Libby and Novick (1982), Jones (2001) and Olkin and Liu (2003)),

$$f(z_1, z_2) = \frac{z_1^{a_1-1} z_2^{a_2-1} (1 - z_1)^{a_2+b-1} (1 - z_2)^{a_1+b-1}}{B(a_1, a_2, b)(1 - z_1 z_2)^{a_1+a_2+b}}, \quad 0 < z_1, z_2 < 1 \tag{22}$$

where $B(a_1, a_2, b) = \Gamma(a_1)\Gamma(a_2)\Gamma(b)/\Gamma(a_1 + a_2 + b)$ and the marginal distributions are $Z_1 \sim \mathcal{B}e(a_1, b)$ and $Z_2 \sim \mathcal{B}e(a_2, b)$. Note that (22) belongs to the three-parametric exponential family, where sufficient statistics for (a_1, a_2, b) are given by,

$$\left(\prod_{i=1}^{n} \frac{y_{1i}(1-y_{2i})}{1-y_{1i}y_{2i}}, \prod_{i=1}^{n} \frac{y_{2i}(1-y_{1i})}{1-y_{1i}y_{2i}}, \prod_{i=1}^{n} \frac{(1-y_{1i})(1-y_{2i})}{1-y_{1i}y_{2i}}\right).$$

For $n = 1$, the distributions of the sufficient statistics are:

$$\frac{1-Z_1Z_2}{Z_1(1-Z_2)} \sim \mathcal{B}2(a_2+b, a_1)+1,$$

$$\frac{1-Z_1Z_2}{Z_2(1-Z_1)} \sim \mathcal{B}2(a_1+b, a_2)+1,$$

and

$$\frac{1-Z_1Z_2}{(1-Z_1)(1-Z_2)} \sim \mathcal{B}2(a_1+a_2, b)+1,$$

where $\mathcal{B}2(a, b)$ denotes beta distribution of the second kind.

The log-moments of (22) are:

$$E\left\{\log\frac{Z_1(1-Z_2)}{1-Z_1Z_2}\right\} = \psi(a_1) - \psi(a_1+a_2+b),$$

$$E\left\{\log\frac{Z_2(1-Z_1)}{1-Z_1Z_2}\right\} = \psi(a_2) - \psi(a_1+a_2+b),$$

$$E\left\{\log\frac{(1-Z_1)(1-Z_2)}{1-Z_1Z_2}\right\} = \psi(b) - \psi(a_1+a_2+b).$$

The joint PDF of the bivariate beta density (8) is givenby (see El-Bassiouny and Jones (2009)),

$$f(z_1, z_2) = k\frac{z_1^{a_1-1}(1-z_1)^{A-a_1-1}z_2^{a_2-1}(1-z_2)^{A-a_2-1}}{(1-z_1z_2)^A} \times {}_2F_1\left[A, a_4; A-a_2; \frac{z_1(1-z_2)}{1-z_1z_2}\right], \tag{23}$$

where $A = a_1+a_2+a_3+a_4$, $k^{-1} = B(a_1, a_3)B(a_2, a_1+a_3+a_4)$ and ${}_2F_1[..;.;]$ denote the Gauss hypergeometric function.

The expression for the $f_{V,W}(v, w)$ function is given by (see Arnold and Ng 2011),

$$f_{V,W}(v, w) = \int_0^\infty \int_0^\infty \int_{u_4/(w-u_5)}^{(u_4+u_5)v} f(v, w, u_3, u_4, u_5)\,du_3du_4du_5, \quad u, w > 0, \tag{24}$$

where

$$f(v, w, u_3, u_4, u_5) = \frac{(u_3+u_5)(u_4+u_5)}{\prod_{i=1}^5 \Gamma(a_i)}[(v(u_4+u_5)-u_3)]^{a_1-1}$$
$$\times [w(u_3+u_5)-u_4]^{a_2-1}\prod_{i=3}^{5} u_i^{a_i-1}\exp\{-[u_3w+u_4v+u_5(v+w+1)]\},$$

where $u_4/w - u_5 < u_3 < (u_4+u_5)v$, $u_4, u_5, v, w > 0$.

Description of the data set

The list of countries used in the analysis are the following:

Afghanistan, Guatemala, Pakistan, Albania, Guyana, Panama, Algeria, Haiti, Papua New Guinea, Argentina, Honduras, Paraguay, Armenia, Hong Kong, China (SAR), Peru, Australia, Hungary, Philippines, Austria, Iceland, Poland, Bahrain, India, Portugal, Bangladesh, Indonesia, Qatar, Belgium, Iran (Islamic Republic of), Romania, Belize, Ireland, Russian Federation, Benin, Israel, Rwanda, Bolivia (Plurinational State of),

Italy, Saudi Arabia, Botswana, Jamaica, Senegal, Brazil, Japan, Sierra Leone, Brunei Darussalam, Jordan, Slovakia, Bulgaria, Kenya, Slovenia, Burundi, Korea (Republic of), South Africa, Cameroon, Kuwait, Spain, Canada, Lao PDR, Sri Lanka, Central African Republic, Latvia, Sudan, Chile, Lesotho, Swaziland, China, Liberia, Sweden, Colombia, Lithuania, Switzerland, Congo, Luxembourg, Syrian Arab Republic, Congo (Democratic Republic of), Malawi, Tajikistan, Costa Rica, Malaysia, Tanzania (United Republic of), Cote D'ivoire, Mali, Thailand, Cuba, Malta, Togo, Cyprus, Mauritania, Tonga, Denmark, Mauritius, Trinidad and Tobago, Dominican Republic, Mexico, Tunisia, Ecuador, Moldova (Republic of), Turkey, Egypt, Mongolia, Uganda, El Salvador, Morocco, Ukraine, Estonia, Mozambique, United Arab Emirates, Fiji, Myanmar, United Kingdom, Finland, Namibia, United States, France, Nepal, Uruguay, Gabon, Netherlands, Venezuela (Bolivarian R.), Gambia, New Zealand, VietNam, Germany, Nicaragua, Yemen, Ghana, Niger, Zambia, Greece, Norway, Zimbabwe.

Data on the health index can be retrieved from https://data.undp.org/dataset/Health-index/9v27-i7ic, data on the education index can be drawn from https://data.undp.org/dataset/Expected-Years-of-Schooling-of-children-years-/qnam-f624 for the variable expected years of schooling and https://data.undp.org/dataset/Mean-years-of-schooling-of-adults-years-/m67k-vi5c for the mean years of schooling. Finally, income data come from https://data.undp.org/dataset/Income-index/qt4g-yea9.

Competing interests
The authors declare that they have no competing interests.

Authors' contributions
All the authors have contributed equally. All authors read and approved the final manuscript.

Acknowledgements
The authors thank to Ministerio de Economía y Competitividad (project ECO2010-15455) for partial support of this work. The authors thank also the comments by the attendants of the first ICOSDA meeting celebrated at Mount Pleasant, MI USA. We are grateful to the Editors-in-Chief and the reviewers for careful reading and for their comments and suggestions which greatly improved the paper.

References
Alexander, C, Cordeiro, GM, Ortega, EMM, Sarabia, JM: Generalized beta-generated distributions. Comput. Stat. Data Anal. **56**, 1880–1897 (2012)

Alexander, C, Sarabia, JM: Generalized Beta-Generated Distributions, ICMA Centre Discussion Papers in Finance DP2010-09, ICMA Centre. The University of Reading, Witheknights, PO Box 242, Reading RG6 6BA, UK (2010)

Alzaatreh, A, Lee, C, Famoye, F: A new method for generating families of continuous distributions. Metron **71**, 63–79 (2013)

Alzaatreh, A, Lee, C, Famoye, F: The gamma-normal distribution: properties and applications. Comput. Stat. Data Anal. **69**, 67–80 (2014)

Akaike, H: A new look at the statistical model identification. IEEE Trans. Automatic Control **19**, 716–723 (1974)

Apostolakis, FJ, Moieni, P: The foundations of models of dependence in probabilistic safety assessment. Reliability Eng. **18**, 177–195 (1987)

Arnold, BC, Ng, HKT: Flexible bivariate beta distributions. J. Multivariate Anal. **102**, 1194–1202 (2011)

Arnold, BC, Castillo, E, Sarabia, JM: Conditional Specification of Statistical Models, Springer Series in Statistics. Springer Verlag, New York (1999)

Arnold, BC, Castillo, E, Sarabia, JM: Conditionally specified distributions: an introduction (with discussion). Stat. Sci. **16**, 249–274 (2001)

Arnold, BC, Castillo, E, Sarabia, JM: Families of multivariate distributions involving the Rosenblatt construction. J. Am. Stat. Assoc. **101**, 1652–1662 (2006)

Balakrishnan, N, Lai, C-D: Continuous Bivariate Distributions. 2nd ed. Springer, New York (2009)

Barndorff-Nielsen, O, Kent, J, Sorensen, M, mixtures, Normalvariance-mean, z distributions, Int. Stat. Rev. **50**, 145–159 (1982)

Barro, RJ, Lee, JW: A new data set of educational attainment in the world, 1950–2010. J. Dev. Econ. **104**, 184–198 (2013)

Cordeiro, GM, de Castro, M: A new family of generalized distributions. J. Stat. Comput. Simul. **81**, 883–893 (2011)

El-Bassiouny, AH, Jones, MC: A bivariate F distribution with marginals on arbitrary numerator and denominator degrees of freedom, and related bivariate beta and t distributions. Stat. Methods Appl. **18**, 465–481 (2009)

Esary, JD, Proschan, F, Walkup, DW: Association of random variables, with applications. Ann. Math. Stat. **38**, 1466–1474 (1967)

Eugene, N, Lee, C, Famoye, F: The beta-normal distribution and its applications. Commun. Stat. Theory Methods **31**, 497–512 (2002)

Fritsch, FN, Carlson, RE: Monotone piecewise cubic interpolation. SIAM J. Numerical Anal. **17**, 238–246 (1980)

Holland, PW, Wang, YJ: Dependence function for continuous bivariate densities. Commun. Stat. Theory Methods **16**, 863–876 (1987)

Jones, MC: Multivariate t and beta distributions associated with the multivariate F distribution. Metrika **54**, 215–231 (2001)

Jones, MC: Families of distributions arising from distributions of order statistics. Test **13**, 1–43 (2004)

Jones, MC, Larsen, PV: Multivariate distributions with support above the diagonal. Biometrika **91**, 975–986 (2004)

Kotz, S, Balakrishnan, N: Johnson, N L. 2nd ed., Vol. 1. John Wiley and Sons, New York (2000)

Lee, C, Famoye, F, Alzaatreh, A: Methods for generating families of continuous distribution in the recent decades. Wiley. Interdiscip. Rev.: Comput. Stat. **5**, 219–238 (2013)

Libby, DL, Novick, MR: Multivariate generalized beta distributions with applications to utility assessment. J. Educ. Stat. **7**, 271–294 (1982)

Marshall, AW, Olkin, I: Life Distributions. Structure of Nonparametrics, Semiparametric and Parametric Families. Springer, New York (2007)

McDonald, JB: Some generalized functions for the size distribution of income. Econometrica **52**, 647–663 (1984)

Nadarajah, S, Kotz, S: The beta exponential distribution. Reliability Eng. Syst. Safety **91**, 689–697 (2006)

Olkin, I, Liu, R: A bivariate beta distribution. Stat. Probability Lett. **62**, 407–412 (2003)

R Development Core Team: R: A language and environment for statistical computing. R Foundation for Statistical Computing, Vienna, Austria (2011). http://www.R-project.org/

Rosenblatt, M: Remarks on a multivariate transformation. Ann. Math. Stat. **23**, 470–472 (1952)

Sarabia, JM, Gómez-Déniz, E: Construction of multivariate distributions: a review of some recent results (with discussion), SORT. Stat. Oper. Res. Trans. **32**, 3–36 (2008)

SAS Institute, Inc.: SAS/STAT, version 9.2, Cary, NC, USA (2010)

Shaked, M: A family of concepts of dependence for bivariate distributions. J. Am. Stat. Assoc. **72**, 642–650 (1977)

The Mathworks, Inc. Matlab, release 2011, Novi, MI, USA (2011)

UNDP: International Human Development Indicators (2012). Retrieved from http://hdr.undp.org/en/statistics/. Last Accessed 10 Nov 2012

Venter, G: Transformed beta and gamma distributions and aggregate losses. Proc. Casualty Actuarial Soc. **LXX**, 156–193 (1983)

Wolfram Research, Inc.: Mathematica, version 8.0, Champaign, IL, USA (2010)

Zografos, K: Generalized beta generated-II distributions. In: Pardo, L, Balakrishnan, N, Angeles Gil, M (eds.) Modern Mathematical Tools and Techniques in Capturing Complexity. Springer, Berlin, (2011)

8

Nonlinear mixed-effects HIV dynamic models with considering left-censored measurements

Tao Lu[1*] and Min Wang[2]

*Correspondence: stat.lu11@gmail.com
[1] Department of Epidemiology and Biostatistics, State University of New York, Albany, New York, USA
Full list of author information is available at the end of the article

Abstract

HIV dynamic model offers a different perspective of studying HIV pathogenesis and developing treatment strategies for AIDS patients. Many HIV dynamic models have recently been developed to characterize short-term AIDS treatment, whereas in long-term HIV dynamics, viral load often rebounds in the later stage of treatment primarily due to reduced drug efficacy. Although time-varying drug efficacy can be incorporated into the ordinary differential equations (ODE) model, such a system has no analytical solution, and the measurement of viral load is usually censored at the detection limit due to technological constraints. We consider nonlinear mixed-effects ODE model with stochastic approximation EM algorithm to overcome these difficulties. The performance of the proposed method is illustrated by means of a simulation study and a real-data application. Numerical evidence shows that the HIV infection is generally more severe when considering left-censored data. The T cell production rate from human body source varies, but the death rate of infected T cells, infection rate of virus, and other dynamic parameters do not have much difference among patients. We hope these findings inspire more research on clarifying biological mechanism of HIV infection and developing better treatment.

Keywords: Below detection limit; Drug efficacy; HIV dynamic model; Long-term treatment; Nonlinear mixed-effect model

1 Introduction

HIV dynamic model, a set of ordinary differential equations (ODE) that describes the interaction between HIV virus and human body cells, has been proven useful for understanding the pathogenesis of HIV infection and developing treatment strategies. In order to estimate biologically/clinically meaningful parameters in the HIV dynamic model, many statistical models have been developed in the last decade, ranging from the simple nonlinear least squares (NLS) approach to more general nonlinear mixed-effect modeling approaches (Nowak and May 2000; Perelson and Nelson 1999; Perelson et al. 1996; Perelson et al. 1997; Wu and Ding 1999; Wu et al. 1998; Wu et al. 2005). However, in more complex scenarios, the ODE system has no close-form solution and thus needs to be solved numerically. Note that data from viral dynamic studies usually consist of repeated viral load measurements taken over time for each subject. In addition, the viral dynamic processes share certain similar patterns among patients while still having distinct individual characteristics. These properties indicate that the nonlinear mixed-effects (NLME) models appear to be reasonable for modeling HIV dynamics.

Disparate from short-term (a few hours) AIDS treatment, viral load remains sustained, whereas in long-term treatment (a few weeks to years), viral load usually decreases at the beginning of treatment and is followed by rebounds later, which leads to the treatment failure. There are various reasons that could contribute to the failure; one of which is the non-constant drug efficacy due to time-varying drug resistance, drug adherence or/and pharmacokinetical factors, such as median inhibitory concentration (IC_{50}), and so on. Huang et al. (2006) incorporated these clinical/PK factors into a time-varying drug efficacy model to characterize viral load resurgence featured in the long-term (24 weeks) AIDS treatment trajectory. This approach is capable of capturing the long-term treatment effect, but the difficulty may arise for the case in which the solution of ODE can not be linearized, because analytical formula does not exist. Accordingly, the classical nonlinear mixed-effects modeling approach becomes futile under this scenario.

It is remarkable that only a few methods have recently been proposed to tackle this problem in the longitudinal setting. For example, Huang et al. (2006) combined the Bayesian approach with the mixed-effects model to estimate both population and individual parameters within a framework of hierarchical Bayesian nonlinear models. From a frequentist point of view, Guedj et al. (2007) adapted a Newton-like approach using only first derivative to deal with the problem. Later on, in the presence of left-censoring, Huang and Getachew (2012) investigated the NLME model with a skew-t (ST) distribution for the response process and the ST nonparametric mixed-effects model for covariate measurement error process. However, the first approach ignores below detection limit viral load data but simply imputes the missing data with detection limit. The second approach requires reiteratively evaluating integral through an adaptive Gaussian quadrature approach, which quickly becomes intractable when the dimension of random effects is high. In addition, the computation of the approach is very complicated. The third approach is only applicable to short-term HIV dynamics and cannot be directly applied for long-term HIV dynamics. In particular, we may face severe bias in parameter estimation if we simply ignore censoring (below detection limit) or do not carefully handle it. As an illustration, in the simulation study performed later, we show that either imputing censored viral load with detection limit or completely removing those censored data will lead to large bias.

A classical way to deal with the incomplete or missing data problem is the Expectation-maximization (EM) algorithm, originally proposed by Dempster et al. (1977). For the E-step of the EM algorithm, one needs to compute the expectation of the complete log-likelihood with respect to missing data distribution at current estimates. The M-step includes the maximization of the expected log-likelihood function to update unknown parameters. One then follows iteration method until convergence is achieved. Within the mixed-effects modeling framework, the random effects can be treated as missing data so that the EM algorithm is capable of estimating fixed effects after averaging random effects in the E-step. Also, the below detection limit data are missing and can thus be averaged along with random effects at the E-step. Nevertheless, the E-step usually fails quickly when the number of censored data and/or the dimension of random effects increases since numerical integration becomes intractable in high dimensional situations.

To circumvent this difficulty, within linear mixed-effects model setting, Hughes (1999) used Monte Carlo integration to evaluate the E-step by repeatedly sampling from the conditional distribution of below detection data. Wu (2002, 2004) extended the Monte Carlo version of EM (MCEM) for the nonlinear mixed-effects model. However, sampling the missing data is often time consuming and the algorithm is also slow to converge. Delyon et al. (1999) proposed a stochastic approximation version of EM (SAEM) to obtain the maximum likelihood estimate (MLE) of the unknown parameter. Kuhn and Lavielle (2005) coupled the Markov chain Monte Carlo (MCMC) procedure to the SAEM so that only one simulation of missing data at each iteration is required. The algorithm has been proven to converge quickly toward the MLE under general conditions. For theoretical proof, we refer readers to Delyon et al. (1999) and Kuhn and Lavielle (2005). In this paper, we couple the SAEM algorithm with the nonlinear mixed-effects model under the scenarios in which the ODE system has no analytical solution and the data from long-term AIDS treatment are left-censored. The performance of the proposed method is illustrated by means of a simulation study and a real-data application. Numerical evidence shows that the HIV infection is generally more severe when considering left-censored data.

The remainder of the paper is organized as follows. In Section 2, we introduce HIV dynamic model and time-varying drug efficacy model, which are used to model long-term AIDS treatment. In Section 3, the statistical model and method of estimating parameters are presented. In Section 4, we carry out a simulation study to examine the performance of the proposed method and in Section 5, we evaluate the performance of our method via analysis of the real AIDS clinical trial study. We conclude with a brief discussion in Section 6.

2 Model specification

2.1 Antiviral response model

One of the commonly used HIV dynamic models that describes the interaction between human body cells and virus is given by (Huang et al. 2006):

$$\begin{aligned}
\frac{d}{dt}T(t) &= \lambda - \rho T(t) - [1-\gamma(t)]\,kT(t)V(t),\\
\frac{d}{dt}T^*(t) &= [1-\gamma(t)]\,kT(t)V(t) - \delta T(t)^*, \qquad (1)\\
\frac{d}{dt}V(t) &= N\delta T^*(t) - cV(t),
\end{aligned}$$

where $T(t)$, $T^*(t)$, and $V(t)$ stand for uninfected T cells, infected T cells, and virus at time t, respectively, λ represents the rate at which new T cells are generated from body source like thymus, ρ is the death rate of uninfected T cells, k is the infection rate of T cells, δ is the death rate of infected T cells, N is the number of new virions produced from each infected T cells during their life-time, and c is the clearance rate of free virions. The time-varying parameter $\gamma(t)$ is the antiviral drug efficacy as defined below. In this paper, we assume that the system of equations (1) is in a steady-state before initiating antiretroviral treatment, and then the initial conditions for (1) are given by

$$T_0 = \frac{c}{kN}, \quad T_0^* = \frac{cV_0}{\delta N}, \quad \text{and} \quad V_0 = \frac{\lambda N}{c} - \frac{\rho}{k}. \qquad (2)$$

The above ordinary differential equations have no analytical solutions and thus need to be solved numerically. One feature that distinguishes this dynamic model from others is that by considering time-varying drug efficacy, we are capable of characterizing rebounds of virus in long-term antiviral treatment, which is not seen typically in the short-term period. On the other hand, more complexity may be added to the model by incorporating various clinical or/and pharmacokinetical (PK) factors such as drug susceptibility, drug resistance, and drug adherence etc. Thereafter, it becomes more challenging to carry out statistical inference.

2.2 Antiviral drug efficacy model

In many studies, drug efficacy is often assumed to be either a constant or perfect over treatment time; see, for example, Perelson et al. (1997) Perelson and Nelson (1999), Ding and Wu (2000), among others. Such an assumption may be reasonable during short-term period, whereas in long-term treatment, drug efficacy varies across time due to drug resistance, imperfect drug adherence or/and other clinical/pharmacological factors, which could lead to treatment failure. To deal with this issue, we advocate the modified E_{max} model to account for time-varying drug efficacy (Huang et al. 2006). In this paper, we consider two antiretroviral drugs given by

$$\gamma(t) = \frac{C_1 A_1(t)/IC_{50}^1(t) + C_2 A_2(t)/IC_{50}^2(t)}{\phi + C_1 A_1(t)/IC_{50}^1(t) + C_2 A_2(t)/IC_{50}^2(t)}, \tag{3}$$

where C_1 and C_2 represent drug concentration or any PK parameters such as C_{12h}. The median inhibitory concentrations $IC_{50}^1(t)$ and $IC_{50}^2(t)$ are used to quantify drug susceptibility (Molla et al. 1996) and are written as

$$IC_{50}(t) = \begin{cases} I_0 + \frac{I_r - I_0}{t_r} t & \text{for } 0 < t < t_r, \\ I_r & \text{for } t \geq t_r, \end{cases} \tag{4}$$

where I_0 and I_r are respective values of $IC_{50}(t)$ at the baseline and time point t_r, which is the time of virological failure. Let ϕ be a conversion factor between $IC_{50}(t)$ *in vitro* and $IC_{50}(t)$ *in vivo* and it can be estimated from data. Here, $A_1(t)$ and $A_2(t)$ are adherence profiles of the two drugs, which are measured by pill counts and are modeled as

$$A(t) = \begin{cases} 1 & \text{if all doses are taken in } (T_k, T_{k+1}], \\ R & \text{if } 100R\% \text{ doses are taken in } (T_k, T_{k+1}], \end{cases} \tag{5}$$

where T_k is the time at the kth clinical visit of patients, R_k represents the percentage of drugs that is consumed by patients during the interval of two visits. In particular, if all of the prescribed drugs are taken, R_k is one. This suggests that poor adherence is one of the major causes of treatment failure, and thus it is definitely necessary to take this factor into account; see, for example, Besch (1995).

3 Statistical model and parameter estimation

3.1 Nonlinear mixed-effects ODE model

Viral load data from an HIV/AIDS clinical trial are composed of repeated measurements on a group of patients so that a hierarchical modeling approach is necessary to account for within subject as well as between subject variation simultaneously. We are interested in estimating biologically/clinically meaningful parameters in (1) and conducting statistical inferences while taking below detection limit measurements for all patients into account. In the area of longitudinal data analysis, the mixed-effects model is often used to characterize both within and between subjects variation. Let y_{ij} be logarithmic measurement of viral load for subject i at time t_{ij} for $i = 1, \cdots, n$ and $j = 1, \cdots, n_i$; population parameter $\boldsymbol{\mu} = (\log c, \log \delta, \log \lambda, \log \rho, \log N, \log k, \log \phi)^T$; individual parameter $\boldsymbol{\theta}_i = (\log c_i, \log \delta_i, \log \lambda_i, \log \rho_i, \log N_i, \log k_i, \log \phi_i)^T$. Let $g(\boldsymbol{\theta}_i, t_{ij}) = \log_{10}(V_{ij}(\boldsymbol{\theta}_i, t_{ij}))$ with $V_{ij}(\boldsymbol{\theta}_i, t_{ij})$ being the true amount of viral load based on (1) for subject i at time t_{ij}. Following Davidian and Giltinan (1995), a natural Nonlinear Mixed-effects ODE Model (NLME-ODE) is modeled as

(i) Within-subject variation: $y_{ij} = g(\boldsymbol{\theta}_i, t_{ij}) + \boldsymbol{\varepsilon}_{ij}$. Measurement error $\boldsymbol{\varepsilon}_{ij}$ is assumed to follow a normal distribution with mean zero and variance σ^2.

(ii) Between-subject variation: $\boldsymbol{\theta}_i = \boldsymbol{\mu} + \mathbf{b}_i$. Random effect $\mathbf{b}_i$ characterizes the deviation of individual parameters from population level and we assume $\mathbf{b}_i \sim \mathcal{N}(\mathbf{0}, \mathbf{D})$.

It deserves mentioning that the model described here is different from the classical NLME model in that $g(\cdot)$ has no explicit form in the long-term HIV dynamic setting, which leads to additional challenges for estimating parameters. A few methods such as the Bayesian approach and a Newton-like method were proposed to attack these challenges; see, for example, Huang et al. (2006), Guedj et al. (2007). However, these procedures either ignore missing (under detection limit) mechanism or are not applicable in the high dimensional setting. We here advocate the SAEM algorithm to take below detection limit data for parameter estimation into account in the longitudinal HIV dynamic model.

3.2 Parameter estimation

Both random effects from NLME-ODE and below detection limit (left-censoring) viral load can be treated as missing data. EM algorithm is able to deal with missing data where the log-likelihood with regard to missing components distribution is obtained at the E-step and the parameter estimates are updated through maximization (M-step). In light of high dimensionality of random effects and censored data, SAEM algorithm coupled with MCMC provides a convenient way of drawing samples at the E-step (Delyon et al. 1999; Kuhn and Lavielle 2005). The detail of this method is described below.

3.2.1 *A. Expectation*

From one point of view, both individual parameter $\boldsymbol{\theta}_i$ and below detection limit data can be treated as missing data. A classical way to cope with missing data is the EM algorithm proposed by Dempster et al. (1977). Let $\boldsymbol{\theta} = (\boldsymbol{\theta}_1, \cdots, \boldsymbol{\theta}_n)$. Denote $\mathbf{Y}^o$ and $\mathbf{Y}^m$ as observations beyond and below detection limit for all subjects across the study period, respectively. Let $L(\mathbf{Y}^o, \mathbf{Y}^m, \boldsymbol{\theta}; \boldsymbol{\mu}, \mathbf{D}, \sigma^2)$ represent the complete data likelihood. The MLE of $(\boldsymbol{\mu}, \mathbf{D}, \sigma^2)$ is determined by the marginal likelihood of the observed data $L(\mathbf{Y}^o; \boldsymbol{\mu}, \mathbf{D}, \sigma^2)$, whereas this quantity is often intractable. As an alternative, the EM algorithm calculates the expected value of the complete log-likelihood function, with respect to the joint conditional distribution of $\mathbf{Y}^m, \boldsymbol{\theta}$ given $\mathbf{Y}^o$ under the current estimates of the parameter $(\boldsymbol{\mu}^{(k)}, \mathbf{D}^{(k)}, \sigma^{2\,(k)})$

$$\mathbf{Q}\left(\boldsymbol{\mu}, \mathbf{D}, \sigma^2 \mid \boldsymbol{\mu}^{(\mathbf{k})}, \mathbf{D}^{(\mathbf{k})}, \sigma^{2\,(\mathrm{k})}\right) = \mathbf{E}_{\mathbf{Y}^m, \theta \mid \mathbf{Y}^o, \mu^{(k)}, \mathbf{D}^{(k)}, \sigma^{2\,(k)}} \left[\log L(\boldsymbol{\mu}, \mathbf{D}, \sigma^2; \mathbf{Y}^o, \mathbf{Y}^m, \boldsymbol{\theta})\right]. \tag{6}$$

For simplicity of notation, let $I_o = \{(i,j); y_{ij} \geq \mathrm{DL}\}$ with DL being the detection limit, and y^o_{ij} be the corresponding observation such that $I_m = \{(i,j); y_{ij} < \mathrm{DL}\}$, and y^m_{ij} be the corresponding missing data. Also, let n_t be the total number of observations and n_s be the number of subjects. In the long-term HIV treatment, it follows

$$L(\boldsymbol{\mu}, \mathbf{D}, \sigma^2; \mathbf{Y}^o, \mathbf{Y}^m, \boldsymbol{\theta}) \propto (\sigma^2)^{-\frac{n_t}{2}} \times |\mathbf{D}|^{-\frac{n_s}{2}} \exp\left\{-\frac{1}{2\sigma^2}\Sigma_{i,j\in I_o}\left[y_{ij}^o - g(\boldsymbol{\theta}_i, t_{ij})\right]^2 - \frac{1}{2\sigma^2}\Sigma_{i,j\in I_m}\left[y_{ij}^m - g(\boldsymbol{\theta}_i, t_{ij})\right]^2 - \frac{1}{2}\Sigma_i(\boldsymbol{\theta}_i - \boldsymbol{\mu})^T\mathbf{D}^{-1}(\boldsymbol{\theta}_i - \boldsymbol{\mu})\right\}. \tag{7}$$

Of particular note is that when $g(\cdot)$ is a linear function of $\boldsymbol{\theta}$, it is easy to obtain that equation (6) follows a normal distribution. For our problem, $g(\cdot)$ is the result of numerically integrating the ODE system in (1), which is not only a nonlinear function of $\boldsymbol{\theta}$, but the close-form expression does not exist as well. Accordingly, we follow the idea of a stochastic version of EM algorithm (SAEM) (Delyon et al. 1999) and evaluate equation (6) as follows.

3.2.2 B. Gibbs sampler for incomplete data

It has been long known that Gibbs sampler is useful for simulating data from the joint posterior distribution (Gelfand et al. 1990; Wakefield 1996). In our case, at the kth iteration, $\boldsymbol{\theta}$ and $\mathbf{Y}^m$ can be alternatively generated from the joint posterior distribution $P\left(\boldsymbol{\theta}, \mathbf{Y}^m \mid \mathbf{Y}^o, \boldsymbol{\mu}^{(k-1)}, \mathbf{D}^{(k-1)}, \sigma^{2(k-1)}\right)$ summarized in the following two steps.

Step 1 Simulate $\mathbf{Y}^{m(k)}$ from the marginal conditional posterior distribution $P\left(\mathbf{Y}^m \mid \boldsymbol{\theta}^{(k-1)}, \mathbf{Y}^o, \boldsymbol{\mu}^{(k-1)}, \mathbf{D}^{(k-1)}, \sigma^{2(k-1)}\right)$ which follows a normal distribution truncated at the detection limit. Each $y_{ij}^{m(k)}$ is then centered at $g\left(\boldsymbol{\theta}_i^{(k-1)}, t_{ij}\right)$ with variance $\sigma^{2(k-1)}$ and can thus be simulated as follows (Breslaw 1994):

- (a) calculate the cumulative probability of the detection limit under the same distribution as $y_{ij}^{m(k)}$ and denote as $\mathbf{P}_{\mathrm{DL}}$;
- (b) draw u from the uniform distribution $\mathcal{U}(0, 1)$; and
- (c) obtain a sample of $y_{ij}^{m(k)}$ as $y_{ij}^{m(k)} = g(\boldsymbol{\theta}_i^{(k-1)}, t_{ij}) + \sigma^{(k-1)}\Phi^{-1}[u \times \mathbf{P}_{\mathrm{DL}}]$, where Φ is the standard normal cumulative distribution function.

It should be noted that this sampling algorithm requires only one draw at each iteration therefore is efficient.

Step 2 Simulate $\boldsymbol{\theta}^k$ from the conditional posterior distribution $P\left(\boldsymbol{\theta} \mid \mathbf{Y}^{m(k)}, \mathbf{Y}^o, \boldsymbol{\mu}^{(k-1)}, \mathbf{D}^{(k-1)}, \sigma^{2(k-1)}\right)$, which has no close-form formula but is proportional to (7) with all the parameters given at current values. The Metropolis-Hastings (M-H) algorithm is capable of generating samples from this distribution. Indeed, one choice of the proposal distribution is $q_{k-1} \sim \mathcal{N}(\boldsymbol{\mu}^{(k-1)}, \mathbf{D}^{(k-1)})$. Then the procedure proceeds as follows

- (a) Calculate acceptance probability $\alpha(\varphi|\boldsymbol{\theta}^{(k-1)})$ as

$$\min\left\{1, \frac{P\left(\varphi \mid \mathbf{Y}^{\mathbf{m(k)}}, \mathbf{Y}^{\mathbf{o}}, \boldsymbol{\mu}^{(k-1)}, \mathbf{D}^{(k-1)}, \sigma^{2(k-1)}\right)}{P\left(\boldsymbol{\theta}^{(k-1)} \mid \mathbf{Y}^{\mathbf{m(k)}}, \mathbf{Y}^{\mathbf{o}}, \boldsymbol{\mu}^{(k-1)}, \mathbf{D}^{(k-1)}, \sigma^{2(k-1)}\right)} \frac{q_{(k-1)}\left(\boldsymbol{\theta}^{(k-1)} \mid \boldsymbol{\mu}^{(k-1)}, \mathbf{D}^{(k-1)}\right)}{q_{(k-1)}\left(\varphi \mid \boldsymbol{\mu}^{(k-1)}, \mathbf{D}^{(k-1)}\right)}\right\}, \tag{8}$$

where φ is a candidate simulated from q_{k-1}. If we assume that each $\boldsymbol{\theta}_i$ is independent, then $\mathbf{D}$ is diagonal. We may thus simulate φ for each i (denote as φ_i) separately. After some arrangements, the acceptance probability α is simplified as

$$\alpha(\varphi^i \mid \boldsymbol{\theta}^{(k-1)}) = \min\left\{1,\ \mathbf{R}^i\right\}, \tag{9}$$

where

$$\mathbf{R}^i = \exp\left\{\frac{1}{2\sigma^{2(k-1)}}\left(\Sigma_{j;i,j\in I_o}[y_{ij}^o - g(\boldsymbol{\theta}_i^{(k-1)}, t_{ij})]^2 + \Sigma_{j;i,j\in I_m}\left[y_{ij}^{m(k)} - g\left(\boldsymbol{\theta}_i^{(k-1)}, t_{ij}\right)\right]^2 - \Sigma_{j;i,j\in I_o}\left[y_{ij}^o - g\left(\varphi_i, t_{ij}\right)\right]^2 - \Sigma_{j;i,j\in I_m}\left[y_{ij}^{m(k)} - g\left(\varphi_i, t_{ij}\right)\right]^2\right)\right\}. \tag{10}$$

(b) For each i, draw u from the uniform distribution $U(0,\ 1)$. If $u \le \alpha(\varphi^i|\boldsymbol{\theta}^{(k-1)})$, then accept φ^i as new $\boldsymbol{\theta}_i^{(k)}$; otherwise keep $\boldsymbol{\theta}_i^{(k-1)}$ as $\boldsymbol{\theta}_i^{(k)}$.

Notice that, unlike the implementation of M-H algorithm in other cases in which the choice of variance for proposal density q is essential to the efficiency of the algorithm. Herein, there is no need to choose an appropriate variance manually. Since variance $\mathbf{D}$ is always estimated from the last iteration, the algorithm updates the proposal variance automatically to make itself adaptive. Often, the candidate parameter φ simulated from q makes the integration of (1) unstable, which is the so-called stiffness. To handle the stiff ODEs, we apply a Rosenbrock method, which is relatively easy to implement and also provides a good accuracy. We refer the interested readers to Kaps and Rentrop (1979) for more details.

3.2.3 C. Maximization

Once $\boldsymbol{\theta}$ and $\mathbf{Y}^m$ are simulated, it is straightforward to update $(\boldsymbol{\mu}, \mathbf{D}, \sigma^2)$ by maximizing equation (6) and we obtain

$$\begin{aligned}
\boldsymbol{\mu}^{(k)} &= \frac{1}{n_s}\Sigma_i\boldsymbol{\theta}_i^{(k)},\\
\operatorname{diag}\left(\mathbf{D}^{(k)}\right) &= \frac{1}{n_s}\Sigma_i\boldsymbol{\theta}_i^{2(k)} - \boldsymbol{\mu}^{2(k)},\\
\sigma^{2(k)} &= \frac{1}{n_t}\left\{\Sigma_{i,j\in I_o}[y_{ij}^o - g(\boldsymbol{\theta}_i^{(k)}, t_{ij})]^2 + \Sigma_{i,j\in I_m}[y_{ij}^{m(k)} - g(\boldsymbol{\theta}_i^{(k)}, t_{ij})]^2\right\}.
\end{aligned} \tag{11}$$

Note that the above estimates are composed of minimum sufficient statistics of $(\boldsymbol{\mu}, \mathbf{D}, \sigma^2)$. Denote $s_1 = \Sigma_i\boldsymbol{\theta}_i$, $s_2 = \Sigma_i\boldsymbol{\theta}_i^2$, and $s_3 = \Sigma_{i,j\in I_o}[y_{ij}^o - g(\boldsymbol{\theta}_i, t_{ij})]^2 + \Sigma_{i,j\in I_m}[y_{ij}^m - g(\boldsymbol{\theta}_i, t_{ij})]^2$. Stochastic approximation step of SAEM is composed of updating s_1, s_2, and s_3 with a sequence $\gamma^{(k)}$ at the kth iteration $s_i^{(k)} = (1-\gamma^{(k)})s_i^{(k-1)} + \gamma^{(k)}s_i^{(k)}$. Kuhn and Lavielle (2005) recommended to use $\gamma^{(k)} = 1$ for the first K1 iterations followed by diminishing $\gamma^{(k)} = 1/(k - K1)$ for another K2 iterations in order to satisfy the assumptions of SAEM and to ensure the convergence of the algorithm. It deserves mentioning that the estimates of variance can be easily obtained by the inverse of the observed fisher information matrix of (7).

4 Simulation

We illustrate the effectiveness of our method via analysis of an example in the long-term HIV dynamic study with considering below detection limit measurements. Note that the simulation experiment is similar to an actual AIDS clinical trial. Basically, 40 patients were involved in a long-term antiretroviral treatment and then followed up to 200 days. Some patients were loss to follow up due to various reasons as in the real study. Consequently, the simulated viral load data are unbalanced longitudinal measurements. All data for the pharmacokinetic factor (C_1, C_2), phenotype marker (baseline and failure IC_{50}s) and adherence as well as the baseline viral load (V_0) were taken from an AIDS clinical trial study.

In order to avoid parameter identifiability problem, ρ, N, and K are fixed at 0.1, 980, and 0.0001, respectively. These values are chosen from previous studies in the literature (Ding and Wu 2000; Nowak and May 2000; Perelson and Nelson 1999). In addition, we assume the system is at steady state at the beginning of the treatment. This indicates that only three parameters in (1) need to be estimated. Denote $\boldsymbol{\mu} = (\log c, \log \delta, \log \phi)^T$ and $\boldsymbol{\theta}_i = (\log c_i, \log \delta_i, \log \phi_i)^T$. The diagonal covariance matrix $\mathbf{D}$ has a vector of diagonal elements $d^2 = (d_c^2, d_\delta^2, d_\phi^2)^T$. The value of $\log \lambda$ is obtained from (2). Altogether, there are totally seven parameters in the model.

The true values for $\boldsymbol{\mu} = (\log c, \log \delta, \log \phi)^T$ and $d^2 = (d_c^2, d_\delta^2, d_\phi^2)^T$ are set to be $(1.1, -1, 3)^T$ and $(0.25, 0.04, 0.0625)^T$, respectively. These values are similar to the ones used in the simulation study of Huang et al. (2006). The true value for σ^2 is 0.1 and the detection limit is 25 as in the real study. The true values of parameters $\boldsymbol{\theta}_i$ for $i = 1, \cdots, 40$ are simulated from $\mathcal{N}(\boldsymbol{\mu}, \mathbf{D})$ independently. Given PK/clinial data $(C, IC_{50}, A(t))$ and true parameters $(\boldsymbol{\theta}_i)$ as well as initial viral load (V_0), viral load V_{ij} are generated by numerically solving the ODE system in (1) for each individual. A measurement error term ε_{ij} is then added to $\log V_{ij}$ to mimic observation data $\log y_{ij}$ for the ith subject at time t_{ij}, where $\varepsilon_{ij} \sim N(0, \sigma^2)$. We then censor the viral load data at detection limit 25, producing totally 23 percent of below detection data which are imputed at 25 as observed in a real experiment.

We examine the effect of taking into account censored data or not on parameter estimates. First, the data below detection limit is completely hold out before fitting model, named as 'case A'; second, the below detection limit data are retained but assumed 25 as the true value, named as 'case B'; third, the below detection limit data are considered as censored data, named as 'case C'. For each case, we fit the nonlinear mixed-effects ODE model coupled with the SAEM algorithm. However, in cases A and B, the data are fit as they are, i.e., the data are considered as complete while censoring mechanisms are accounted for in case C. In this simulation study, K1 is set to be 10,000 to make sure the estimates converging to the neighborhood of MLE, then a decreasing step size γ is used for another 2,000 steps to fine-tune the final estimates. The same generating and model fitting processes are repeated 100 times. The performance of the proposed method on three cases is measured with respect to the bias and square root of MSE (RMSE).

The simulation results are presented in Table 1. In all three cases, the biases for inter-subject variance components are significantly larger than the ones of fixed effects. Biases for d_c^2 in all the three cases exceed 45%. The biases for variance estimates are especially large (almost 500% for 'case A' and 'case B'; 83% for 'case C'). For fixed effects, case A is superior to case B in terms of both bias and RMSE. Among all the three cases, case C

Table 1 The true values and mean estimates of population dynamic parameters with 100 replications as well as the corresponding bias, RMSE

		$\log c$	$\log \delta$	$\log \phi$	d_c^2	d_δ^2	d_ϕ^2	σ^2
Case A								
	True value	1.1	-1	3	0.25	0.04	0.0625	0.01
	Mean estimate	0.9	-0.8	3.07	0.13	0.06	0.11	0.05
	Bias(%)	15.2	23.7	2.4	47.3	46.4	69.5	484
	RMSE(%)	78	69	4.6	85	198	147	658
Case B								
	True value	1.1	-1	3	0.25	0.04	0.0625	0.01
	Mean estimate	1.9	-0.8	3.6	0.03	0.067	0.669	0.06
	Bias(%)	71	22	19	90	68	970	490
	RMSE(%)	141	58	23	105	125	1013	687
Case C								
	True value	1.1	-1	3	0.25	0.04	0.0625	0.01
	Mean estimate	1.2	-1.0	3.02	0.12	0.036	0.048	0.02
	Bias(%)	9.9	0.2	0.9	53	9.5	23.3	83
	RMSE(%)	54	5	3	75	39	211	783

In case A, the below detection limit data are hold out before fitting the model; In case B, the below detection limit data are retained but assumed 25 as the true value; In case C, the below detection limit data are considered as censored data and censoring mechanism is introduced by the proposed method.

has the least bias and RMSE for all of the seven parameters, thus performs the best. In general, imputing censored viral load with the detection limit (25) leads to larger bias for both fixed effects and random effects variance estimates. Our simulation results show that simply omitting censored observations (case A) is even better than retaining the censored values (case B) in terms of parameter estimating accuracy. Handling below detection limit data with the proposed method gives us the least bias in the current setting.

We also compared the fitting results with the Bayesian method and Gaussian quadrature method. We observed that for cases A and B, all the three methods provide similar estimates (Table 2). However, the Bayesian method fails to handle with below detection data so that it can not deal with case C. In addition, we also tried to fit case C with the Gaussian quadrature method. Unfortunately, this method does not converge in our case.

To test the robustness of parameter estimates with regard to initial values, we carry out a sensitivity analysis. Three sets of initial settings are examined. One sets the mean vector $\mu = (\log c, \log \delta, \log \phi)^T$ at the true values $(1.1, -1, 3)^T$, the other two set μ at the two extremes $(3, -0.5, 5)^T$ and $(0.2, -0.2, 0.6)^T$, separately. Variance components $d^2 = (d_c^2, d_\delta^2, d_\phi^2)^T$ are also increased to $(5, 5, 5)^T$ and $(10, 10, 10)^T$, respectively. We then

Table 2 Comparison of bias(%) on parameter estimates by three methods: Bayesian, Gaussian quadrature and NLME-SAEM methods

Case	Method	$\log c$	$\log \delta$	$\log \phi$	d_c^2	d_δ^2	d_ϕ^2	σ^2
A	Bayesian	16.1	24.2	2.1	49.6	47.7	65.1	543
	Gaussian quadrature	15.6	24	2.2	53.1	44.3	75.5	512
	NLME-SAEM	15.2	23.7	2.4	47.3	46.4	69.5	484
B	Bayesian	78	20	22	78	75	998	657
	Gaussian quadrature	72	25	16	94	61	932	586
	NLME-SAEM	71	22	19	90	68	970	490

In case A, the below detection limit data are hold out before fitting each model; In case B, the below detection limit data are retained but assumed 25 as the true value. The number of simulation run is 100.

estimate parameters using the same method as above. The first 10,000 steps are followed by additional 2,000 steps. Figure 1 depicts the graphical diagnostic of convergence of SAEM algorithm for the three cases. As shown clearly, after the first 10,000 steps, all three chains converge to the neighborhood of true values and 2,000 more steps make the estimate more precise. Based on these results, the different initial values give similar parameter estimates, therefore the method is quite robust. We may thus conclude that parameter estimates using the proposed method are insensitive to the choices of initial values.

5 A real AIDS clinical study

In this section, we apply the proposed method to a real AIDS clinical trial study. This study was a Phase I/II, randomized, open-label, 24-week comparative study of the PK, tolerability and antiretroviral effects of two regimens of indinavir (IDV), ritonavir (RTV), plus two nucleoside analogue reverse transcriptase inhibitors (NRTIs) on HIV-1-infected subjects failing protease inhibitor (PI) containing antiretroviral therapies. The 44 subjects were randomly assigned to the two treatment regimens, Arm A (IDV 800 mg q12h+RTV 200 mg q12 h), and Arm B (IDV 400 mg q12h+RTV 400 mg q12h). Study visits occurred at pre-entry, entry (within 14 days of pre-entry) and days 7, 14, 28, 56, 84, 112, 140 and 168 of follow-up. Plasma HIV RNA testing was conducted at each study visit. The detection limit is 25 cp/ml. Clinical assessment and laboratory parameters were performed at all visit weeks with the exception of week 1. Phenotypic determination (IC_{50}) of antiretroviral drug resistance was performed at baseline and at the time of virological failure. The PK parameters of IDV and RTV were determined using noncompartmental methods. Pill counts were measured to monitor adherence at each study visit from weeks 2 to 24. Of the 44 subjects, 42 subjects were included in this analysis; of the remaining two subjects, one was excluded from the analysis because the PK parameters were not obtained and the other was excluded because the phenotype assay could not be completed on this subject. Acosta et al. (2004) gives more detailed descriptions of this study.

The analysis here is to investigate the effect of missing data on the parameter estimation of HIV dynamic model (1). There are totally 14 percent (45 out of 328) of below detection observations in this real study. We compare the parameter estimate results based on the full data, with and without taking censoring mechanism into account. To avoid parameter identifiability problem, we fix $c = 3.0$ and $N = 980$; see Stafford et al. (2000). There are

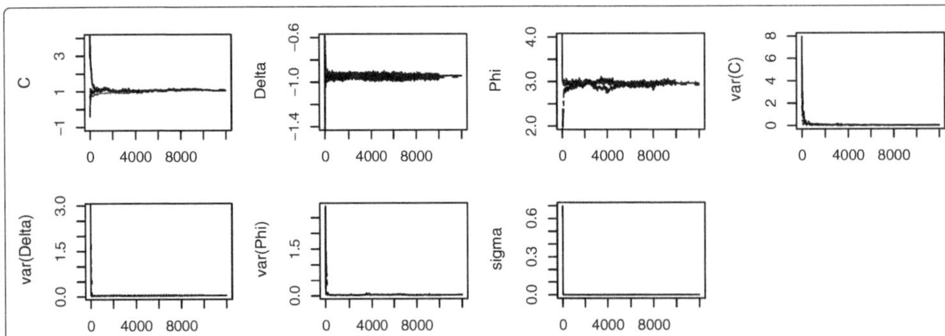

Figure 1 Convergence diagnostics with respect to three different initial values from a simulation study.

totally five fixed effects and six variance parameters (five from random effects and one from measurement error). We tried different initial values which all gave similar results. Table 3 presents parameter estimates and corresponding standard deviation for the 11 parameters.

The biological and clinical interpretation of estimating results are quite interesting. Compared with the estimate with considering censoring mechanism, the death rate of infected cell (T^*) is overestimated ($\log \delta$ = -1.613 vs -1.801); the death rate of uninfected cell (T) is underestimated ($\log \rho$ = -3.798 vs -3.563); the infection rate of T cells by virus is underestimated ($\log k$ = -11.87 vs -9.397); conversion factor is overestimated ($\log \phi$ = 1.395 vs 1.02); the production rate of T cells from body source is overestimated ($\log \lambda$ = 5.68 vs 4.876) from the model without taking censoring into account. As a result, the comparison above consistently indicates that the degree of infection is underestimated by model without considering censoring if we define high degree of infection as low death rate of infected cell (T^*), high death rate of uninfected cell (T), high infection rate of T cells by virus, low production rate of T cells from body source. In other words, the degree of infection is actually more severe if censored data are taken care of. As one can see, these results are logical and meaningful. Moreover, as suggested by (3), conversion factor ϕ is negatively associated with drug efficacy. By considering censoring, the drug efficacy is higher, which is suggested by lower ϕ. Alternatively, the model without considering censoring is more likely to underestimate drug efficacy.

In addition, the estimates from the model that incorporates censoring mechanism are more precise in terms of smaller standard error except for $\log \lambda$. The inter-subject variance components are smaller for the ones based on the model with censoring mechanism, especially for k and ϕ. These results indicate most of parameter estimates are not much different among patients. On the contrary, the T cell production rate has a larger variation (d_λ^2) among patients when taking censoring into account. These results inform us that the dynamic parameters are not varying much from one subject to another, but each subject do present different health conditions as indexed by self production rate of T cells (λ). Again, these estimates are more precise compared to those from the model without considering censoring, as reflected by smaller standard errors. The variance for measurement error (σ^2) is not comparable since in the model with censoring, the censored data are replaced by viral load simulated from the model.

Table 3 Parameter estimates and standard deviation of parameters from a real AIDS clinical study using models with and without considering censoring mechanism, respectively

	W/o censor mechanism		With censor mechanism	
Parameter	Estimate	SE	Estimate	SE
$\log \delta$	-1.613	0.086	-1.801	0.07
$\log \rho$	-3.798	0.17	-3.564	0.02
$\log k$	-11.87	0.274	-9.397	0.001
$\log \phi$	1.395	0.094	1.02	0.001
$\log \lambda$	5.68	0.17	4.876	0.21
d_δ^2	0.312	0.068	0.203	0.044
d_ρ^2	1.219	0.266	0.016	0.004
d_k^2	3.146	0.686	7×10^{-4}	2×10^{-5}
d_ϕ^2	0.37	0.081	2×10^{-4}	3×10^{-6}
d_λ^2	1.215	0.265	1.787	0.39
σ^2	0.14	0.01	0.29	0.02

We also compared the model fitting from the two models with and without taking censoring for each patient into account. Individual parameter estimates are from the posterior mode of the Gibbs samples. The model fits the data very well. Figure 2 depicted the model fitting results of four patients. Clearly, the model not only describes consistently decreasing viral load trajectory (subject 13 and 42), but also rebounds (subject 1 and 12), which is a character of long-term treatment failure and a challenge to short-term models as explained in Section 1. When no below detection limit data are present, i.e. subject 12, the two models behave similarly. Even when censoring data occur, the two models fit the data beyond detection limit approximately the same way. However, without considering censoring, the model does not provide any information below detection limit. After accounting for censoring mechanism, we have a better idea of what is going on underneath the line. For example, in subject 1, viral load rebound has already occurred before it is detectable (solid line below detection limit). In subject 13, viral load keeps decreasing until diminished below detection limit.

Based on the analysis, we conclude that handling below detection data carefully is crucial to obtain more accurate estimates and also correct interpretation of the results.

6 Discussion

HIV dynamic model offers an opportunity to study HIV infection mechanism and treatment development from a perspective, other than laboratory experiments. However, research on this area is sparse mainly due to the absence of advanced statistical methods. The issue is further complicated by the lack of closed-form expression of the model and the numerical solution always incurs errors. In contrast to other methods introduced in Section 1, the SAEM algorithm is very efficient and fast, especially in the setting of

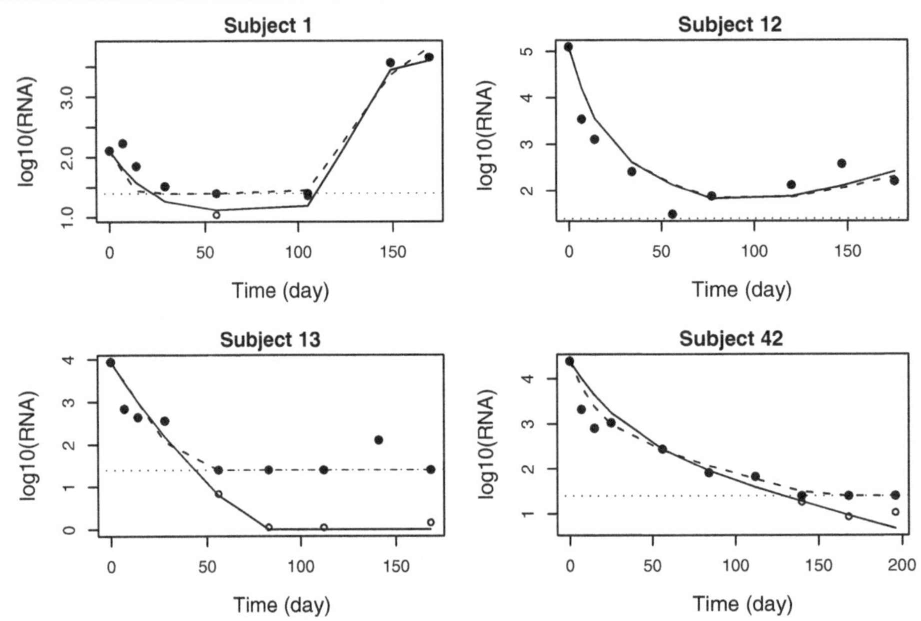

Figure 2 Viral load fit for four patients. Filled circles represent observation values; circles stand for simulated below detection limit viral load by the proposed model; Logarithm of detection limit (1.398) is represented by horizontal dotted lines; Dashed lines are viral load estimated from the model without considering censoring mechanism, while solid lines are estimated from the proposed model accounting for censoring.

nonlinear mixed-effects ODE models. Numerical evidence shows that the algorithm converges very fast, usually in a few minutes whereas other algorithms, such as the Bayesian method, seem to be computationally expensive and also slow to converge. Note that the simulation step of SAEM employs a similar MCMC procedure, such as the Gibbs sampling and M-H step as in the Bayesian method. Nevertheless, the MCMC algorithm converges much faster when coupled with the SAEM as compared to the employment in a traditional Bayesian posterior sampling.

One disadvantage of the proposed method is that not all the parameters can be identified due to limitation of data on hand. We may use CD4 T cell counts in the model so that the constraints on some parameters can be released. However, the measurements of CD4 T cells are not reliable so that the use of these data may induce significant bias. The ODE model identifiability is an interesting topic but beyond the scope of this article. Some references for nonlinear ODE model identifiability, including HIV dynamic models, can be found in Jeffrey and Xia (2005), Wu et al. (2008), and Xia and Moog (2003).

In spite of this limitation, we compared the models with and without handling below detection data and found that ignoring these data is very dangerous for interpreting the results. The biological and clinical interpretation of the results accounting for censoring mechanism is inspiring. We hope the study results in this paper may stimulate more research in this area and contribute to the research on identifying the biological mechanism of HIV infection as well as clinical treatment development.

Competing interests

The authors declare that they have no competing interests.

Authors' contributions

TL initiated and carried out the study. TL drafted the manuscript. MW participated in the discussion and proofread the article. Both authors read and approved the final manuscript.

Acknowledgments

We would like to thank the editor and two anonymous referees whose valuable suggestions led to major improvements in the overall clarity and presentation of this paper.

Author details

[1] Department of Epidemiology and Biostatistics, State University of New York, Albany, New York, USA. [2] Department of Mathematical Sciences, Michigan Technological University, Houghton, Michigan, USA.

References

Acosta, EP, Wu, H, Hammer, SM, Yu, S, Kuritzkes, DR, Walawander, A, Eron, JJ, Fichtenbaum, CJ, Pettinelli, C, Neath, D, Ferguson, E, Saah, AJ, Gerber, JG: Comparison of two indinavir/ritonavir regimens in the treatment of HIV-infected individuals. JAIDS. **9**, 1358–1366 (2004)

Besch, CL: Compliance in clinical trials. AIDS. **9**, 1–10 (1995)

Breslaw, JA: Random sampling from a truncated multivariate normal distribution. Appl. Math. Lett. **7**, 1–6 (1994)

Davidian, M, Giltinan, DM: Nonlinear Models for Repeated Measurement Data. Chapman & Hall, London (1995)

Delyon, B, Lavielle, M, Moulines, E: Convergence of a stochastic approximation version of the EM algorithm. Ann. Stat. **27**, 94–128 (1999)

Dempster, AP, Laird, NM, Rubin, DB: Maximum likelihood from incomplete data via the EM algorithm. J. Roy. Stat. Soc. B. **39**, 1–38 (1977)

Ding, AA, Wu, H: A comparison study of models and fitting procedures for biphasic viral decay rates in viral dynamic models. Biometrics. **56**, 16–23 (2000)

Gelfand, AE, Hills, SE, Racine-Poon, A, Smith, AFM: Illustration of Bayesian inference in normal data models using Gibbs sampling. J. Am. Stat. Assoc. **85**, 972–985 (1990)

Guedj, J, Thiebaut, R, Commenges, D: Maximum likelihood estimation in dynamical models of HIV. Biometrics. **63**, 1198–1206 (2007)

Huang, Y, Getachew, D: Simultaneous Bayesian inference for longitudinal data with asymmetry, left-censoring and covariates measured with errors. J. Janpanes Stat. Soc. **42**, 1–22 (2012)

Huang, Y, Liu, D, Wu, H: Hierarchical Bayesian methods for estimation of parameters in a longitudinal HIV dynamic system. Biometrics. **62**, 413–423 (2006)

Hughes, J: Mixed-effectss models with censored data with applications to HIV RNA levels. Biometrics. **55**, 625–629 (1999)

Jeffrey, AM, Xia, X: Identifiability of HIV/AIDS model. In: Tan, WY, Wu, H (eds.) *Deterministic and Stochastic Models of AIDS Epidemics and HIV Infections with Intervention*. World Scientific, Singapore, (2005)

Kaps, P, Rentrop, P: Generalized Runge-Kutta methods of order four with stepsize control for stiff ordinary differential equations. Numer. Math. **33**, 55–68 (1979)

Kuhn, E, Lavielle, M: Coupling a stochastic approximation version of EM with a MCMC procedure. ESAIM: P. S. **8**, 115–131 (2005)

Molla, A, Korneyeva, M, Gao, Q, Vasavanonda, S, Schipper, P, Mo, H, Markowitz, M, Chernyavskiy, T, Niu, P, Lyons, N, Hsu, A, Granneman, G, Ho, DD, Boucher, C, Leonard, J, Norbeck, D, Kempf, D: Ordered accumulation of mutations in HIV protease confers resistance to ritonavir. Nat. Med. **2**, 760–766 (1996)

Nowak, MA, May, RM: Virus Dynamics: Mathematical Principles of Immunology and Virology. Oxford University Press, Oxford (2000)

Perelson, AS, Nelson, PW: Mathematical analysis of HIV-1 dynamics *in vivo*. SIAM Rev. **41**(1), 3–44 (1999)

Perelson, AS, Neumann, AU, Markowitz, M, Leonard, JM, Ho, DD: HIV-1 dynamics *in vivo*: virion clearance rate, infected cell life-span, and viral generation time. Science. **271**, 1582–1586 (1996)

Perelson, AS, Essunger, P, Cao, Y, Vesanen, M, Hurley, A, Saksela, K, Markowitz, M, Ho, DD: Decay characteristics of HIV-1-infected compartments during combination therapy. Nature. **387**, 188–191 (1997)

Stafford, MA, Corey, L, Cao, Y, Daar, ES, Ho, DD, Perelson, AS: Modeling plasma virus concentration during primary HIV infection. J. Theor. Biol. **203**, 285–301 (2000)

Wakefield, JC: The Bayesian analysis to population Pharmacokinetic models. J. Am. Stat. Assoc. **91**, 62–75 (1996)

Wu, L: A joint model for nonlinear mixed-effects models with censoring and covariates measured with error, with application toAIDS studies. J. Am. Stat. Assoc. **97**, 955–964 (2002)

Wu, L: Exact and approximate inferences for nonlinear mixed-effects models with missing covariates. J. Am. Stat. Assoc. **99**, 700–709 (2004)

Wu, H, Ding, AA: Population HIV-1 dynamics *in vivo*: applicable models and inferential tools for virological data from AIDS clinical trials. Biometrics. **55**, 410–418 (1999)

Wu, H, Ding, AA, de Gruttola, V: Estimation of HIV dynamic parameters. Stat. Med. **17**, 2463–2485 (1998)

Wu, H, Huang, Y, Acosta, EP, Rosenkaranz, S, Kuritzkes, D, Eron, JJ, Perelson, AS, Gerber, JG: Modeling long-term HIV dynamics and antiretroviral response: effects of drug potency, pharmacokinetics, adherence and drug resistance. J. Acquir. Immune Defic. Syndr. **39**, 272–283 (2005)

Wu, H, Zhu, H, Miao, H, Perelson, AS: Parameter identifiability and estimation of HIV/AIDS dynamic models. Bull. Math. Biol. **70**, 785–799 (2008)

Xia, X, Moog, CH: Identifiability of nonlinear systems with applications to HIV/AIDS models. IEEE Trans. Automat. Contr. **48**, 330–336 (2003)

The odd generalized exponential family of distributions with applications

Muhammad H Tahir[1*†], Gauss M Cordeiro[2†], Morad Alizadeh[3†], Muhammad Mansoor[4†], Muhammad Zubair[4†] and Gholamhossein G Hamedani[5†]

*Correspondence: mtahir.stat@gmail.com
†Equal contributors
[1] Department Statistics, Baghdad Campus, The Islamia University of Bahawalpur, 63100 Bahawalpur, Pakistan
Full list of author information is available at the end of the article

Abstract

We propose a new family of continuous distributions called the odd generalized exponential family, whose hazard rate could be increasing, decreasing, J, reversed-J, bathtub and upside-down bathtub. It includes as a special case the widely known exponentiated-Weibull distribution. We present and discuss three special models in the family. Its density function can be expressed as a mixture of exponentiated densities based on the same baseline distribution. We derive explicit expressions for the ordinary and incomplete moments, quantile and generating functions, Bonferroni and Lorenz curves, Shannon and Rényi entropies and order statistics. For the first time, we obtain the generating function of the Fréchet distribution. Two useful characterizations of the family are also proposed. The parameters of the new family are estimated by the method of maximum likelihood. Its usefulness is illustrated by means of two real lifetime data sets.

AMS Subject Classification: Primary 60E05; secondary 62N05; 62F10

Keywords: Generalized exponential; Hazard function; Maximum likelihood; Moments; Order statistic; Rényi entropy

1 Introduction

The art of proposing generalized classes of distributions has attracted theoretical and applied statisticians due to their flexible properties. Most of the generalizations are developed for one or more of the following reasons: a physical or statistical theoretical argument to explain the mechanism of the generated data, an appropriate model that has previously been used successfully, and a model whose empirical fit is good to the data. The last decade is full on new classes of distributions that become precious for applied statisticians. There are two main approaches for adding new shape parameter(s) to a baseline distribution.

1.1 Approach 1: Adding one shape parameter

The first approach of generalization was suggested by Marshall and Olkin (1997) by adding one parameter to the survival function $\overline{G}(x) = 1 - G(x)$, where $G(x)$ is the cumulative distribution function (cdf) of the baseline distribution. Gupta et al. (1998) added one parameter to the cdf, $G(x)$, of the baseline distribution to define the *exponentiated-G* ("exp-G" for short) class of distributions based on Lehmann-type alternatives (see

Lehmann 1953). Following Gupta et al. 's class, Gupta and Kundu (1999) studied the two-parameter *generalized-exponential* (GE) distribution as an extension of the exponential distribution based on Lehmann type I alternative. The GE distribution is also known as the exponentiated exponential (EE) distribution. Since it is the most attractive generalization of the exponential distribution, the GE model has received increased attention and many authors have studied its various properties and also proposed comparisons with other distributions. Some significant references are: Gupta and Kundu (2001a; 2001b; 2002; 2003; 2004; 2006; 2007; 2008; 2011), Kundu et al. (2005), Nadarajah and Kotz (2006), Dey and Kundu (2009), Pakyari (2010) and Nadarajah (2011). In fact, the GE model has been proven to be a good alternative to the gamma, Weibull and log-normal distributions, all of them with two-parameters. The GE distribution can be used quite effectively for analyzing lifetime data which have monotonic hazard rate function (hrf) but unfortunately it cannot be used if the hrf is upside-down, J or reversed-J shapes. Differently, the new family can have increasing, decreasing, J, reversed-J, bathtub and upside-down bathtub and, therefore, it can be used effectively for analyzing lifetime data of various types. The generalization of the exp-G class to other distributions is beyond the scope of the paper.

1.2 Approach 2: Adding two or more shape parameters

A second approach of generalization was pioneered by Eugene et al. (2002) and Jones (2004) by defining the beta-generated (beta-G) class from the logit of the beta distribution. Further works on generalized distributions were the Kumaraswamy-G (Kw-G) by Cordeiro and de Castro (2011), McDonald-G (Mc-G) by Alexander et al. (2012), gamma-G type 1 by Zografos and Balakrishanan (2009) and Amini et al. (2014), gamma-G type 2 by Ristić and Balakrishanan (2012) and Amini et al. (2014), odd-gamma-G type 3 by Torabi and Montazari (2012), logistic-G by Torabi and Montazari (2014), odd exponentiated generalized (odd exp-G) by Cordeiro et al. (2013), transformed-transformer (T-X) (Weibull-X and gamma-X) by Alzaatreh et al. (2013), exponentiated T-X by Alzaghal et al. (2013), odd Weibull-G by Bourguignon et al. (2014), exponentiated half-logistic by Cordeiro et al. (2014a), T-X{Y}-quantile based approach by Aljarrah et al. (2014) and T-R{Y} by Alzaatreh et al. (2014), Lomax-G by Cordeiro et al. (2014b), logistic-X by Tahir et al. (2015a), a new Weibull-G by Tahir et al. (2015b) and Kumaraswamy odd log-logistic-G by Alizadeh et al. (2015).

We propose a new class of distributions called the *odd generalized exponential* ("OGE" for short) family, which is flexible because of the hazard rate shapes: increasing, decreasing, J, reversed-J, bathtub and upsidedown bathtub. In Section 2, we define the OGE family of distributions. Two special cases of this family are presented in Section 3. The density and hazard rate functions are described analytically in Section 4. A useful mixture representation for the pdf of the new family is derived in Section 5. In Section 6, we obtain explicit expressions for the moments, generating function, mean deviations and entropies. In Section 7, we determine a power series for the quantile function (qf). In Section 8, we investigate the order statistics. Section 9 refers to some characterizations of the OGE family. In Section 10, the parameters of the new family are estimated by the method of maximum likelihood. In Section 11, we illustrate its performance by means of two applications to real life data sets. The paper is concluded in Section 12.

2 The new family

Consider a parent distribution depending on a parameter vector $\boldsymbol{\xi}$ with cdf $G(x;\boldsymbol{\xi})$, survival function $\overline{G}(x;\boldsymbol{\xi}) = 1 - G(x;\boldsymbol{\xi})$ and probability density function (pdf) $g(x;\boldsymbol{\xi})$. The cdf of the GE model with positive parameters λ and α is given by $\Pi(x) = \left(1 - \mathrm{e}^{-\lambda x}\right)^{\alpha}$ (for $x > 0$). For $\alpha = 1$, it reduces to the exponential distribution with mean λ^{-1}. In fact, the density and hazard functions of the GE and gamma distributions are quite similar. The GE density function is always right-skewed and can be used quite effectively to analyze skewed data. Its hrf can be increasing, decreasing and constant depending on the shape parameter in a similar manner of the gamma distribution. Its applications have been wide-spread as a model to power system equipment, rainfall data, software reliability and analysis of animal behavior, among others.

We define the cdf of the OGE family by replacing x by $G(x;\boldsymbol{\xi})/\overline{G}(x;\boldsymbol{\xi})$ in the cdf $\Pi(x)$ leading to

$$F(x) = F(x;\alpha,\lambda,\boldsymbol{\xi}) = \left(1 - \mathrm{e}^{-\lambda \frac{G(x,\boldsymbol{\xi})}{\overline{G}(x,\boldsymbol{\xi})}}\right)^{\alpha}, \tag{1}$$

where $\alpha > 0$ and $\lambda > 0$ are two additional parameters.

The pdf corresponding to (1) is given by

$$f(x) = f(x;\alpha,\lambda,\boldsymbol{\xi}) = \frac{\lambda\,\alpha\,g(x,\boldsymbol{\xi})}{\overline{G}(x,\boldsymbol{\xi})^2}\,\mathrm{e}^{-\lambda \frac{G(x,\boldsymbol{\xi})}{\overline{G}(x,\boldsymbol{\xi})}} \left(1 - \mathrm{e}^{-\lambda \frac{G(x,\boldsymbol{\xi})}{\overline{G}(x,\boldsymbol{\xi})}}\right)^{\alpha-1}, \tag{2}$$

where $g(x;\boldsymbol{\xi})$ is the baseline pdf. We can omit the dependence on the vector of parameters $\boldsymbol{\xi}$ and write simply $G(x) = G(x;\boldsymbol{\xi})$. Equation 2 will be most tractable when the cdf $G(x)$ and pdf $g(x)$ have explicit expressions. Hereafter, a random variable X with density function (2) is denoted by $X \sim \mathrm{OGE}(\alpha,\lambda,\boldsymbol{\xi})$. The main motivations for using the OGE family are to make the kurtosis more flexible (compared to the baseline model) and possible to construct heavy-tailed distributions that are not long-tailed for modeling real data. The family (1) contains some special models as those listed in Table 1. In particular, it includes as a special case the widely known *exponentiated Weibull* (EW) distribution pioneered by Mudholkar and Srivastava (1993) and Mudholkar et al. (1995). For a detail survey on the EW model, the reader is referred to Nadarajah et al. (2013). It has been well-established in the literature that the EW distribution gives significantly better fits than the exponential, gamma, Weibull and lognormal distributions.

We offer a physical interpretation of X when α is an integer. Consider a system formed by α independent components following the *odd exponential-G class* (Bourguignon et al. 2014) given by

$$H(x;\lambda,\boldsymbol{\xi}) = 1 - \mathrm{e}^{-\lambda \frac{G(x;\boldsymbol{\xi})}{\overline{G}(x;\boldsymbol{\xi})}}.$$

Suppose the system fails if all α components fail and let X denote the lifetime of the entire system. Then, the cdf of X is $F(x;\alpha,\lambda,\boldsymbol{\xi}) = H(x;\lambda,\boldsymbol{\xi})^{\alpha}$, which is identical to (1).

Table 1 Some OGE models

SN	α	λ	$G(x;\xi)$	Reduced distribution
1	1	-	$G(x;\xi)$	Odd exponential-G family (Bourguignon et al. 2014)
2	-	-	$\frac{x}{1+x}$	Generalized exponential distribution (Gupta and Kundu 1999)
3	1	-	$\frac{x^c}{1+x^c}$	Exponentiated Weibull distribution (Mudholkar and Srivastava 1993)
4	-	-	$1-\mathrm{e}^{-ax}$	Generalized Gompertz distribution (El-Gohary et al. 2013)

In Table 1, we provide some members of the OGE family.
The hrf of X is given by

$$h(x) = h(x;\alpha,\lambda,\boldsymbol{\xi}) = \frac{\lambda\,\alpha\,g(x;\boldsymbol{\xi})\,e^{-\lambda\frac{G(x;\boldsymbol{\xi})}{\overline{G}(x;\boldsymbol{\xi})}}}{\overline{G}(x;\boldsymbol{\xi})^2\left\{1-\left(1-e^{-\lambda\frac{G(x;\boldsymbol{\xi})}{\overline{G}(x;\boldsymbol{\xi})}}\right)^{\alpha}\right\}}\left(1-e^{-\frac{\lambda\,G(x;\boldsymbol{\xi})}{\overline{G}(x;\boldsymbol{\xi})}}\right)^{\alpha-1}. \tag{3}$$

Theorem 1 provides some relations of the OGE family with other distributions.

Theorem 1. *Let* $X \sim OGE(\alpha,\lambda,\boldsymbol{\xi})$.
(a) If $Y = G(X;\boldsymbol{\xi})$, *then* $F_Y(y) = \left(1-e^{-\lambda\frac{y}{1-y}}\right)^{\alpha}$, $0 < y < 1$, *and*
(b) If $Y = \dfrac{G(X;\boldsymbol{\xi})}{\overline{G}(X;\boldsymbol{\xi})}$, *then* $Y \sim GE(\alpha,\lambda)$.

The OGE family is easily simulated by inverting (1) as follows: if U has a uniform $U(0,1)$ distribution, then

$$x = Q_G\left[\frac{-\log\left(1-u^{\frac{1}{\alpha}}\right)}{\lambda-\log\left(1-u^{\frac{1}{\alpha}}\right)}\right] \tag{4}$$

has the density function (2), where $Q_G(\cdot) = G^{-1}(\cdot)$ is the baseline qf.

3 Special OGE distributions

Lifetime distributions play a fundamental role in survival analysis, biomedical science, engineering and social sciences. Typically, lifetime refers to human life length, the life span of a device before it fails or the survival time of a patient with serious disease from the diagnosis to the death. Here, we present three special OGE distributions that can be useful in applied survival analysis.

3.1 The OGE-Weibull (OGE-W) distribution

The OGE-W distribution is defined from (2) by taking $G(x;\boldsymbol{\xi}) = 1-e^{-\theta\,x^{\beta}}$ and $g(x;\boldsymbol{\xi}) = \beta\,\theta\,x^{\beta-1}\,e^{-\theta\,x^{\beta}}$ to be the cdf and pdf of the Weibull distribution with positive parameters β and θ, respectively, and $\boldsymbol{\xi} = (\beta,\theta)$.

The cdf and pdf of the OGE-W distribution are given by (for $x > 0$)

$$F(x;\alpha,\lambda,\beta,\theta) = \left(1-e^{-\lambda(e^{\theta\,x^{\beta}}-1)}\right)^{\alpha}$$

and

$$f(x;\alpha,\lambda,\beta,\theta) = \alpha\lambda\theta\beta\,x^{\beta-1}\,e^{\theta\,x^{\beta}}\,e^{-\lambda(e^{\theta\,x^{\beta}}-1)}\left(1-e^{-\lambda\left(e^{\theta\,x^{\beta}}-1\right)}\right)^{\alpha-1}, \tag{5}$$

respectively, where $\alpha > 0$ and $\beta > 0$ are shape parameters and $\lambda > 0$ and $\theta > 0$ are scale parameters.

The applications of the OGE-W and EW distributions can be directed to model extreme value observations in floods, software reliability, insurance data, tree diameters, carbon fibrous composites, firmware system failure, reliability prediction and fracture toughness, among others.

3.2 The OGE-Fréchet (OGE-Fr) distribution

The OGE-Fr distribution is defined from (2) by taking $G(x;\boldsymbol{\xi}) = \mathrm{e}^{-(b/x)^a}$ and $g(x;\boldsymbol{\xi}) = a\,b^a\,x^{-(a+1)}\,\mathrm{e}^{-(b/x)^a}$ to be the cdf and pdf of the Fréchet distribution with parameters a and b, respectively, and $\boldsymbol{\xi} = (a,b)$.

The cdf and pdf of the OGE-Fr distribution are given by (for $x > 0$)

$$F(x;\alpha,\lambda,a,b) = \left\{1 - \mathrm{e}^{-\lambda\left(\mathrm{e}^{(b/x)^a}-1\right)^{-1}}\right\}^{\alpha}$$

and

$$f(x;\alpha,\lambda,a,b) = \alpha\,\lambda\,a\,b^a\,x^{-(a+1)}\,\mathrm{e}^{-(b/x)^a}\,\mathrm{e}^{-\lambda\left(\mathrm{e}^{(b/x)^a}-1\right)^{-1}}\left\{1-\mathrm{e}^{-(b/x)^a}\right\}^{-2}$$
$$\times\left\{1 - \mathrm{e}^{-\lambda\left[\mathrm{e}^{(b/x)^a}-1\right]^{-1}}\right\}^{\alpha-1}, \qquad (6)$$

respectively, where $\alpha > 0$ and $a > 0$ are shape parameters and $\lambda > 0$ and $b > 0$ are scale parameters.

The OGE-Fr distribution has applications ranging from accelerated life testing, rainfall and floods, queues in supermarkets, sea currents, wind speeds, and track race records, among others.

3.3 The OGE-Normal (OGE-N) distribution

The OGE-N distribution is defined from (2) by taking $G(x;\boldsymbol{\xi}) = \Phi\left[(x-\mu)/\sigma\right]$ and $g(x;\boldsymbol{\xi}) = \sigma^{-1}\phi\left[(x-\mu)/\sigma\right]$ to be the cdf and pdf of the normal distribution with parameters μ and σ, respectively, and $\boldsymbol{\xi} = (\mu,\sigma)$.

The cdf and pdf of the OGE-N distribution are given by (for $x \in \Re$)

$$F(x;\alpha,\lambda,\mu,\sigma) = \left[1 - \mathrm{e}^{-\lambda\left\{\frac{\Phi[(x-\mu)/\sigma]}{1-\Phi[(x-\mu)/\sigma]}\right\}}\right]^{\alpha}$$

and

$$f(x;\alpha,\lambda,\mu,\sigma) = \frac{\alpha\,\lambda}{\sigma}\,\frac{\phi\left[(x-\mu)/\sigma\right]}{\left\{1-\Phi\left[(x-\mu)/\sigma\right]\right\}^2}\,\mathrm{e}^{-\lambda\left\{\frac{\Phi[(x-\mu)/\sigma]}{1-\Phi[(x-\mu)/\sigma]}\right\}}$$
$$\times\left[1 - \mathrm{e}^{-\lambda\left\{\frac{\Phi[(x-\mu)/\sigma]}{1-\Phi[(x-\mu)/\sigma]}\right\}}\right]^{\alpha-1}, \qquad (7)$$

respectively, where $\alpha > 0$ is a shape parameter, $\mu \in \Re$ is a location parameter, and $\lambda > 0$ and $\sigma > 0$ are scale parameters.

In Figures 1, 2, 3, 4 and 5, we display some plots of the pdf and hrf of the OGE-W, OGE-Fr and OGE-N distributions for selected parameter values. Figures 1, 2 and 3 indicate that the OGE family generates distributions with various shapes such as symmetrical, left-skewed, right-skewed and reversed-J. Further, Figures 4 and 5 reveal that this family can produce hazard rate shapes such as increasing, decreasing, J, reversed-J, bathtub and upside-down bathtub. This fact implies that the OGE family can be very useful to fit data sets under various shapes.

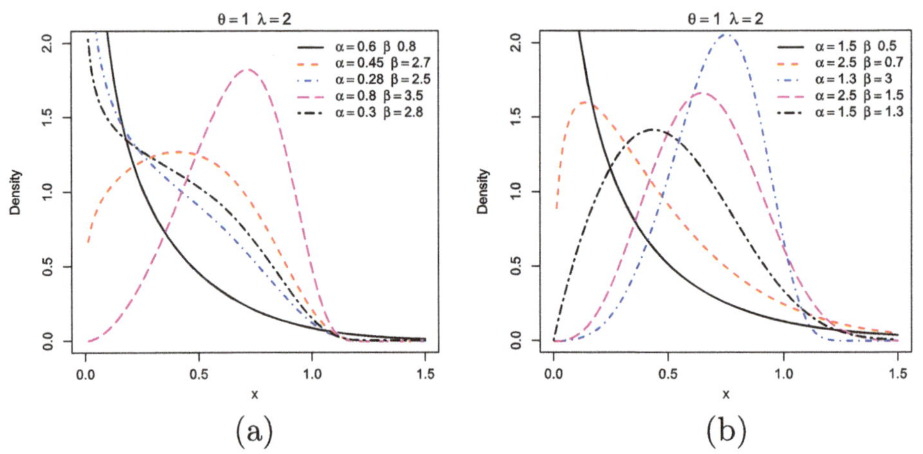

Figure 1 Density function plots (a)-(b) of the OGE-W distribution.

4 Asymptotics and shapes

Proposition 1. *The asymptotics of equations (1), (2) and (3) as $G(x) \to 0$ are given by*

$$F(x) \sim [\lambda G(x)]^{\alpha} \qquad \text{as} \quad G(x) \to 0,$$
$$f(x) \sim \alpha\, \lambda^{\alpha}\, g(x)\, G(x)^{\alpha-1} \qquad \text{as} \quad G(x) \to 0,$$
$$h(x) \sim \alpha\, \lambda^{\alpha}\, g(x)\, G(x)^{\alpha-1} \qquad \text{as} \quad G(x) \to 0.$$

Proposition 2. *The asymptotics of equations (1), (2) and (3) as $x \to \infty$ are given by*

$$1 - F(x) \sim \alpha\, e^{\frac{-\lambda}{\bar{G}(x)}} \qquad \text{as} \quad x \to \infty,$$
$$f(x) \sim \frac{\alpha \lambda g(x)}{\bar{G}(x)^2}\, e^{\frac{-\lambda}{\bar{G}(x)}} \qquad \text{as} \quad x \to \infty,$$
$$h(x) \sim \frac{\lambda g(x)}{\bar{G}(x)^2} \qquad \text{as} \quad x \to \infty.$$

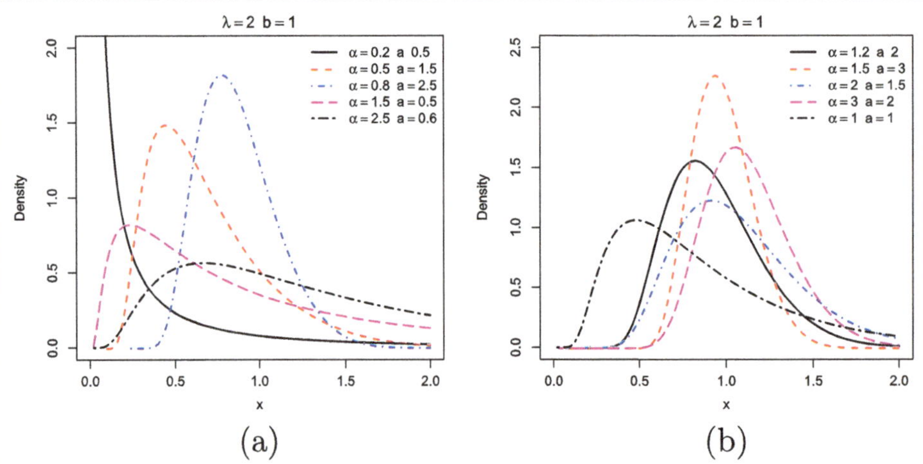

Figure 2 Density function plots (a)-(b) of the OGE-Fr distribution.

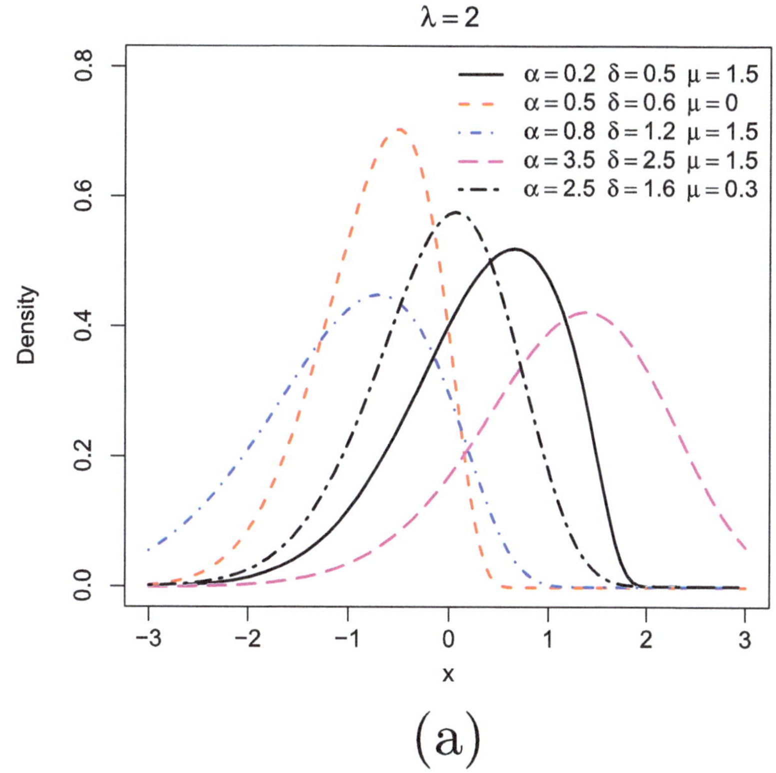

Figure 3 Density function plot (a) of the OGE-N distribution.

The shapes of the density and hazard rate functions can be described analytically. The critical points of the OGE-G density functions are the roots of the equation:

$$\frac{g'(x)}{g(x)} + \frac{g(x)}{\overline{G}(x)} - \frac{\lambda\, g(x)}{\overline{G}(x)^2} + \frac{(\alpha-1)\lambda\, g(x) \mathrm{e}^{-\lambda \frac{G(x)}{\overline{G}(x)}}}{\overline{G}(x)^2 \left(1 - \mathrm{e}^{-\lambda \frac{G(x)}{\overline{G}(x)}}\right)} = 0. \tag{8}$$

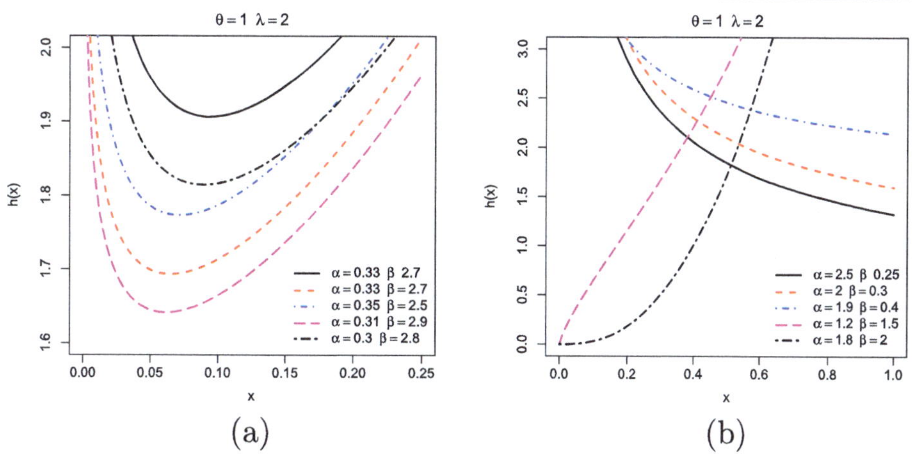

Figure 4 Hazard rate plots (a)-(b) of the OGE-W distribution.

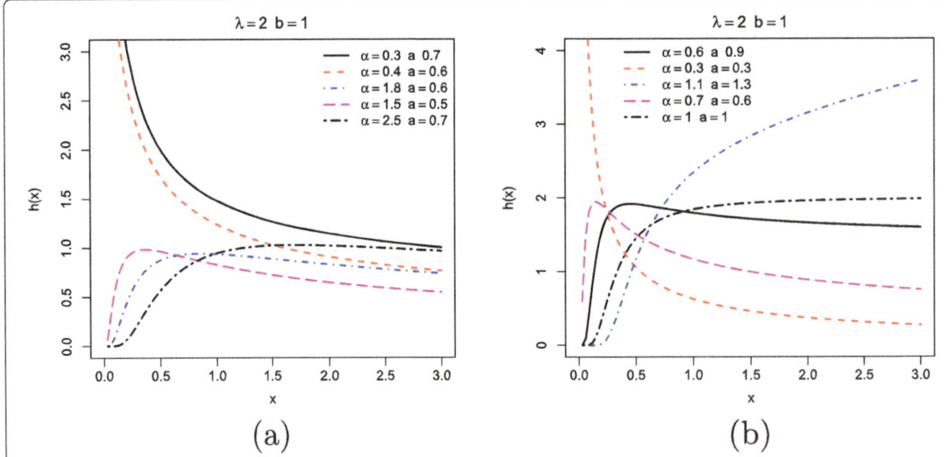

Figure 5 Hazard rate plot (a)-(b) of the OGE-Fr distribution.

There may be more than one root to (8). Let $\lambda(x) = \frac{d^2 \log[f(x)]}{dx^2}$. We have

$$\begin{aligned}\lambda(x) &= \frac{g''(x)g(x) - [g'(x)]^2}{g(x)^2} + \frac{g'(x)\overline{G}(x) + g(x)^2}{\overline{G}(x)^2} - \lambda\,\frac{g'(x)\overline{G}(x) + 2g(x)^2}{\overline{G}(x)^3} \\ &\quad + \frac{(\alpha-1)\lambda g'(x) e^{-\lambda\frac{G(x)}{\overline{G}(x)}}}{\overline{G}(x)^2\left(1 - e^{-\lambda\frac{G(x)}{\overline{G}(x)}}\right)} + \frac{(\alpha-1)\lambda g(x)^2 e^{-\lambda\frac{G(x)}{\overline{G}(x)}}}{\overline{G}(x)^3\left(1 - e^{-\lambda\frac{G(x)}{\overline{G}(x)}}\right)} \\ &\quad + (\alpha-1)\left\{\frac{\lambda g(x) e^{-\lambda\frac{G(x)}{\overline{G}(x)}}}{\overline{G}(x)^2\left(1 - e^{-\lambda\frac{G(x)}{\overline{G}(x)}}\right)}\right\}^2.\end{aligned}$$

If $x = x_0$ is a root of (8) then it corresponds to a local maximum if $\lambda(x) > 0$ for all $x < x_0$ and $\lambda(x) < 0$ for all $x > x_0$. It corresponds to a local minimum if $\lambda(x) < 0$ for all $x < x_0$ and $\lambda(x) > 0$ for all $x > x_0$. It refers to a point of inflexion if either $\lambda(x) > 0$ for all $x \neq x_0$ or $\lambda(x) < 0$ for all $x \neq x_0$.

The critical point of $h(x)$ are obtained from the equation

$$\begin{aligned}&\frac{g'(x)}{g(x)} + \frac{g(x)}{\overline{G}(x)} - \frac{\lambda g(x)}{\overline{G}(x)^2} + \frac{(\alpha-1)\lambda g(x) e^{-\lambda\frac{G(x)}{\overline{G}(x)}}}{\overline{G}(x)^2\left(1 - e^{-\lambda\frac{G(x)}{\overline{G}(x)}}\right)} \\ &\qquad + \frac{\lambda\alpha g(x)\, e^{-\lambda\frac{G(x)}{\overline{G}(x)}}}{\overline{G}(x)^2\left[1 - \left(1 - e^{-\lambda\frac{G(x)}{\overline{G}(x)}}\right)^{\alpha}\right]}\left(1 - e^{-\lambda\frac{G(x)}{\overline{G}(x)}}\right)^{\alpha-1} = 0. \qquad (9)\end{aligned}$$

There may be more than one root to (9). Let $\tau(x) = d^2 \log[h(x)]/dx^2$, then we have

$$\begin{aligned}\tau(x) &= \frac{g''(x)g(x) - [g'(x)]^2}{g(x)^2} + \frac{g'(x)\overline{G}(x) + g(x)^2}{\overline{G}(x)^2} - \lambda\,\frac{g'(x)\overline{G}(x) + 2g(x)^2}{\overline{G}(x)^3} \\ &\quad + \frac{(\alpha-1)\lambda g'(x) e^{-\lambda\frac{G(x)}{\overline{G}(x)}}}{\overline{G}(x)^2\left(1 - e^{-\lambda\frac{G(x)}{\overline{G}(x)}}\right)} + \frac{(\alpha-1)\lambda g(x)^2 e^{-\lambda\frac{G(x)}{\overline{G}(x)}}}{\overline{G}(x)^3\left(1 - e^{-\lambda\frac{G(x)}{\overline{G}(x)}}\right)}\end{aligned}$$

$$+(\alpha-1)\left\{\frac{\lambda g(x)e^{-\lambda\frac{G(x)}{\overline{G}(x)}}}{\overline{G}(x)^2\left(1-e^{-\lambda\frac{G(x)}{\overline{G}(x)}}\right)}\right\}^2$$

$$+\frac{\lambda\,\alpha\,g'(x)\,e^{-\lambda\frac{G(x)}{\overline{G}(x)}}}{\overline{G}(x)^2\left[1-\left(1-e^{-\lambda\frac{G(x)}{\overline{G}(x)}}\right)^{\alpha}\right]}\left(1-e^{-\lambda\frac{G(x)}{\overline{G}(x)}}\right)^{\alpha-1}$$

$$-\frac{\lambda^2\,\alpha\,g(x)^2\,e^{-\lambda\frac{G(x)}{\overline{G}(x)}}}{\overline{G}(x)^4\left[1-\left(1-e^{-\lambda\frac{G(x)}{\overline{G}(x)}}\right)^{\alpha}\right]}\left(1-e^{-\lambda\frac{G(x)}{\overline{G}(x)}}\right)^{\alpha-1}$$

$$+\frac{2\lambda\,\alpha\,g(x)^2\,e^{-\lambda\frac{G(x)}{\overline{G}(x)}}}{\overline{G}(x)^4\left[1-\left(1-e^{-\lambda\frac{G(x)}{\overline{G}(x)}}\right)^{\alpha}\right]}\left(1-e^{-\lambda\frac{G(x)}{\overline{G}(x)}}\right)^{\alpha-1}$$

$$+\frac{\lambda^2\,\alpha(\alpha-1)\,g(x)^2\,e^{-2\lambda\frac{G(x)}{\overline{G}(x)}}}{\overline{G}(x)^4\left[1-\left(1-e^{-\lambda\frac{G(x)}{\overline{G}(x)}}\right)^{\alpha}\right]}\left(1-e^{-\lambda\frac{G(x)}{\overline{G}(x)}}\right)^{\alpha-2}$$

$$+\frac{\lambda^2\,\alpha^2\,g(x)^2\,e^{-2\lambda\frac{G(x)}{\overline{G}(x)}}}{\overline{G}(x)^4\left[1-\left(1-e^{-\lambda\frac{G(x)}{\overline{G}(x)}}\right)^{\alpha}\right]^2}\left(1-e^{-\lambda\frac{G(x)}{\overline{G}(x)}}\right)^{2\alpha-2}.$$

If $x=x_0$ is a root of (9) then it refers to a local maximum if $\tau(x)>0$ for all $x<x_0$ and $\tau(x)<0$ for all $x>x_0$. It corresponds to a local minimum if $\tau(x)<0$ for all $x<x_0$ and $\tau(x)>0$ for all $x>x_0$. It gives an inflexion point if either $\tau(x)>0$ for all $x\neq x_0$ or $\tau(x)<0$ for all $x\neq x_0$.

5 Useful expansions

First, we consider the quantity $M=1-e^{-\lambda\frac{G(x;\xi)}{\overline{G}(x;\xi)}}$. By expanding the exponential function in power series and using the generalized binomial expansion, we obtain

$$M=\sum_{i=1}^{\infty}\frac{(-1)^{i+1}\,\lambda^i}{i!}\left[\frac{G(x;\xi)}{\overline{G}(x;\xi)}\right]^i=\sum_{i=1,j=0}^{\infty}\frac{(-1)^{i+j+1}\,\lambda^i}{i!}\binom{-i}{j}G(x;\xi)^{i+j}.$$

The quantity M can be expressed as

$$M=\sum_{k=1}^{\infty}a_k\,G(x;\xi)^k=G(x;\xi)\sum_{k=0}^{\infty}b_k\,G(x;\xi)^k, \tag{10}$$

where, for $k\geq 0$, $b_k=a_{k+1}$, and, for $k\geq 1$, $a_k=\sum_{(i,j)\in I_k}\frac{(-1)^{i+j+1}\lambda^i}{i!}\binom{-i}{j}$, $I_k=\{(i,j)|i+j=k;i=1,2,\ldots,j=0,1,2,\ldots\}$.

Second, we can expand z^{α} in Taylor series as

$$z^{\alpha}=\sum_{l=0}^{\infty}(\alpha)_l\,(z-1)^l/l!=\sum_{m=0}^{\infty}f_m\,z^m, \tag{11}$$

where $f_m=f_m(\alpha)=\sum_{l=m}^{\infty}\frac{(-1)^{l-m}}{l!}\binom{l}{m}(\alpha)_l$. Here and from now on, we use the notation $(\alpha)_l=\alpha(\alpha-1)\ldots(\alpha-l+1)$ for the descending factorial with $(\alpha)_0=1$.

We use throughout the paper a result of Gradshteyn and Ryzhik (2000, Section 0.314) for a power series raised to a positive integer n (for $n \geq 1$)

$$\left(\sum_{i=0}^{\infty} b_i\, u^i\right)^n = \sum_{i=0}^{\infty} c_{n,i}\, u^i, \tag{12}$$

where the coefficients $c_{n,i}$ (for $n = 1, 2, \ldots$) can be determined from the recurrence equation $c_{n,i} = (i\, b_0)^{-1} \sum_{m=1}^{i} [m(n+1) - i]\, b_m\, c_{n,i-m}$ (for $i \geq 1$), and $c_{n,0} = b_0^n$.

Third, based on Eqs. (10)-(12), the cdf (1) can be expressed as

$$\begin{aligned} F(x;\alpha,\lambda,\boldsymbol{\xi}) &= G(x;\boldsymbol{\xi})^{\alpha} \left[\sum_{k=0}^{\infty} b_k\, G(x;\boldsymbol{\xi})^k\right]^{\alpha} \\ &= G(x;\boldsymbol{\xi})^{\alpha} \sum_{m=0}^{\infty} f_m \left[\sum_{k=0}^{\infty} b_k\, G(x;\boldsymbol{\xi})^k\right]^{m} \\ &= G(x;\boldsymbol{\xi})^{\alpha} \sum_{m=0}^{\infty} f_m \sum_{k=0}^{\infty} c_{m,k}\, G(x;\boldsymbol{\xi})^k, \end{aligned} \tag{13}$$

where $c_{m,k}$ can be obtained from the quantities $b_0, \ldots, b_k$ as in equation (12).

Then, we have

$$F(x) = F(x;\alpha,\lambda,\boldsymbol{\xi}) = \sum_{k=0}^{\infty} d_k\, G(x;\boldsymbol{\xi})^{\alpha+k} = \sum_{k=0}^{\infty} d_k\, H_{\alpha+k}(x), \tag{14}$$

where $d_k = \sum_{m=0}^{\infty} f_m\, c_{m,k}$ (for $k \geq 0$), $H_{\alpha+k}(x) = G(x)^{\alpha+k}$ denotes the cdf of the exp-G distribution with power parameter $\alpha + k$.

Finally, the density function of X can be expressed in the mixture form

$$f(x) = f(x;\alpha,\lambda,\boldsymbol{\xi}) = \sum_{k=0}^{\infty} d_k\, h_{\alpha+k}(x), \tag{15}$$

where $h_{\alpha+k}(x) = (\alpha + k)\, G(x;\boldsymbol{\xi})^{\alpha+k-1}\, g(x;\boldsymbol{\xi})$ is the exp-G density function with power parameter $\alpha + k$. Equation (15) reveals that the OGE density function is a linear combination of exp-G densities. Thus, some mathematical properties of the new model such as the ordinary and incomplete moments, and moment generating function (mgf) can be derived from those properties of the exp-G distribution. Some exp-G structural properties are studied by Mudholkar and Srivastava (1993), Mudholkar et al. (1995), Mudholkar and Hutson (1996), Gupta et al. (1998), Gupta and Kundu (2001a; 2001b), Nadarajah and Kotz (2006) and Nadarajah (2011).

6 Mathematical properties

6.1 Moments

The need for necessity and the importance of moments in Statistics especially in applications is obvious. Some of the most important features and characteristics of a distribution can be studied through moments. Let Y_k be a random variable having the exp-G pdf $h_{\alpha+k}(x)$ with power parameter $\alpha + k$.

A first formula for the nth moment of X follows from (15) as

$$\mu'_n = E(X^n) = \sum_{k=0}^{\infty} d_k\, E(Y_k^n). \tag{16}$$

Expressions for moments of several exp-G distributions are given by Nadarajah and Kotz (2006), which can be used to obtain μ'_n.

We provide two applications of (16) for the OGE-W and OGE-Fr models discussed in Section 3. First, the moments of the OGE-W model can be derived from closed-forms moments of the EW distribution given by Choudhury (2005). We obtain

$$\mu'_n = \theta^{-n/\beta}\,\Gamma(n/\beta+1)\sum_{k=0}^{\infty}(\alpha+k)\,d_k\left[1+\sum_{j=0}^{\infty}A_{j,k}\,(j+1)^{-n/\beta-1}\right], \quad (17)$$

where the quantities $A_{j,k}$ (for $j,k \geq 0$) are given by

$$A_{j,k} = \frac{(-1)^j\,(\alpha+k)}{j!\,(\alpha+k)_{j+1}}.$$

The double infinite series on the right-hand side converges for all parameter values.

Second, the moments of the OGE-Fr model (for $n < a$) follow immediately from the moments of the Fréchet distribution as

$$\mu'_n = b^n\,\Gamma(1-n/a)\sum_{k=0}^{\infty} d_k\,(\alpha+k)^{n/a}.$$

A second alternative formula for μ'_n is obtained from (16) in terms of the baseline qf as

$$\mu'_n = \sum_{k=0}^{\infty}(\alpha+k)\,d_k\,\tau(n,\alpha+k-1), \quad (18)$$

where (for $a > 0$) $\tau(n,a) = \int_0^1 Q_G(u)^n\,u^a du$. Cordeiro and Nadarajah (2011) obtained $\tau(n,a)$ for some well-known distribution such as the normal, beta, gamma and Weibull distributions, which can be applied to obtain raw moments of the corresponding OGE distributions.

The central moments (μ_s) and cumulants (κ_s) of X can be determined from the ordinary moments using the recurrence equations

$$\mu_s = \sum_{k=0}^{s}(-1)^k\binom{s}{k}\mu_1'^{s}\,\mu'_{s-k}, \qquad \kappa_s = \mu'_s - \sum_{k=1}^{s-1}\binom{s-1}{k-1}\kappa_k\,\mu'_{s-k},$$

where $\kappa_1 = \mu'_1$. The skewness $\rho_1 = \kappa_3/\kappa_2^{3/2}$ and kurtosis $\rho_2 = \kappa_4/\kappa_2^2$ can be calculated from the third and fourth standardized cumulants.

For empirical purposes, the shapes of many distributions can be usefully described by what we call the first incomplete moment which plays an important role for measuring inequality, for example, income quantiles and Lorenz and Bonferroni curves.

The nth incomplete moment of X can be expressed as

$$m_n(y) = \sum_{k=0}^{\infty}(\alpha+k)\,d_k\int_0^{G(y)} Q_G(u)^n\,u^{\alpha+k-1}du. \quad (19)$$

The last integral can be computed for most G distributions.

6.2 Generating function

The generating function provides the basis of an alternative route to analytical results compared with working directly with the pdf and cdf and it is widely used in the characterization of the distributions, and the application of goodness-of-fit tests. Let $M(t) = E(e^{tX})$ be the mgf of X. Then, we can write from (15)

$$M(t) = \sum_{k=0}^{\infty} d_k\, M_k(t), \tag{20}$$

where $M_k(t)$ is the mgf of Y_k. Hence, $M(t)$ can be determined from the exp-G generating function.

Now, we provide two applications of (20) for the OGE-W and OGE-Fr models. Nadarajah et al. (2015) derived a formula for the mgf of Y_k which holds for the OGE-W model depending on the complex parameter Wright generalized hypergeometric function with p numerator and q denominator parameters (Kilbas et al. 2006, Equation (1.9)) given by

$${}_p\Psi_q\left[\begin{matrix}(\alpha_1, A_1), \ldots, (\alpha_p, A_p)\\ (\beta_1, B_1), \ldots, (\beta_q, B_q)\end{matrix}; z\right] = \sum_{n=0}^{\infty} \frac{\prod_{j=1}^{p} \Gamma\left(\alpha_j + A_j n\right)}{\prod_{j=1}^{q} \Gamma\left(\beta_j + B_j n\right)} \frac{z^n}{n!}$$

for $z \in \mathbb{C}$, where $\alpha_j,\ \beta_k \in \mathbb{C},\ A_j,\ B_k \neq 0,\ j = 1, \ldots, p,\ k = 1, \ldots, q$, which converges for $1 + \sum_{j=1}^{q} B_j - \sum_{j=1}^{p} A_j > 0$.

Nadarajah et al. (2015) demonstrated that the mgf of Y_k can be expressed as

$$M_k(t) = (\alpha + k) \sum_{j=0}^{\infty} \frac{(-1)^j}{(j+1)} \binom{\alpha + k - 1}{j}\, {}_1\Psi_0\left[\begin{matrix}\left(1, \beta^{-1}\right)\\ -\end{matrix}; t\,[(j+1)^{1/\beta}\,\theta^{1/\beta}]^{-1}\right].$$

Combining (20) and the last equation gives the mgf of the OGE-W distribution.

The mgf of the Fréchet distribution can be obtained by

$$M(t) = a\, b^a \int_0^{\infty} x^{-a-1}\, e^{-b^a x^{-a} + tx}\, dx. \tag{21}$$

The calculations of the integral in (21) involve the generalized hypergeometric function defined in equation (2.3.1.14) (Prudnikov et al. 1986, p. 322).

For $a > 0$ and $s > 0$, if $b = p/q$ (with $p \geq 1$ and $q \geq 1$ are co-primes integers), we can write

$$\begin{aligned} J(a, \gamma, b, s) &= \int_0^{\infty} x^{a-1}\, e^{-\gamma\, x^{-b} - sx}\, dx \\ &= \sum_{j=0}^{q-1} \frac{(-\gamma)^j}{j!}\, \Gamma\left(1 + \frac{p(1+j)}{q}\right) s^{-\left(1 + \frac{p(1+j)}{q}\right)} \\ &\quad \times {}_1F_{p+q}\left(1; \Delta\left(p, 1 + \frac{p(1+j)}{q}\right), \Delta(q, 1+j); z\right) \\ &\quad + \sum_{j=0}^{p-1} \frac{q\,(-s)^j}{p\, j!}\, \Gamma\left(1 + \frac{q(1+j)}{p}\right) \gamma^{-\left(1 + \frac{q(1+j)}{p}\right)} \\ &\quad \times {}_1F_{p+q}\left(1, \Delta\left(q, 1 + \frac{q(1+j)}{p}\right), \Delta(p, 1+j); z\right), \end{aligned} \tag{22}$$

where $z = (-1)^{p+q}\, s^p\, a^q / (p^p\, q^q)$, $\Delta(k, a)$ denote the sequence

$$\Delta(k, a) = (a/k, (a+1)/k, \ldots, (a+k-1)/k),$$

${}_mF_n$ denotes the generalized hypergeometric function defined by

$$ {}_mF_n\ (\alpha_1, \alpha_2 \ldots, \alpha_m; \beta_1, \beta_2 \ldots, \beta_n; z) = \sum_{k=0}^{\infty} \frac{(\alpha_1)_k, (\alpha_2)_k \ldots, (\alpha_m)_k}{(\beta_1)_k, (\beta_2)_k \ldots, (\beta_n)_k} \frac{z^k}{k!} $$

and $(c)_k = c(c+1)\ldots(c+k-1)$ denotes the ascending factorial.

Numerical routines for computing the generalized hypergeometric function are available in most mathematical packages, e.g., Maple, Mathematica and Matlab. Nadarajah and Kotz (2005), and Nadarajah (2007) used also this result to obtain the properties of the distribution of the difference between two independent Gumbel variates, and to Iacbellis and Fiorentino (2000), and Fiorentino et al. (2006)'s model for peak streamflow.

By using (22), the mgf of the Fréchet distribution follows as

$$ M(t) = a\, b^a\, J(-a, b^a, b, -t). \tag{23} $$

Using (15), the mgf of the OGE-Fr model can be expressed as

$$ \begin{aligned} M(t) &= \sum_{k=0}^{\infty} a\, b^a\, d_k \int_0^{\infty} h_{(\alpha+k)}(z)\, e^{tz}\, dz \\ &= \sum_{k=0}^{\infty} a(\alpha+k)\, b^a\, d_k \int_0^{\infty} z^{-a-1}\, e^{-(b/z)^a}\, e^{-(b/z)^{a(\alpha+k-1)}}\, e^{tz}\, g(z)\, dz \end{aligned} $$

and then

$$ M(t) = \sum_{k=0}^{\infty} a\,(\alpha+k)\, b^a\, d_k \int_0^{\infty} z^{-a-1}\, e^{-b^{a(\alpha+k)}\, z^{-a(\alpha+k)}+tz}\, dz. $$

By using (23), $M(t)$ reduces to

$$ M(t) = \sum_{l=0}^{\infty} a(\alpha+k)\, b^a\, d_k\, J\left(-a, b^{a(\alpha+k)}, b(\alpha+k), -t\right). $$

A second alternative formula for $M(t)$ can be derived from (15) as

$$ M(t) = \sum_{k=0}^{\infty} (\alpha+k)\, d_k\, \rho(t, \alpha+k-1), \tag{24} $$

where (for $a > 0$) $\rho(t,a) = \int_{-\infty}^{\infty} e^{tx}\, G(x)^a\, g(x) dx = \int_0^1 \exp[t\, Q_G(u)]\, u^a\, du$.

We can obtain the mgf's of several OGE distributions directly from equation (24). Equations (20) and (24) are the main results of this section.

6.3 Mean deviations

The mean deviations about the mean ($\delta_1 = E\left(|X - \mu_1'|\right)$) and about the median ($\delta_2 = E(|X - M|)$) of X can be expressed as

$$ \delta_1(X) = 2\mu_1' F\left(\mu_1'\right) - 2m_1\left(\mu_1'\right) \quad \text{and} \quad \delta_2(X) = \mu_1' - 2m_1(M), \tag{25} $$

respectively, where $\mu_1' = E(X)$, $M = Median(X) = Q(0.5)$ is the median which comes from (4), $F\left(\mu_1'\right)$ is easily calculated from the cdf (1) and $m_1(z)$ is the first incomplete moment determined from (19) with $n = 1$.

Now, we provide two alternative ways to compute δ_1 and δ_2. A general formula for $m_1(z)$ can be derived from equation (15) as

$$m_1(z)=\sum_{k=0}^{\infty} d_k J_k(z), \tag{26}$$

where $J_k(z)=\int_{-\infty}^{z} x\, h_{\alpha+k}(x)dx$ is the first incomplete moment of Y_k. Hence, the mean deviations in (25) depend only on quantity $J_k(z)$.

We give two applications of $J_k(z)$ for the the OGE-W and OGE-Fr distributions. For the first model, the EW pdf (for $x>0$) with power parameter $\alpha+k$ is given by

$$h_{\alpha+k}(x)=\lambda\,(\alpha+k)\,\beta^{\lambda}\,x^{\lambda-1}\exp\left\{-(\beta x)^{\lambda}\right\}\left[1-\exp\left\{-(\beta x)^{\lambda}\right\}\right]^{\alpha+k-1}$$

and then

$$\begin{aligned} J_k(z) &= \lambda\,(\alpha+k)\,\beta^{\lambda}\int_0^z x^{\lambda}\exp\left\{-(\beta x)^{\lambda}\right\}\left[1-\exp\left\{-(\beta x)^{\lambda}\right\}\right]^{\alpha+k-1}dx \\ &= \lambda\,(\alpha+k)\,\beta^{\lambda}\sum_{r=0}^{\infty}(-1)^r\binom{\alpha+k-1}{r}\int_0^z x^{\lambda}\exp\left\{-(r+1)(\beta x)^{\lambda}\right\}dx. \end{aligned}$$

The last integral is equal to the incomplete gamma function and then the mean deviation for the OGE-W distribution is given by

$$m_1(z)=\beta^{-1}\sum_{k,r=0}^{\infty}\frac{(-1)^r\,(\alpha+k)\,d_k}{(r+1)^{1+\lambda^{-1}}}\binom{\alpha+k-1}{r}\gamma(1+\lambda^{-1},(r+1)(\beta z)^{\lambda}),$$

where $\gamma(a,z)=\int_0^z w^{a-1}\,\mathrm{e}^{-\mathrm{w}}\mathrm{dw}$ is the incomplete gamma function.

The first incomplete moment of the Fréchet distribution is given by

$$\mu'_{(1,Y)}(z)=b\,\Gamma\left(1-\tfrac{1}{a},\left(\frac{b}{x}\right)^{a}\right).$$

So, the first incomplete moment of the OGE-Fr model can be derived from the last equation and (15) as

$$m_1(z)=b\sum_{k=0}^{\infty}(\alpha+k)^{1/a}\,d_k\,\Gamma\left(1-\tfrac{1}{a},\left[\tfrac{b\,(\alpha+k)^{1/a}}{x}\right]^{a}\right),$$

where $\Gamma(a,z)=\int_z^{\infty} w^{a-1}\,\mathrm{e}^{-\mathrm{w}}\mathrm{dw}$ is the complementary incomplete gamma function.

A second general formula for $m_1(z)$ can be derived by setting $u=G(x)$ in (15)

$$m_1(z)=\sum_{k=0}^{\infty}(\alpha+k)\,d_k\,T_k(z), \tag{27}$$

where $T_k(z)=\int_0^{G(z)} Q_G(u)\,u^{\alpha+k-1}du$ depends on the baseline qf.

Applications of equations (26) and (27) can be directed to the Bonferroni and Lorenz curves defined for a given probability π by $B(\pi)=m_1(q)/(\pi\mu'_1)$ and $L(\pi)=m_1(q)/\mu'_1$, respectively, where $\mu'_1=E(X)$ and $q=Q(\pi)$ is the qf of X at π.

6.4 Entropies

An entropy is a measure of variation or uncertainty of a random variable X. Two popular entropy measures are the Rényi and Shannon entropies (Rényi 1961; Shannon 1948). The Rényi entropy of a random variable with pdf $f(x)$ is defined by

$$I_R(\gamma)=\frac{1}{1-\gamma}\log\left(\int_0^{\infty} f^{\gamma}(x)dx\right),$$

for $\gamma > 0$ and $\gamma \neq 1$. The Shannon entropy of a random variable X is defined by $\mathbb{E}\{-\log[f(X)]\}$. It is the special case of the Rényi entropy when $\gamma \uparrow 1$. Direct calculation yields

$$\begin{aligned}\mathbb{E}\{-\log[f(X)]\} = &-\log(\alpha\lambda) - \mathbb{E}\{\log[g(X;\xi)]\} + 2\mathbb{E}\{\log[\bar{G}(X;\xi)]\}\\ &+\lambda\,\mathbb{E}\left\{\frac{G(X)}{\bar{G}(X)}\right\} + (1-\alpha)\,\mathbb{E}\left\{\log\left[1-\mathrm{e}^{\frac{-\lambda G(X)}{\bar{G}(X)}}\right]\right\}.\end{aligned}$$

First, we define

$$A(a_1,a_2,a_3;\lambda) = \int_0^1 \frac{x^{a_1}\,\mathrm{e}^{-\lambda\left(\frac{x}{1-x}\right)}\left[1-\mathrm{e}^{-\lambda\left(\frac{x}{1-x}\right)}\right]^{a_3}}{(1-x)^{a_2}}dx.$$

Using Taylor and generalized binomial expansions, and after some algebraic manipulations, we obtain

$$A(a_1,a_2,a_3;\lambda) = \sum_{i,j,k=0}^{\infty} \frac{(-1)^{i+j+k}\,[\lambda(1+i)]^j}{j!\,(a_1+j+k+1)}\binom{a_3}{i}\binom{-a_2-j}{k}.$$

So, we can write:

Proposition 3. *Let X be a random variable with pdf* (2). *Then,*

$$\mathbb{E}\{\log[\bar{G}(X)]\} = \alpha\lambda\,\frac{\partial}{\partial t}A(0,2-t,\alpha-1;\lambda)\Big|_{t=0},$$

$$\mathbb{E}\left\{\frac{G(X)}{\bar{G}(X)}\right\} = \alpha\lambda\,A(1,3,\alpha-1;\lambda),$$

$$\mathbb{E}\left\{\log\left[1-\mathrm{e}^{\frac{-\lambda G(X)}{\bar{G}(X)}}\right]\right\} = \alpha\lambda\,\frac{\partial}{\partial t}A(0,2,t+\alpha-1;\lambda)\Big|_{t=0}.$$

The simplest formula for the entropy of X can be reduced to

$$\begin{aligned}\mathbb{E}\{-\log[f(X)]\} = &-\log(\alpha\lambda) - \mathbb{E}\{\log[g(X;\xi)]\}\\ &+2\alpha\lambda\,\frac{\partial}{\partial t}A(0,2-t,\alpha-1;\lambda)\Big|_{t=0}\\ &+\alpha\lambda^2 A(1,3,\alpha-1;\lambda)\\ &+(1-\alpha)\alpha\lambda\,\frac{\partial}{\partial t}A(0,2,t+\alpha-1;\lambda)\Big|_{t=0}.\end{aligned}$$

After some algebraic developments, we obtain an expression for $I_R(\gamma)$

$$I_R(\gamma) = \frac{\gamma}{1-\gamma}\log(\alpha\lambda) + \frac{1}{1-\gamma}\log\left\{\sum_{j,k=0}^{\infty} w_{j,k}\,\mathbb{E}_{Y_{j,k}}[g^{\gamma-1}[G^{-1}(Y)]]\right\}, \tag{28}$$

where $Y_{j,k}$ has a beta distribution with parameters $j+k+1$ and one, and

$$w_{j,k} = \sum_{i=0}^{\infty} \frac{(-1)^{i+j+k}\,[\lambda(\gamma+i)]^j}{j!\,(j+k+1)}\binom{\gamma(\alpha-1)}{i}\binom{-2\gamma-j}{k}.$$

7 Quantile power series

Power series methods are at the heart of many aspects of applied mathematics and statistics. Here, we derive a power series for the qf $x = Q(u) = F^{-1}(u)$ of X by expanding (4). If the baseline qf $Q_G(u)$ does not have a closed-form expression, it can usually be expressed in terms of a power series as

$$Q_G(u) = \sum_{i=0}^{\infty} a_i\, u^i, \tag{29}$$

where the coefficients a_i's are suitably chosen real numbers which depend on the parameters of the G distribution. For several important distributions, such as the normal, Student t, gamma and beta distributions, $Q_G(u)$ does not have explicit expressions but it can be expanded as in equation (29). As a simple example, for the normal $N(0, 1)$ distribution, $a_i = 0$ for $i = 0, 2, 4, \ldots$ and $a_1 = 1$, $a_3 = 1/6$, $a_5 = 7/120$ and $a_7 = 127/7560, \ldots$

Next, we derive an expansion for the argument of $Q_G(\cdot)$ in equation (4)

$$A = \frac{-\log(1 - u^{\frac{1}{\alpha}})}{\lambda - \log(1 - u^{\frac{1}{\alpha}})}.$$

For $|u| < 1$, we have $-\log(1 - u) = \sum_{i=0}^{\infty} u^{i+1}/(i + 1)$. Then, using the generalized binomial expansion, we obtain

$$-\log(1 - u^{\frac{1}{\alpha}}) = \sum_{k=0}^{\infty} \alpha_k^*\, u^k,$$

where (for $k \geq 0$)

$$\alpha_k^* = \sum_{i=0}^{\infty} \sum_{j=k}^{\infty} \frac{(-1)^{j+k}}{(i+1)} \binom{\frac{i+1}{\alpha}}{j} \binom{j}{k}.$$

Using the generalized binomial expansion again, we can write

$$A = \frac{\sum_{k=0}^{\infty} \alpha_k^*\, u^k}{\sum_{k=0}^{\infty} b_k^*\, u^k} = \sum_{k=0}^{\infty} \delta_k\, u^k,$$

where $b_0^* = \lambda + \alpha_0^*$, $b_k^* = \alpha_k^*$ for $k \geq 1$ and the coefficient δ_k (for $k \geq 0$) can be determined from the recurrence equation (for $k \geq 1$ and $\delta_0 = \alpha_0^*/b_0^*$)

$$\delta_k = \frac{1}{b_0^*} \left(\alpha_k^* - \frac{1}{b_0^*} \sum_{r=1}^{k} b_r^*\, \delta_{k-r} \right).$$

Then, the qf of X can be expressed from (4) as

$$Q(u) = Q_G \left(\sum_{k=0}^{\infty} \delta_k\, u^k \right). \tag{30}$$

Combining (29) and (30), we obtain

$$Q(u) = \sum_{i=0}^{\infty} a_i \left(\sum_{m=0}^{\infty} \delta_m\, u^m \right)^i,$$

and then

$$Q(u) = \sum_{m=0}^{\infty} e_m\, u^m, \tag{31}$$

where $e_m = \sum_{i=0}^{\infty} a_i\, d_{i,m}$ and, for $i \geq 0$, $d_{i,0} = \delta_0^i$ and, for $m > 1$, $d_{i,m} = (m\,\delta_0)^{-1} \sum_{n=1}^{m} [\,n(i+1) - m]\ \delta_n\, d_{i,m-n}$.

Let $W(\cdot)$ be any integrable function in the positive real line. We can write

$$\int_{-\infty}^{\infty} W(x) f(x) dx = \int_{0}^{1} W\left(\sum_{m=0}^{\infty} e_m \, u^m\right) du. \tag{32}$$

Equations (31) and (32) are the main results of this section since from them we can obtain various OGE mathematical quantities. In fact, various of these properties can follow by using the second integral for special $W(\cdot)$ functions, which are usually more simple than if they are based on the first integral. Established algebraic expansions to determine mathematical quantities of the OGE distributions based on these equations can be more efficient then using numerical integration of the pdf (2), which can be prone to rounding off errors among others. For the great majority of these quantities, we can adopt twenty terms in the power series (31).

The effects of the shape parameter α on the skewness and kurtosis of X can be studied using quantile measures. The shortcomings of the classical kurtosis measure are well-known. The Bowley skewness is given by

$$B = \frac{Q(3/4) + Q(1/4) - 2\,Q(2/4)}{Q(3/4) - Q(1/4)}.$$

Since only the middle two quartiles are considered and the outer two quartiles are ignored, this adds robustness to the measure. The Moors kurtosis is based on octiles

$$M = \frac{Q(3/8) - Q(1/8) + Q(7/8) - Q(5/8)}{Q(6/8) - Q(2/8)}.$$

These measures are less sensitive to outliers and they exist even for distributions without moments. In Figures 6 and 7, we plot the measures B and M for the OGE-W and OGE-Fr distributions. These plots indicate how both measures B and M vary on the parameter α.

8 Order statistics

Order statistics make their appearance in many areas of statistical theory and practice. Suppose $X_1, \ldots, X_n$ is a random sample from the OGE-G distribution. Let $X_{i:n}$ denote the ith order statistic. Using (14) and (15), the pdf of $X_{i:n}$ can be expressed as

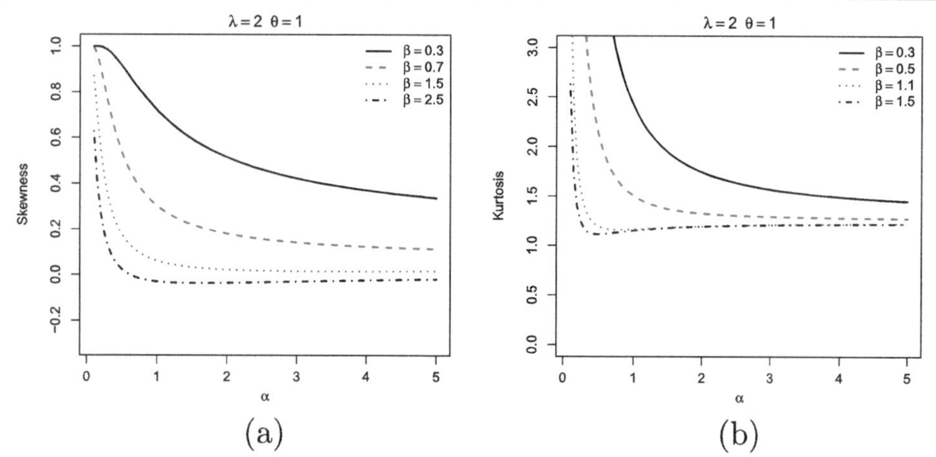

Figure 6 Skewness (a) and kurtosis (b) plots for the OGE-W distribution.

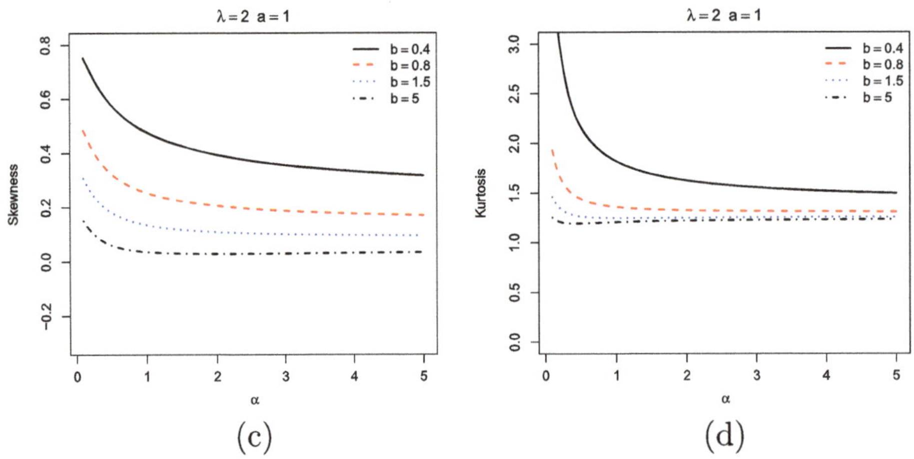

Figure 7 Skewness (c) and kurtosis (d) plots for the OGE-Fr distribution.

$$\begin{aligned} f_{i:n}(x) &= K f(x)\, F(x)^{i-1}\, \{1-F(x)\}^{n-i} \\ &= K \sum_{j=0}^{n-i} (-1)^j \binom{n-i}{j} f(x)\, F(x)^{j+i-1} \\ &= K \sum_{j=0}^{n-i} (-1)^j \binom{n-i}{j} \left[\sum_{r=0}^{\infty} (\alpha+r) d_r\, G(x)^{\alpha+r-1} g(x)\right] \\ &\times \left[\sum_{k=0}^{\infty} d_k\, G(x)^{\alpha+k}\right]^{j+i-1}, \end{aligned}$$

where $K = n!/[\,(i-1)!\,(n-i)!\,]$ and then

$$\left(\sum_{k=0}^{\infty} d_k\, G(x)^{\alpha+k}\right)^{j+i-1} = \sum_{k=0}^{\infty} \varphi_{j+i-1,k}\, G(x)^{\alpha(j+i-1)+k},$$

where $\varphi_{j+i-1,0} = d_0^{j+i-1}$ and (for $k \geq 1$)

$$\varphi_{j+i-1,k} = (k\, d_0)^{-1} \sum_{m=1}^{k} \left[m(j+i)-k\right] d_m\, \varphi_{j+i-1,k-m}.$$

Hence,

$$f_{i:n}(x) = \sum_{j=0}^{n-i} \sum_{r,k=0}^{\infty} m_{i,j,k}\, h_{\alpha(i+j)+k+r}(x), \tag{33}$$

where

$$m_{i,j,k} = \frac{(-1)^j \binom{n-i}{j} n!}{(i-1)!\,(n-i)!} \frac{(\alpha+r)\, d_r\, \varphi_{j+i-1,k}}{\alpha(j+i)+k+r}.$$

Equation (33) is the main result of this section. It reveals that the pdf of the OGE order statistics is a triple linear combination of exp-G distributions. So, several mathematical quantities of these order statistics like ordinary and incomplete moments, factorial moments, mgf, mean deviations and several others come from these quantities of the OGE distributions.

9 Characterization of the OGE family

Characterizations of distributions are important to many researchers in the applied fields. An investigator will be vitally interested to know if their model fits the requirements of a particular distribution. To this end, one will depend on the characterizations of this distribution which provide conditions under which the underlying distribution is indeed that particular distribution. Various characterizations of distributions have been established in many different directions. In this section, two main characterizations of the OGE family are presented. These characterizations are based on: (i) a simple relationship between two truncated moments; (ii) a single function of the random variable.

9.1 Characterizations based on truncated moments

In this subsection we present characterizations of the OGE family in terms of a simple relationship between two truncated moments. Our characterization results presented here will employ an interesting result due to Glänzel (1987) (Theorem 2, below). The advantage of these characterizations is that, the cdf $F(x)$ does not require to have a closed-form and are given in terms of an integral whose integrand depends on the solution of a first order differential equation, which can serve as a bridge between probability and differential equation.

Theorem 2. *Let $(\Omega, \Sigma, \mathbf{P})$ be a given probability space and let $H = [a, b]$ be an interval for some $a < b$ $\left(a = -\infty, b = \infty\right.$ might as well be allowed$\left.\right)$. Let $X : \Omega \to H$ be a continuous random variable with the distribution function F and let q_1 and q_2 be two real functions defined on H such that $\mathbf{E}\left[q_1(X) \mid X \geq x\right] = \mathbf{E}\left[q_2(X) \mid X \geq x\right] \eta(x)$, $x \in H$, is defined with some real function η. Consider that $q_1, q_2 \in C^1(H)$, $\eta \in C^2(H)$ and $F(x)$ is twice continuously differentiable and strictly monotone function on the set H. Finally, assume that the equation $q_2 \eta = q_1$ has no real solution in the interior of H. Then, F is uniquely determined by the functions q_1, q_2 and η, particularly*

$$F(x) = \int_a^x C \left| \frac{\eta'(u)}{\eta(u) q_2(u) - q_1(u)} \right| \exp(-s(u))\, du,$$

where the function s is a solution of the differential equation $s' = \frac{\eta' q_2}{\eta q_2 - q_1}$ and C is a constant chosen to make $\int_H dF = 1$.

Clearly, Theorem 2 can be stated in terms of two functions q_1 and η by taking $q_2(x) \equiv 1$, which will reduce the condition given in Theorem 2 to $E\left[q_1(X) \mid X \geq x\right] = \eta(x)$. However, adding an extra function will give a lot more flexibility, as far as its application is concerned.

Proposition 4. *Let $X : \Omega \to (0, \infty)$ be a continuous random variable and let $q_2(x) = \left[1 - e^{-\lambda\left(\frac{G(x;\xi)}{\bar{G}(x;\xi)}\right)}\right]^{1-\alpha}$ and $q_1(x) = q_2(x)\, e^{-\lambda\left(\frac{G(x;\xi)}{\bar{G}(x;\xi)}\right)}$ for $x > 0$. The pdf of X is (2) if and only if the function η defined in Theorem 2 has the form*

$$\eta(x) = \frac{1}{2}\left\{1 + e^{-\lambda\left(\frac{G(x;\xi)}{\bar{G}(x;\xi)}\right)}\right\}, \quad x > 0.$$

Proof. Let X has density (2), then

$$[1-F(x)]\,\mathbf{E}\left[q_2(X)\,|X \geq x\right] = \alpha\left\{1-e^{-\lambda\left(\frac{G(x;\xi)}{\overline{G}(x;\xi)}\right)}\right\}, \quad x>0,$$

and

$$[1-F(x)]\,\mathbf{E}\left[q_1(X)\,|X \geq x\right] = \frac{\alpha}{2}\left\{1-e^{-2\lambda\left(\frac{G(x;\xi)}{\overline{G}(x;\xi)}\right)}\right\}, \quad x>0.$$

Finally,

$$\eta(x)\,q_2(x) - q_1(x) = \frac{1}{2}q_2(x)\left\{1-e^{-\lambda\left(\frac{G(x;\xi)}{\overline{G}(x;\xi)}\right)}\right\} > 0 \quad \text{for} \quad x>0.$$

Conversely, if η is given as above, then

$$s'(x) = \frac{\eta'(x)\,q_2(x)}{\eta(x)\,q_2(x)-q_1(x)} = \frac{-\lambda\left(\frac{g(x;\xi)}{\left(\overline{G}(x;\xi)\right)^2}\right)e^{-\lambda\left(\frac{G(x;\xi)}{\overline{G}(x;\xi)}\right)}}{\left\{1-e^{-\lambda\left(\frac{G(x;\xi)}{\overline{G}(x;\xi)}\right)}\right\}}, \quad x>0$$

and hence

$$s(x) = -\ln\left\{1-e^{-\lambda\left(\frac{G(x;\xi)}{\overline{G}(x;\xi)}\right)}\right\} \quad x>0.$$

Now, in view of Theorem 2, X has density (2). □

Corollary 1. *Let $X:\Omega\to\mathbb{R}$ be a continuous random variable and let $q_2(x)$ be as in Proposition 4. The pdf of X is (2) if and only if there exist functions q_1 and η defined in Theorem 2 satisfying the differential equation*

$$\frac{\eta'(x)\,q_2(x)}{\eta(x)\,q_2(x)-q_1(x)} = \frac{-\lambda\left(\frac{g(x;\xi)}{\left(\overline{G}(x;\xi)\right)^2}\right)e^{-\lambda\left(\frac{G(x;\xi)}{\overline{G}(x;\xi)}\right)}}{\left\{1-e^{-\lambda\left(\frac{G(x;\xi)}{\overline{G}(x;\xi)}\right)}\right\}}, \quad x>0.$$

Remark 1. **(a)** The general solution of the differential equation in Corollary 1 is

$$\eta(x) = \left\{1-e^{-\lambda\left(\frac{G(x;\xi)}{\overline{G}(x;\xi)}\right)}\right\}^{-1}\left[\int \lambda q_1(x)\left(\frac{g(x;\xi)}{\left(\overline{G}(x;\xi)\right)^2}\right)e^{-\lambda\left(\frac{G(x;\xi)}{\overline{G}(x;\xi)}\right)}\left[q_2(x)\right]^{-1}dx + D\right],$$

where D is a constant. One set of appropriate functions is given in Proposition 4 with $D=\frac{1}{2}$.

(b) Clearly there are other triplets of functions (q_2, q_1, η) satisfying the conditions of Theorem 2. We presented one such triplet in Proposition 4.

9.2 Characterizations based on single function of the random variable

In this subsection we employ a single function ψ of X and state characterization results in terms of $\psi(X)$.

Proposition 5. *Let $X:\Omega\to(0,\infty)$ be a continuous random variable with cdf $F(x)$. Let $\psi(x)$ be a differentiable function on $(0,\infty)$ with $\lim_{x\to\infty}\psi(x)=1$. Then for $\delta\neq 1$,*

$$E\left[\psi(X)\,|X<x\right] = \delta\psi(x),\, x\in(0,\infty)$$

if and only if

$$\psi(x) = (F(x))^{\frac{1}{\delta}-1}, x \in (0,\infty).$$

Proof. Proof is straightforward. □

Remark 2. For $\psi(x) = \left[1 - e^{-\lambda\left(\frac{G(x;\xi)}{\overline{G}(x;\xi)}\right)}\right]^{\frac{\alpha(1-\delta)}{\delta}}$, $x \in (0,\infty)$, Proposition 5 will give a cdf $F(x)$ given by (1).

10 Estimation

Inference can be carried out in three different ways: point estimation, interval estimation and hypothesis testing. Several approaches for parameter point estimation were proposed in the literature but the maximum likelihood method is the most commonly employed. The maximum likelihood estimates (MLEs) enjoy desirable properties and can be used when constructing confidence intervals and also in test-statistics. Large sample theory for these estimates delivers simple approximations that work well in finite samples. Statisticians often seek to approximate quantities such as the density of a test-statistic that depend on the sample size in order to obtain better approximate distributions. The resulting approximation for the MLEs in distribution theory is easily handled either analytically or numerically.

Here, consider the estimation of the unknown parameters of the OGE family by the method of maximum likelihood from complete samples only. Let $x_1, \ldots, x_n$ be observed values from this family with parameters α, λ and $\boldsymbol{\xi}$. Let $\Theta = (\alpha, \lambda, \boldsymbol{\xi})^\top$ be the $(r \times 1)$ parameter vector. The total log-likelihood function for Θ is given by

$$\begin{aligned}\ell_n = \ell_n(\Theta) &= n\log(\alpha\,\lambda) + \sum_{i=1}^{n}\log\left[g(x_i;\boldsymbol{\xi})\right] - 2\sum_{i=1}^{n}\log\left[\overline{G}(x_i;\boldsymbol{\xi})\right] \\ &\quad -\lambda\sum_{i=1}^{n}V(x_i;\boldsymbol{\xi}) + (\alpha-1)\sum_{i=1}^{n}\log\left\{1-\mathrm{e}^{-\lambda\,V(x_i;\xi)}\right\},\end{aligned} \tag{34}$$

where $V(x_i;\boldsymbol{\xi}) = G(x_i;\boldsymbol{\xi})/\overline{G}(x_i;\boldsymbol{\xi})$.

The components of the score function $U_n(\Theta) = \left(U_\alpha, U_\lambda, U_{\xi_\mathbf{k}}\right)^\top$ are

$$\begin{aligned}U_\alpha &= \frac{n}{\alpha} + \sum_{i=1}^{n}\log\left\{1-\mathrm{e}^{-\lambda V(x_i;\xi)}\right\}, \\ U_\lambda &= \frac{n}{\lambda} - \sum_{i=1}^{n}V(x_i;\boldsymbol{\xi}) + (\alpha-1)\sum_{i=1}^{n}\left\{\frac{V(x_i;\boldsymbol{\xi})\ \mathrm{e}^{-\lambda\,V(x_i;\xi)}}{1-\mathrm{e}^{-\lambda\,V(x_i;\xi)}}\right\}, \\ U_{\xi_\mathbf{k}} &= \sum_{i=1}^{n}\frac{g^{(\xi_k)}(x_i,\boldsymbol{\xi})}{g(x_i,\boldsymbol{\xi})} - 2\sum_{i=1}^{n}\frac{\overline{G}^{(\xi_k)}(x_i,\boldsymbol{\xi})}{\overline{G}(x_i,\boldsymbol{\xi})} - \lambda\sum_{i=1}^{n}V^{(\xi_k)}(x_i,\boldsymbol{\xi}) \\ &\quad +\lambda(\alpha-1)\sum_{i=1}^{n}\left\{\frac{V^{(\xi_k)}(x_i;\boldsymbol{\xi})\ \mathrm{e}^{-\lambda\,V(x_i;\xi)}}{1-\mathrm{e}^{-\lambda\,V(x_i;\xi)}}\right\},\end{aligned}$$

where $v^{(\xi)}(\cdot)$ means the derivative of the function v with respect to ξ.

Setting these equations to zero and solving them simultaneously yields the MLEs $\widehat{\Theta} = (\widehat{\alpha}, \widehat{\lambda}, \widehat{\boldsymbol{\xi}})^\top$ of $\Theta = (\alpha, \lambda, \boldsymbol{\xi})^\top$. These equations cannot be solved analytically, and analytical softwares are required to solve them numerically.

For interval estimation of the parameters, we obtain the 3 × 3 observed information matrix $J(\Theta) = \{U_{rs}\}$ (for $r, s = \alpha, \lambda, \xi_k$), whose elements are listed in Appendix A. Under standard regularity conditions, the multivariate normal $N_3(0, J(\widehat{\Theta})^{-1})$ distribution is used to construct approximate confidence intervals for the parameters. Here, $J(\widehat{\Theta})$ is the total observed information matrix evaluated at $\widehat{\Theta}$. Then, the $100(1-\gamma)\%$ confidence intervals for α, λ and ξ_k are given by $\hat{\alpha} \pm z_{\gamma^*/2} \times \sqrt{var(\hat{\alpha})}$, $\hat{\lambda} \pm z_{\gamma^*/2} \times \sqrt{var(\hat{\lambda})}$ and $\hat{\xi}_k \pm z_{\gamma^*/2} \times \sqrt{var(\hat{\xi}_k)}$, respectively, where the $var(\cdot)$'s denote the diagonal elements of $J(\widehat{\Theta})^{-1}$ corresponding to the model parameters, and $z_{\gamma^*/2}$ is the quantile $(1-\gamma^*/2)$ of the standard normal distribution.

10.1 Simulation study

We evaluate the performance of the maximum likelihood method for estimating the OGE-W parameters using Monte Carlo simulation for a total of twenty four parameter combinations and the process is repeated 200 times. Two different sample sizes $n = 100$ and 300 are considered. The MLEs and their standard deviations of the parameters are listed in Table 2. The estimates of α, β and λ are determined by solving the nonlinear equations $U(\Theta) = \mathbf{0}$. Based on Table 2, we note that the ML method performs well for estimating the model parameters. Also, as the sample size increases, the biases and the standard deviations of the MLEs decrease as expected.

Table 2 MLEs and standard deviations for various parameter values

Sample size	Actual values			Estimated values			Standard deviations		
n	α	β	λ	$\tilde{\alpha}$	$\tilde{\beta}$	$\tilde{\lambda}$	$\tilde{\alpha}$	$\tilde{\beta}$	$\tilde{\lambda}$
100	0.5	0.5	1	0.5264	0.5493	1.0198	0.0205	0.0188	0.0284
	0.5	1	2	0.4894	1.3254	2.1484	0.0254	0.0709	0.0461
	0.5	1.5	1	0.5214	1.7122	1.0097	0.0249	0.0602	0.0284
	0.5	2	2	0.4897	2.6385	2.2290	0.0252	0.1421	0.0619
	1	0.5	1	1.0695	0.5210	1.0321	0.0429	0.0124	0.0244
	1	1	2	1.0063	1.1701	2.0607	0.0451	0.0470	0.0327
	1	1.5	1	1.0838	1.5895	1.0074	0.0476	0.0403	0.0267
	1	2	2	1.0753	2.2144	2.0797	0.0538	0.0765	0.0346
	1.5	0.5	1	1.6511	0.5250	1.0207	0.0853	0.0111	0.0264
	1.5	1	2	1.7591	1.0921	2.0900	0.1136	0.0355	0.0323
	1.5	1.5	1	1.6315	1.5700	1.0269	0.0670	0.0325	0.0247
	1.5	2	2	1.7981	2.2429	2.1356	0.1037	0.1227	0.0437
300	0.5	0.5	1	0.4980	0.5261	0.9972	0.0064	0.0052	0.0086
	0.5	1	2	0.4996	1.0557	2.0265	0.0070	0.0125	0.0111
	0.5	1.5	1	0.5014	1.5597	0.9960	0.0066	0.0154	0.0089
	0.5	2	2	0.4844	2.2030	2.0532	0.0079	0.0299	0.0123
	1	0.5	1	1.0319	0.5014	1.0132	0.0130	0.0036	0.0083
	1	1	2	0.9856	1.0539	2.0084	0.01450	0.01112	0.0097
	1	1.5	1	1.0277	1.5324	1.0063	0.0127	0.0116	0.0079
	1	2	2	1.0053	2.0929	2.0217	0.01536	0.0216	0.0097
	1.5	0.5	1	1.5457	0.5078	1.0068	0.0182	0.0038	0.0077
	1.5	1	2	1.5414	1.0310	2.0300	0.0250	0.0096	0.0094
	1.5	1.5	1	1.5356	1.5182	1.0085	0.0185	0.0098	0.0077
	1.5	2	2	1.5377	2.0670	2.0177	0.02377	0.02022	0.0085

11 Applications

In this section, we provide two applications to real data to illustrate the importance of the OGE family by means of the OGE-W, OGE-Fr and OGE-N models presented in Section 3. We consider $\theta = 1$, $b = 1$ and $\mu = 0$ for the OGE-W, OGE-Fr and OGE-N models, respectively, due to the fact that one scale parameter is enough for fitting these univariate models. The MLEs of the parameters for the these models are calculated and four goodness-of-fit statistics are used to compare the new family with its sub-models.

The first real data set represents the survival times of 121 patients with breast cancer obtained from a large hospital in a period from 1929 to 1938 (Lee 1992). The data examined by Ramos et al. (2013) are: 0.3, 0.3, 4.0, 5.0, 5.6, 6.2, 6.3, 6.6, 6.8, 7.4, 7.5, 8.4, 8.4, 10.3, 11.0, 11.8, 12.2, 12.3, 13.5, 14.4, 14.4, 14.8, 15.5, 15.7, 16.2, 16.3, 16.5, 16.8, 17.2, 17.3, 17.5, 17.9, 19.8, 20.4, 20.9, 21.0, 21.0, 21.1, 23.0, 23.4, 23.6, 24.0, 24.0, 27.9, 28.2, 29.1, 30.0, 31.0, 31.0, 32.0, 35.0, 35.0, 37.0, 37.0, 37.0, 38.0, 38.0, 38.0, 39.0, 39.0, 40.0, 40.0, 40.0, 41.0, 41.0, 41.0, 42.0, 43.0, 43.0, 43.0, 44.0, 45.0, 45.0, 46.0, 46.0, 47.0, 48.0, 49.0, 51.0, 51.0, 51.0, 52.0, 54.0, 55.0, 56.0, 57.0, 58.0, 59.0, 60.0, 60.0, 60.0, 61.0, 62.0, 65.0, 65.0, 67.0, 67.0, 68.0, 69.0, 78.0, 80.0,83.0, 88.0, 89.0, 90.0, 93.0, 96.0, 103.0, 105.0, 109.0, 109.0, 111.0, 115.0, 117.0, 125.0, 126.0, 127.0, 129.0, 129.0, 139.0, 154.0.

The second data set consists of 63 observations of the strengths of 1.5 cm glass fibres, originally obtained by workers at the UK National Physical Laboratory. Unfortunately, the units of measurement are not given in the paper. The data are: 0.55, 0.74, 0.77, 0.81, 0.84, 0.93, 1.04, 1.11, 1.13, 1.24, 1.25, 1.27, 1.28, 1.29, 1.30, 1.36, 1.39, 1.42, 1.48, 1.48, 1.49, 1.49, 1.50, 1.50, 1.51, 1.52, 1.53, 1.54, 1.55, 1.55, 1.58, 1.59, 1.60, 1.61, 1.61, 1.61, 1.61, 1.62, 1.62, 1.63, 1.64, 1.66, 1.66, 1.66, 1.67, 1.68, 1.68, 1.69, 1.70, 1.70, 1.73, 1.76, 1.76, 1.77, 1.78, 1.81, 1.82, 1.84, 1.84, 1.89, 2.00, 2.01, 2.24. These data have also been analyzed by Smith and Naylor (1987).

The MLEs are computed using the Limited-Memory Quasi-Newton Code for Bound-Constrained Optimization (L-BFGS-B) and the log-likelihood function evaluated at the MLEs ($\hat{\ell}$). The measures of goodness of fit including the Akaike information criterion (AIC), Anderson-Darling (A^*), Cramér–von Mises (W^*) and Kolmogrov-Smirnov (K-S) statistics are computed to compare the fitted models. The statistics A^* and W^* are described in details in Chen and Balakrishnan (1995). In general, the smaller the values of these statistics, the better the fit to the data. The required computations are carried out in the R-language.

Tables 3 and 4 list the MLEs and their corresponding standard errors (in parentheses) of the model parameters. The numerical values of the model selection statistics $\hat{\ell}$, AIC, A^*, W^* and K-S are listed in Tables 5 and 6. We note from the figures in Table 5 that both

Table 3 MLEs and their standard errors (in parentheses) for the first data set

Distribution	λ	α	β	a
OGE-W	0.7173	8.1570	0.2208	-
	(0.2019)	(3.6827)	(0.0276)	-
OE-W	0.1114	-	0.3550	-
	(0.0155)	-	(0.0123)	-
OGE-Fr	0.4920	1.9766	-	0.6051
	(0.2704)	(0.7609)	-(0.1314)	
OE-Fr	0.1647	-	-	0.8627
	(0.0303)	-	-	(0.0586)

Table 4 MLEs and their standard errors (in parentheses) for the second data set

Distribution	λ	α	β	σ
OGE-W	0.7173	8.1570	0.2208	-
	(0.2019)	(3.6827)	(0.0276)	-
OE-W	0.0721	-	1.9603	-
	(0.0162)	-	(0.0940)	-
OGE-N	0.1010	3.1230	-	0.9884
	(0.0798)	(1.5660)	-	(0.1484)
OE-N	0.0121	-	-	0.7385
	(0.0043)	-	-	(0.0364)

Table 5 The statistics $\hat{\ell}$, AIC, A^*, W^* K-S and K-S p-value for the first data set

Distribution	$\hat{\ell}$	AIC	A^*	W^*	K-S	p-value (K-S)
OGE-W	-410.8347	827.6693	0.3111	0.0472	0.0460	0.9498
OE-W	-431.1625	866.3251	2.6578	0.4510	0.1430	0.0107
OGE-Fr	-414.2511	834.5023	0.6838	0.1011	0.0651	0.6492
OE-Fr	-416.3539	836.7078	0.8991	0.1367	0.0939	0.2095

Table 6 The statistics $\hat{\ell}$, AIC, A^*, W^*, K-S and K-S p-value for the second data set

Distribution	$\hat{\ell}$	AIC	A^*	W^*	K-S	p-value (K-S)
OGE-N	-14.1653	34.3306	0.8755	0.1548	0.1286	0.2483
OE-N	-17.5979	39.1958	0.9684	0.1557	0.1232	0.2945
OGE-W	-14.2733	34.5465	0.9645	0.1717	0.1334	0.2119
OE-W	-16.4613	36.9227	0.9619	0.1614	0.1377	0.1832

OGE-W and OGE-Fr models have the lowest values of the AIC, A^*, W^* and K-S statistics (for the first data set) as compared to their sub-models, suggesting that the OGE-W and OGE-Fr models provide the best fits. The histogram of the first data and the estimated pdfs and cdfs of the OGE-W and OGE-Fr models and their sub-models are displayed in Figure 8. Similarly, it is also evident from Table 6 that the OGE-N gives the lowest values for the AIC, A^* and W^* statistics while OGE-W models gives the lowest values for the AIC and K-S statistics (for the second data set) as compared to their sub-models,

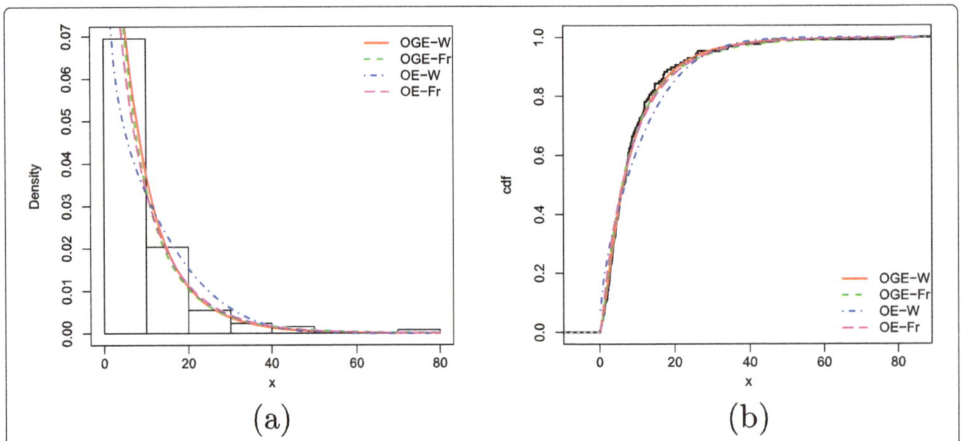

Figure 8 Plots of the estimated (a) pdfs and (b) cdfs for the OGE-W and OGE-Fr, and their sub-models OE-W and OE-Fr for the first data set.

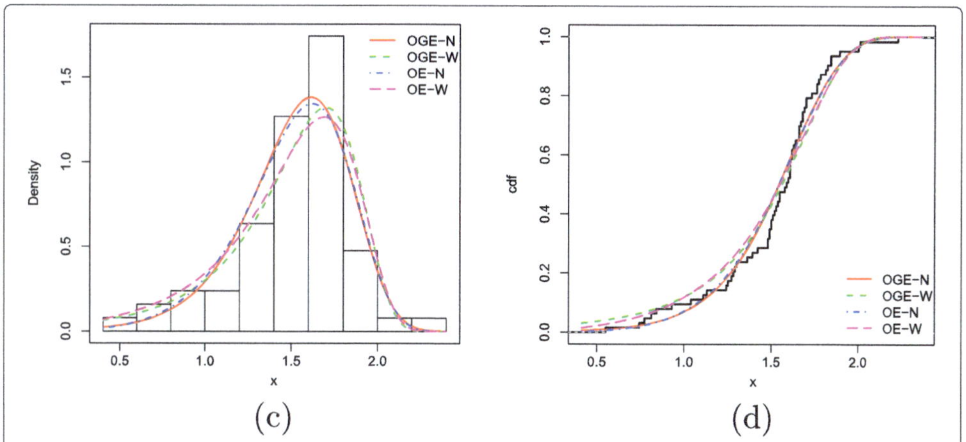

Figure 9 Plots of the estimated (c) pdfs and (d) cdfs for the OGE-N and OGE-W, and their sub-models OE-W and OE-N for the second data set.

and therefore these models can be chosen as the best ones. The histogram of the second data and estimated pdfs and cdfs of the OGE-N and OGE-W distributions and their sub-models are displayed in Figure 9. It is clear from Tables 5 and 6, and Figures 8 and 9 that the OGE-W, OGE-Fr and OGE-N models provide better fits than their sub-models to these two data sets.

12 Concluding remarks

The generalized continuous distributions have been widely studied in the literature. We propose a new class of distributions called the *odd generalized exponential* family. We study some structural properties of the new family including an expansion for its density function. We obtain explicit expressions for the moments, generating function, mean deviations, quantile function and order statistics. The maximum likelihood method is employed to estimate the family parameters. We fit three special models of the proposed family to two real data sets to demonstrate the usefulness of the family. We use four goodness-of-fit statistics in order to verify which distribution provides better fit to these data. We conclude that these three special models provide consistently better fits than other competing models. We hope that the proposed family and its generated models will attract wider applications in several areas such as reliability engineering, insurance, hydrology, economics and survival analysis.

Appendix A

The observed information matrix for the parameter vector $\Theta = (\alpha, \lambda, \boldsymbol{\xi}_k)^\top$ is given by

$$J(\Theta) = -\frac{\partial^2 \ell(\Theta)}{\partial\,\Theta\,\partial\,\Theta^\top} = -\begin{pmatrix} J_{\alpha\alpha} & J_{\alpha\lambda} & J_{\alpha\xi_{\mathbf{k}}} \\ . & J_{\lambda\lambda} & J_{\lambda\xi_{\mathbf{k}}} \\ . & . & J_{\xi_{\mathbf{k}}\xi_{\mathbf{l}}} \end{pmatrix},$$

whose elements are

$$J_{\alpha\alpha} = -\frac{n}{\alpha^2},$$

$$J_{\alpha\lambda} = \sum_{i=1}^{n} \left\{ \frac{V(x_i;\boldsymbol{\xi})\ \mathrm{e}^{-\lambda\ V(x_i;\xi)}}{1-\mathrm{e}^{-\lambda\ V(x_i;\xi)}} \right\},$$

$$J_{\alpha\xi_{\mathbf{k}}} = \lambda \sum_{i=1}^{n} \left\{ \frac{V^{(\xi_{\mathbf{k}})}\ (x_i;\boldsymbol{\xi})\ \mathrm{e}^{-\lambda\,V(x_i;\xi)}}{1-\mathrm{e}^{-\lambda\ V(x_i;\xi)}} \right\},$$

$$J_{\lambda\lambda} = -\frac{n}{\lambda^2} + (\alpha-1)\sum_{i=1}^{n}\left\{\frac{V^2(x_i;\boldsymbol{\xi})\, e^{-\lambda\, V(x_i;\xi)}}{\left\{1-e^{-\lambda\, V(x_i;\xi)}\right\}^2}\right\},$$

$$J_{\lambda\,\xi_k} = -\sum_{i=1}^{n} V^{(\xi_k)}(x_i;\boldsymbol{\xi})$$

$$+(\alpha-1)\sum_{i=1}^{n}\left\{\frac{V^{(\xi_k)}(x_i;\boldsymbol{\xi})\, e^{-\lambda\, V(x_i;\xi)}\left[1-\lambda\, V(x_i;\xi) - e^{-\lambda\, V(x_i;\xi)}\right]}{\left\{1-e^{-\lambda\, V(x_i;\xi)}\right\}^2}\right\},$$

$$J_{\xi_k\xi_l} = \sum_{i=1}^{n}\left\{\frac{g(x_i;\boldsymbol{\xi})\, g''_{kl}(x_i;\boldsymbol{\xi}) - g'_k(x_i;\boldsymbol{\xi})\, g'_l(x_i;\boldsymbol{\xi})}{g^2(x_i;\boldsymbol{\xi})}\right\}$$

$$-2\sum_{i=1}^{n}\left\{\frac{\overline{G}(x_i;\boldsymbol{\xi})\,\overline{G}''_{kl}(x_i;\boldsymbol{\xi}) - \overline{G}'_k(x_i;\boldsymbol{\xi})\,\overline{G}'_l(x_i;\boldsymbol{\xi})}{\overline{G}^2(x_i;\boldsymbol{\xi})}\right\} - \lambda\sum_{i=1}^{n} V''_{kl}(x_i;\boldsymbol{\xi})$$

$$+\lambda(\alpha-1)\sum_{i=1}^{n}\frac{1}{\left\{1-e^{-\lambda\, V(x_i;\xi)}\right\}^2}\quad\times$$

$$\left\{e^{-\lambda\, V(x_i;\xi)}\left[V''_{kl}(x_i;\boldsymbol{\xi}) - \lambda\, V'_k(x_i;\boldsymbol{\xi})\, V'_l(x_i;\boldsymbol{\xi}) - V''_{kl}(x_i;\boldsymbol{\xi})\, e^{-\lambda\, V(x_i;\xi)}\right]\right\},$$

where $t'_k(\cdot,\boldsymbol{\xi}) = \partial\, t(\cdot,\boldsymbol{\xi})/\partial\,\xi_k$ and $t''_{kl}(\cdot,\boldsymbol{\xi}) = \partial^2\, t(\cdot,\boldsymbol{\xi})/\partial\,\xi_k\,\partial\,\xi_l$.

Acknowledgements
The authors would like to thank the Editor and the two referees for careful reading and for their comments which greatly improved the paper.

Competing interests
The authors declare that they have no competing interests.

Authors' contributions
The authors, viz MHT, GMC, MA, MM, MZ and GGH with the consultation of each other carried out this work and drafted the manuscript together. All authors read and approved the final manuscript.

Author details
[1]Department Statistics, Baghdad Campus, The Islamia University of Bahawalpur, 63100 Bahawalpur, Pakistan. [2]Departamento de Estatística, Universidade Federal de Pernambuco, PE 50740-540 Recife, Brazil. [3]Department of Statistics, Persian Gulf University of Bushehr, Bushehr, Iran. [4]Department Statistics, The Islamia University of Bahawalpur, 63100 Bahawalpur, Pakistan. [6]Department of Mathematics, Statistics and Computer Science Marquette University, WI 53201-1881 Milwaukee, USA.

References
Alexander, C, Cordeiro, GM, Ortega, EMM, Sarabia, JM: Generalized beta-generated distributions. Comput. Statist. Data. Anal. **56**, 1880–1897 (2012)
Alizadeh, M, Emadi, M, Doostparast, M, Cordeiro, GM, Ortega, EMM, Pescim, RR: A new family of distributions: the Kumaraswamy odd log-logistic, properties and applications. Hacet. J. Math Stat (2015). forthcoming
Aljarrah, MA, Lee, C, Famoye, F: On generating T-X family of distributions using quantile functions. J. Stat. Dist. Appl. 1, Art. **2** (2014)
Alzaatreh, A, Lee, C, Famoye, F: A new method for generating families of continuous distributions. Metron. **71**, 63–79 (2013)
Alzaatreh, A, Lee, C, Famoye, F: T-normal family of distributions: A new approach to generalize the normal distribution. J. Stat. Dist. Appl. 1, Art, 16 (2014)
Alzaghal, A, Famoye, F, Lee, C: Exponentiated T-X family of distributions with some applications. Int. J. Probab. Statist. **2**, 31–49 (2013)
Amini, M, MirMostafaee, SMTK, Ahmadi, J: Log-gamma-generated families of distributions. Statistics. **48**, 913–932 (2014)
Bourguignon, M, Silva, RB, Cordeiro, GM: The Weibull-G family of probability distributions. J. Data Sci. **12**, 53–68 (2014)
Chen, G, Balakrishnan, N: A general purpose approximate goodness-of-fit test. J. Qual. Tech. **27**, 154–161 (1995)
Choudhury, A: A simple derivation of moments of the exponentiated Weibull distribution. Metrika. **62**, 17–22 (2005)
Cordeiro, GM: de Castro, M: A new family of generalized distributions. J. Stat. Comput. Simul. **81**, 883–893 (2011)
Cordeiro, GM, Nadarajah, S: Closed-form expressions for moments of a class of beta generalized distributions. Braz. J. Probab. Stat. **25**, 14–33 (2011)
Cordeiro, GM, Ortega, EMM, da Cunha, DCC: The exponentiated generalized class of distributions. J. Data Sci. **11**, 1–27 (2013)

Cordeiro, GM, Alizadeh, M, Ortega, EMM: The exponentiated half-logistic family of distributions: Properties and applications. J. Probab. Statist. Art.ID. **864396**, 21 (2014a)

Cordeiro, GM, Ortega, EMM, Popović, BV, Pescim, RR: The Lomax generator of distributions: Properties, minification process and regression model. Appl. Math. Comput. **247**, 465–486 (2014b)

Dey, AK, Kundu, D: Discriminating among the log-normal, Weibull and generalized exponential distributions. IEEE Trans. Reliab. **58**, 416–424 (2009)

El-Gohary, A, Alshamrani, A, Al-Otaibi, AN: The generalized Gompertz distribution. Appl. Math. Model. **37**, 13–24 (2013)

Eugene, N, Lee, C, Famoye, F: Beta-normal distribution and its applications. Commun. Stat. Theory Methods. **31**, 497–512 (2002)

Fiorentino, M, Gioia, A, Iacobellis, V, Manfreda, S: Analysis on flood generation processes by means of a continuous simulation model. Adv. Geosci. **7**, 231–236 (2006)

Glänzel, W: A characterization theorem based on truncated moments and its application to some distribution families. In: Mathematical Statistics and Probability Theory (Bad Tatzmannsdorf, 1986), Vol. B, Reidel, Dordrecht, pp. 75–84, (1987)

Gradshteyn, IS, Ryzhik, IM: Table of Integrals, Series, and Products. 6th eds. Academic Press, San Diego (2000)

Gupta, RC, Gupta, PI, Gupta, RD: Modeling failure time data by Lehmann alternatives. Commun. Stat. Theory Methods. **27**, 887–904 (1998)

Gupta, RD, Kundu, D: Generalized exponential distribution. Aust. N. Z. J. Stat. **41**, 173–188 (1999)

Gupta, RD, Kundu, D: Generalized exponential distribution: An alternative to Gamma and Weibull distributions. Biom. J. **43**, 117–130 (2001a)

Gupta, RD, Kundu, D: Generalized exponential distribution: Different methods of estimations. J. Stat. Comput. Simul. **69**, 315–337 (2001b)

Gupta, RD, Kundu, D: Discriminating between the Weibull and the GE distributions. Comput. Stat. Data Anal. **43**, 179–196 (2002)

Gupta, RD, Kundu, D: Closeness of gamma and generalized exponential distributions. Commun. Stat. Theory Methods. **32**, 705–721 (2003)

Gupta, RD, Kundu, D: Discriminating between the gamma and generalized exponential distributions. J. Stat. Comput. Simul. **74**, 107–121 (2004)

Gupta, RD, Kundu, D: On comparison of the Fisher information of the Weibull and GE distributions. J. Stat. Plann. Inference. **136**, 3130–3144 (2006)

Gupta, RD, Kundu, D: Generalized exponential distribution: Existing results and some recent developments. J. Stat. Plann. Inference. **137**, 3537–3547 (2007)

Gupta, RD, Kundu, D: Generalized exponential distribution: Bayesian Inference. Comput. Stat. Data Anal. **52**, 1873–1883 (2008)

Gupta, RD, Kundu, D: An extension of generalized exponential distribution. Stat. Methodol. **8**, 485–496 (2011)

Iacobellis, V, Fiorentino, M: Derived distribution of floods based on the concept of partial area coverage with a climatic appeal. Water Resour. Res. **36**, 469–482 (2000)

Jones, MC: Families of distributions arising from the distributions of order statistics. Test. **13**, 1–43 (2004)

Kilbas, AA, Srivastava, HM, Trujillo, JJ: Theory and applications of fractional differential equations. Elsevier, Amsterdam (2006)

Kundu, D, Gupta, RD, Manglick, A: Discriminating between the log-normal and the generalized exponential distributions. J. Stat. Plann. Inference. **127**, 213–227 (2005)

Lee, ET: Statistical Methods for Survival Data Analysis. John Wiley, New York (1992)

Lehmann, EL: The power of rank tests. Ann. Math. Statist. **24**, 23–43 (1953)

Marshall, AN, Olkin, I: A new method for adding a parameter to a family of distributions with applications to the exponential and Weibull families. Biometrika. **84**, 641–652 (1997)

Mudholkar, GS, Hutson, AD: The exponentiated Weibull family: Some properties and a flood data application. Commun. Stat. Theory Methods. **25**, 3059–3083 (1996)

Mudholkar, GS, Srivastava, DK: Exponentiated Weibull family for analyzing bathtub failure data. IEEE Trans. Reliab. **42**, 299–302 (1993)

Mudholkar, GS, Srivastava, DK, Freimer, M: The exponentiated Weibull family: A reanalysis of the bus-motor failure data. Technometrics. **37**, 436–445 (1995)

Nadarajah, S: Exact distribution of the peak streamflow. Water Resour. Res. **43**(2), W02501 (2007). doi:10.1029/2006WR005300

Nadarajah, S: The exponentiated exponential distribution: a survey. AStA Adv. Stat. Anal. **95**, 219–251 (2011)

Nadarajah, S, Cordeiro, GM, Ortega, EMM: The exponentiated Weibull distribution: a survey. Stat. Papers. **54**, 839–877 (2013)

Nadarajah, S, Cordeiro, GM, Ortega, EMM: The Zografos-Balakrishnan–G family of distributions: Mathematical properties and applications. Commun. Stat. Theory Methods. **44**, 186–215 (2015)

Nadarajah, S, Kotz, S: A generalized logistic distribution. Int. J. Math. Math. Sci. **19**, 3169–3174 (2005)

Nadarajah, S, Kotz, S: The exponentiated type distributions. Acta Appl. Math. **92**, 97–111 (2006)

Pakyari, R: Discriminating between generalized exponential, geometric extreme exponential and Weibull distribution. J. Stat. Comput. Simul. **80**, 1403–1412 (2010)

Prudnikov, AP, Bruychkov, YA: Marichev, Vol. 1. Gordon and Breach, New York (1986)

Ramos, MWA, Cordeiro, GM, Marinho, PRD, Dias, CRB, Hamedani, GG: The Zografos-Balakrishnan log-logistic distribution: Properties and applications. J. Stat. Theory Appl. **12**, 225–244 (2013)

Rényi, A: On measures of entropy and information. In: Proceedings of the Fourth Berkeley Symposium on Mathematical Statistics and Probability-I, pp. 547–561. University of California Press, Berkeley, (1961)

Ristić, MM, Balakrishnan, N: The gamma-exponentiated exponential distribution. J. Stat. Comput. Simul. **82**, 1191–1206 (2012)

Shannon, CE: A mathematical theory of communication. Bell Sys. Tech. J. **27**, 379–432 (1948)

Smith, RL, Naylor, JC: A comparison of maximum likelihood and Bayesian estimators for the three-parameter Weibull distributuion. Appl. Statist. **36**, 358–369 (1987)

Tahir, MH, Cordeiro, GM, Alzaatreh, A, Mansoor, M, Zubair, M: The Logistic-X family of distributions and its applications. Commun. Stat. Theory Methods (2015a). forthcoming

Tahir, MH, Zubair, M, Mansoor, M, Cordeiro, GM, Alizadeh, M, Hamedani, GG: A new Weibull-G family of distributions. Hacet. J. Math. Stat. (2015b). forthcoming

Torabi, H, Montazari, NH: The gamma-uniform distribution and its application. Kybernetika. **48**, 16–30 (2012)

Torabi, H, Montazari, NH: The logistic-uniform distribution and its application. Commun. Stat. Simul. Comput. **43**, 2551–2569 (2014)

Zografos, K, Balakrishnan, N: On families of beta- and generalized gamma-generated distributions and associated inference. Stat. Methodol. **6**, 344–362 (2009)

10

The multivariate slash and skew-slash student *t* distributions

Fei Tan[1*], Yuanyuan Tang[2] and Hanxiang Peng[1]

*Correspondence:
ftan@math.iupui.edu
[1] Department of Mathematical Sciences, Indiana University Purdue University Indianapolis, N. Blackford Street, IN 46202, Indianapolis, USA
Full list of author information is available at the end of the article

Abstract

In this article, we introduce the multivariate slash and skew-slash *t* distributions which provide alternative choices in simulating and fitting skewed and heavy tailed data. We study their relationships with other distributions and give the densities, stochastic representations, moments, marginal distributions, distributions of linear combinations and characteristic functions of the random vectors obeying these distributions. We characterize the skew *t*, the skew-slash normal and the skew-slash *t* distributions using both the hidden truncation or selective sampling model and the order statistics of the components of a bivariate normal or *t* variable. Density curves and contour plots are drawn to illustrate the skewness and tail behaviors. Maximum likelihood and Bayesian estimation of the parameters are discussed. The proposed distributions are compared with the skew-slash normal through simulations and applied to fit two real datasets. Our results indicated that the proposed skew-slash *t* fitting outperformed the skew-slash normal fitting and is a competitive candidate distribution in analyzing skewed and heavy tailed data.

Mathematics Subject Classification Primary 62E10; Secondary 62P10

Keywords: Heavy tail; Skew distribution; Skew-slash distribution; Slash distribution; Student *t* distribution

1 Introduction

Skewed and heavy tailed data occur frequently in real life and pose challenges to our usual way of thinking. Examples of such data include household incomes, loss data such as crop loss claims and hospital discharge bills, and files transferred through the Internet to name a few. Candidate distributions for simulating and fitting such data are not abundant. One can't simply take the normal or the t distributions or as such as substitutes. Even though Cauchy distribution can be used to simulate and fit such data, its sharp central peak and the fact that its first moment does not exist narrow its applications. Thus additional distributions are needed to study such skewed and heavy tailed data.

Kafadar (1988) introduced the univariate normal slash distribution as the resulting distribution of the ratio of a standard normal random variable (rv) and an independent uniform rv (hereafter referred to the distribution as the slash normal). Generalizing the standard normal by introducing a tail parameter, the slash normal has heavier tails than the standard normal, hence it could be used to simulate and fit heavy tailed data. Wang and Genton (2006) generalized the univariate slash normal to the multivariate slash normal and investigated its properties. They also defined the multivariate skew-slash normal as the resulting distribution of the ratio of a skewed normal rv and an independent uniform rv (hereafter referred to the distribution as the skew-slash normal). They applied it to fit two real datasets.

In this article, we introduce the *slash (student) t distribution* and the *skew-slash (student) t distribution*. They could be used to simulate and fit skewed and heavy tailed data. The slash t distribution generalizes the slash normal distribution of Kafadar (1988) and the multivaraiate slash normal distribution of Wang and Genton (2006), and the skew-slash t distribution generalizes the skew-slash normal distribution of the latter two authors. In the skew-slash t there is one parameter to regulate the skewness of the distribution and another parameter to control the tail behavior. By setting the skewness parameter to zero, the skew-slash t reduces to the slash t. By letting the tail parameter to be infinity, the skew-slash t simplifies to the skew t of Azzalini and Capitanio (2003). As both the slash and skew t take the t and hence the normal as their special cases, so does the skew-slash t. To fit data, one can start with the skew-slash t. If the fitted value of the degrees of freedom is very large, then one takes the simpler skew-slash normal model. This idea of course can be used to perform the hypothesis testing of a skew-slash normal sub-model against a skew-slash t model. We have derived the formulas for the densities, moments, marginal distributions and linear combinations of these distributions. Thus it could be expected that they can be used to analyze skewed and heavy tailed data.

Compared to the slash and skew-slash normal distributions, an additional parameter, the degrees of freedom, is included in the slash and skew-slash t distributions. This parameter gives the latter distributions more flexibility in fitting data than the former. Even though there is a tail parameter in both slash and skew-slash normal and slash and skew-slash t distributions, the degrees of freedom in the t distribution may lend an additional hand to model heavy (fat) tails and joints with the tail parameter to better fit data. This could be used to explain why the skew-slash fitting to the real GAD data in our application outperformed the skew-slash normal fitting. See Figure 1 where the skew-slash t fitting was able to better capture the peak of the histogram, and Table 1 where the AIC values indicated that the skew-slash t fitting was better than that of the skew-slash normal fitting. In fact, one observes that the standard error (SE) of the MLE of the tail parameter q in the table is very big (83.194) for the skew-slash normal fitting and much smaller (2.19) for the skew-slash t fitting. Noticing also the fact that the estimate of the degrees of freedom r is reasonable with a small SE, we would comment that the joint endeavor of q and r had higher fitting capability than that of a single q. Our simulation results in Section 5.1 and in particular in Table 2 also exhibited the superior performance of the proposed skew-slash t distribution to the skew-slash normal.

Azzalini and Dalla Valle (1996) introduced the multivariate skew normal distribution that extends the normal distribution with an additional skewness parameter. It provides an alternative modeling distribution to skewed data that are often observed in many areas such as economics, computer science and life sciences. Many authors have investigated skew t distributions, see e.g. Azzalini and Dalla Valle (1996), Gupta (2003), and Sahu et al. (2003). Azzalini and Capitanio (2003) proposed the multivariate skew t distribution by allowing a skewness parameter in a multivariate t distribution.

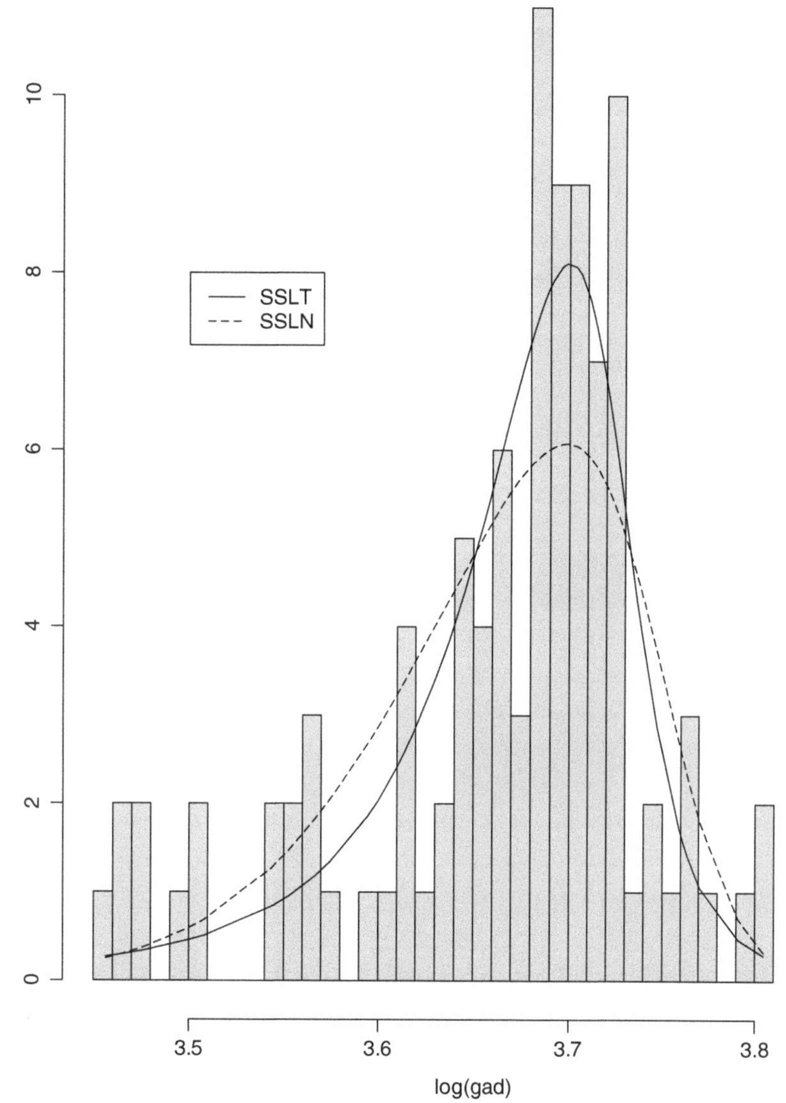

Figure 1 Histogram and fitted density curves.

It is our belief that the proposed slash and skew-slash t distributions throw some additional light on this theory and contribute to the family of candidate distributions for modeling and simulating skewed and heavy tailed data.

The article is organized as follows. In Section 2, we introduce the multivariate slash t distribution, study its relationships with other distributions and derive the density

Table 1 The skew-slash normal and t fitting to the GAD data

	SSLT		SSLN	
	MLE	**SE**	**MLE**	**SE**
AIC	-242.318		-240.687	
μ	3.726	0.0070	3.748	0.036
σ^2	0.002	0.0004	0.009	0.038
λ	-2.056	0.6190	-3.199	3.204
q	3.812	2.1990	7.456	83.194
r	3.044	1.4080		

SSLN=Skew-Slash Normal and SSLT=Skew-Slash t.

Table 2 Skew-slash *t* and skew-slash normal comparison

	True	Estimated		True	Estimated	
	SSLT	SSLT	SSLN	SSLN	SSLT	SSLN
AIC		1524.939	1543.771		1416.485	1425.878
μ_1	0	-0.160	-0.467	0	-0.309	-0.690
μ_2	0	0.032	-0.225	0	0.025	-0.212
σ_{11}^2	1	0.827	0.809	1	0.587	0.651
σ_{12}	0	0.122	0.230	0	0.130	0.304
σ_{22}^2	1	1.154	1.527	1	1.019	1.314
λ_1	-5	-2.830	-1.581	-5	-1.989	-0.860
λ_2	3	1.688	1.422	3	1.363	1.127
q	5	6.383	4.553	5	6.697	5.151
r	8	7.610			7.506	

The comparison of the average AIC and MLE's of the parameters between the skew-slash t and skew-slash normal based on sample size 250 and repetitions 200. SSLT=Skew-Slash t and SSLN=Skew-Slash Normal. A smaller AIC indicates a better fit.

function. We investigate its tractable properties such as heavy tail behavior and closeness of marginal distributions and linear combinations. We give the stochastic representations, moments and characteristic function. We close this section with an example which graphically displays the densities. In Section 3, we define the skew-slash t distribution, derive the densities and characteristic functions, and give the moments and distributions of linear combinations of these distributions. We first define the standard skew-slash distribution in Subsection 3.1 and study their relationships with other distributions. In subsection 3.2, we characterize the skew t, skew-slash normal and skew-slash t distributions using hidden truncation or selective sampling model and the order statistics of the components of a bivariate normal or t variable. In Subsection 3.2, we define the general multivariate skew-slash distribution. An example is presented to illustrate the densities of the proposed distributions. Section 4 covers parameter estimation and statistical inference. Here we briefly discuss the maximum likelihood and Byesian approaches. Section 5 is devoted to simulations as well as applications of the proposed skew-slash t distribution to fit two real datasets. Finally, some concluding remarks are given in Section 6.

2 The multivariate slash *t* distribution

In this section, we define the multivarate slash t distribution, derive the density and study its tail behaviors and relationships with other distributions. We give the stochastic representations, moments, and characteristic function and discuss marginal distributions and linear combinations. We close this section with an example.

Let us first recall the multivariate t distribution. There are several variants of the definitions in the literature and we will adopt the following one. Details can be found in e.g. Kotz and Nadarajah (2004) or Johnson and Kotz (1972, p. 134). A continuous k-variate random vector $\mathbf{T}$ has a t distribution with degrees of freedom r, mean vector $\mathbf{m}$, and correlation matrix $\mathbf{R}$ (covariance matrix Σ), written $\mathbf{T} \sim t_k(r, \mathbf{m}, \mathbf{R})$, if it has the probability density function (pdf) given by

$$t_k(\mathbf{t}; r, \mathbf{m}, \mathbf{R}) = \frac{\Gamma\left(\frac{r+k}{2}\right)}{(r\pi)^{k/2}\Gamma\left(\frac{r}{2}\right)|\mathbf{R}|^{1/2}}\left(1 + \frac{(\mathbf{t}-\mathbf{m})^\top \mathbf{R}^{-1}(\mathbf{t}-\mathbf{m})}{r}\right)^{-(r+k)/2}, \quad \mathbf{t} \in \mathbb{R}^k,$$

where $|\mathbf{M}|$ denotes the determinant of a square matrix $\mathbf{M}$. If $\mathbf{m} = 0$ and $\mathbf{R} = \mathbf{I}_k$ where $\mathbf{I}_k$ denotes the $k \times k$ indentity matrix, it is referred to as the standard k-variate t distribution and denoted by $t_k(r)$. We now introduce the k-variate slash t distribution. Write $U \sim U(0, 1)$ for the rv uniformly distributed over $(0, 1)$.

Definition 1. A k-variate continuous random vector $\mathbf{X}$ is said to have a *slash t distribution* with tail parameter $q > 0$, degrees of freedom r, $\mathbf{m} \in \mathbb{R}^k$, and matrix $\mathbf{R}$, written $\mathbf{X} \sim SLT_k(q, r, \mathbf{m}, \mathbf{R})$, if it can be expressed as the ratio of two independent rv's $\mathbf{T} \sim t_k(r, \mathbf{m}, \mathbf{R})$ and $U \sim U(0, 1)$ as follows:

$$\mathbf{X} = \mathbf{T}/U^{1/q}.$$

When $\mathbf{m} = 0$ and $\mathbf{R} = \mathbf{I}_k$, it is referred to as the *standard (k-variate) slash t* and denoted by $SLT_k(q, r)$. It can be easily seen that the k-variate slash t distribution generalizes the k-variate t distribution as stated below.

Remark 1. *The limiting distribution of the slash t distribution $SLT_k(q, r)$, as $q \to \infty$, is the student t distribution $t_k(r)$.*

Let us now derive the density of the slash t distribution. Note that the joint density of $\mathbf{T}$ and U is

$$t_k(\mathbf{t}; r, m, R)\mathbf{1}_{[0,1]}(u), \quad \mathbf{t} \in \mathbb{R}^k,\ u \in [0, 1]\,.$$

For the substitution $v = u^{1/q}, \mathbf{x} = \mathbf{t}/u^{1/q} = \mathbf{t}/v$, the Jacobian determinant is qv^{k+q-1}. Hence the joint density of $(\mathbf{X}, V)$ is given by

$$qv^{k+q-1} t_k(v\mathbf{x}; r, \mathbf{m}, \mathbf{R})\mathbf{1}_{[0,1]}(v), \quad \mathbf{x} \in \mathbb{R}^k, v \in [\,0, 1]\,.$$

Integrating v out yields the density $f_k(\mathbf{x}; q, r, \mathbf{m}, \mathbf{R})$ of $\mathbf{X}$ as follows:

$$f_k(\mathbf{x}; q, r, \mathbf{m}, \mathbf{R}) = \int_0^1 qv^{k+q-1} t_k(\mathbf{x}v; r, \mathbf{m}, \mathbf{R})\, dv, \quad \mathbf{x} \in \mathbb{R}^k. \tag{2.1}$$

From this density it immediately follows that the standard k-variate slash t distribution $SLT_k(q, r)$ is symmetric about 0 as the standard k-variate t is so.

The heavy-tail behavior The cumulative distribution function (cdf) of the k-variate slash t is given by

$$F_k(\mathbf{x}; q, r, \mathbf{m}, \mathbf{R}) = \int_{-\infty}^{\mathbf{x}} f_k(\mathbf{y}; q, r, \mathbf{m}, \mathbf{R})\, d\mathbf{y} = \int_0^1 qv^{k+q-2} H_k(\mathbf{x}v; r, \mathbf{m}, \mathbf{R})\, dv,$$

for $\mathbf{x} \in \mathbb{R}^k$, where H_k is the cdf of the k-variate $t_k(r, \mathbf{m}, \mathbf{R})$. Denote by $\bar{H}_k = 1 - H_k$ the survival function. The survival function of the k-variate slash t is then given by

$$\bar{F}_k(\mathbf{x}; q, r, \mathbf{m}, \mathbf{R}) = \frac{k-1}{k+q-1} + \int_0^1 qv^{k+q-2} \bar{H}_k(\mathbf{x}v; r, \mathbf{m}, \mathbf{R})\, dv,\ \mathbf{x} \in \mathbb{R}^k. \tag{2.2}$$

Let us focus on the standard univariate case and write $f = f_1, F = F_1, \bar{F} = 1 - F$ and $H = H_1, \bar{H} = 1 - H$. It is well known that the univariate t distribution $t_1(r, 0, 1)$ is a heavy tailed distribution with tail index r. In other words, the survival function decays at the rate of the power function r:

$$\bar{H}(t; r, 0, 1) \propto t^{-r}, \quad t \to \infty,$$

where $a(t) \propto b(t)$ if $\limsup_{t\to\infty} a(t)/b(t) < \infty$. Let c be the constant in the density of the k-variate t. Then by L'Hopital' Rule we have

$$\lim_{x\to\infty} \frac{\bar{H}(x)}{x^{-r}} = c \lim_{x\to\infty} \frac{\int_x^\infty (1+t^2/r)^{-(r+1)/2}\,dt}{x^{-r}}$$
$$= c \lim_{x\to\infty} \frac{-(1+x^2/r)^{-(r+1)/2}\,dt}{-rx^{-r-1}} = cr^{(r-1)/2} < \infty.$$

This shows the above rate holds. Similarly, one can show that if $q > r$ then

$$\bar{F}(x; q, r, 0, 1) \propto x^{-r}, \quad x \to \infty.$$

This manifests that the standard univariate slash t is also heavy tailed.

Further, by (2.2) and in view of $\bar{H}(xv; r, 0, 1) \geq \bar{H}(x; r, 0, 1)$ for $v \in [0, 1]$ one derives

$$\bar{F}(x; q, r, 0, 1) \geq (k-1)/(k+q-1) + \bar{H}(x; r, 0, 1) \int_0^1 qv^{k+q-2}\,dv$$
$$\geq (k-1)/(k+q-1) + q/(k+q-1)\bar{H}(x; r, 0, 1)$$
$$\geq \bar{H}(x; r, 0, 1), \quad x \geq 0.$$

This shows that the standard univariate slash t has heavier tails than the standard univariate t. In fact, the last inequality also holds for k-variate slash t for $\mathbf{x} = (x_1, \ldots, x_k)^\top$ with $x_i \geq 0, i = 1, \ldots, k$ as

$$\begin{aligned}\bar{F}_k(\mathbf{x}; q, r, \mathbf{m}, \mathbf{R}) &= P(X_1 > x_1, \ldots, X_k > x_k)\\ &= \int_0^1 P(T_1 > u^{1/q}x_1, \ldots, T_k > u^{1/q}x_k)\,du\\ &\geq \int_0^1 P(T_1 > x_1, \ldots, T_k > x_k)\,du = \bar{H}_k(\mathbf{x}; r, \mathbf{m}, \mathbf{R}).\end{aligned}$$

Stochastic representations Stochastic Representations not only reveal the relations with other distributions but are very useful, for instance, in calculating moments and random generation. We provide two stochastic representations for the slash t distribution based on the two stochastic representations of the multivariate t distribution. According to Kafadar (1988), a continuous random variable ξ has a slash normal distribution, written $\boldsymbol{\xi} \sim SLN(q, \mathbf{0}, \Sigma)$, if it can be expressed as $\boldsymbol{\xi} = \mathbf{Z}/U^{1/q}$ where $\mathbf{Z} \sim N_k(0, \Sigma)$ and $U \sim U(0, 1)$ are independent.

Note that if a rv $\mathbf{T}$ has a k-variate t distribution $t_k(r, \mathbf{m}, \mathbf{R})$, then it has the stochastic representation

$$\mathbf{T} = S^{-1}\mathbf{Z} + \mathbf{m},$$

where $\mathbf{Z} \sim N_k(0, \mathbf{R})$, rS^2 has the Chi-square distribution χ_r^2 with r degrees of freedom, and $\mathbf{Z}$ and S are independent. This can be easily verified. Let $\mathbf{X} \sim SLT_k(q, r, \mathbf{m}, \mathbf{R})$. Then from the definition of the k-variate slash t it immediately follows that

$$\mathbf{X} = S^{-1}\boldsymbol{\xi} + \boldsymbol{\eta}, \tag{2.3}$$

where $\boldsymbol{\xi} \sim SLN_k(q, 0, \mathbf{R})$, $\boldsymbol{\eta} = \mathbf{m}U^{-1/q}$, and both $\boldsymbol{\xi}$ and $\boldsymbol{\eta}$ are independent of S.

Using another stochastic representation of the k-variate t distribution from page 7 of Kotz and Nadarajah (2004), we obtain the second stochastic representation for the k-variate slash t rv $\mathbf{X} \sim SLT_k(q, r, \mathbf{m}, \mathbf{R})$ as follows:

$$\mathbf{X} = \mathbf{V}^{-1/2}\boldsymbol{\xi} + \boldsymbol{\eta}, \tag{2.4}$$

where $\boldsymbol{\xi} \sim SLN_k(q, 0, r\mathbf{I}_k)$, $\boldsymbol{\eta} = \mathbf{m}U^{-1/q}$, and $\boldsymbol{\xi}, \boldsymbol{\eta}$ are independent of $\mathbf{V}$. Here $\mathbf{V}^{-1/2}$ is the inverse of the symmetric square root $\mathbf{V}^{1/2}$ of $\mathbf{V}$, where $\mathbf{V}$ has a k-variate Wishart distribution with degrees of freedom $r + k - 1$ and covariance matrix $\mathbf{R}^{-1}$.

The moments Let us now calculate the mean vector and covariance matrix of the k-variate slash t. For $\mathbf{X} \sim SLT_k(q, r, \mathbf{m}, \mathbf{R})$ with $\mathbf{R} = (R_{ij})$, by the independence of $\mathbf{T}$ and U we have

$$\boldsymbol{\mu} = \mathbb{E}(\mathbf{X}) = \mathbb{E}(\mathbf{T}/U^{1/q}) = \mathbb{E}(\mathbf{T})\mathbb{E}(U^{-1/q}).$$

It is easy to calculate

$$\mathbb{E}(U^{-1/q}) = q/(q-1), \quad q > 1, \tag{2.5}$$

and

$$\mathbb{E}(U^{-2/q}) = q/(q-2), \quad q > 2. \tag{2.6}$$

For $\mathbf{T} \sim t_k(r, \mathbf{m}, \mathbf{R})$, from page 11 of Kotz and Nadarajah (2004) it follows

$$\mathbb{E}(\mathbf{T}) = \mathbf{m}, \quad \mathrm{Var}(\mathbf{T}) = \mathbf{R}r/(r-2), \quad r > 2. \tag{2.7}$$

Hence by (2.5) and the first equality of (2.7) one has

$$\boldsymbol{\mu} = \mathbf{m}q/(q-1). \tag{2.8}$$

To calculate $\mathrm{Var}(\mathbf{X})$ we use the formula

$$\mathrm{Var}(\mathbf{X}) = \mathrm{Var}(\mathbb{E}(\mathbf{X}|U)) + \mathbb{E}\big(\mathrm{Var}(\mathbf{X}|U)\big). \tag{2.9}$$

It is easy to see for $r > 2$,

$$\mathrm{Var}(\mathbf{X}|U) = \mathrm{Var}(\mathbf{T})/U^{2/q} = \mathbf{R}r/((r-2)U^{2/q}),$$

hence by (2.6) one has

$$\mathbb{E}\big(\mathrm{Var}(\mathbf{X}|U)\big) = \mathbf{R}rq/((r-2)(q-2)), \quad q > 2, r > 2. \tag{2.10}$$

Also since $\mathbb{E}(\mathbf{X}|U) = \mathbb{E}(\mathbf{T})/U^{1/q} = \mathbf{m}/U^{1/q}$ it follows from (2.6) that

$$\mathbb{E}((\mathbb{E}(\mathbf{X}|U))^{\otimes 2}) = \mathbf{m}^{\otimes 2}q/(q-2), \quad \mathbb{E}(\mathbf{X}^{\otimes 2}) = \mathbf{m}^{\otimes 2}q^2/(q-1)^2, \quad q > 2,$$

where $\mathbf{M}^{\otimes 2} = \mathbf{M}\mathbf{M}^{\top}$. Hence for $q > 2$,

$$\begin{aligned}\mathrm{Var}(\mathbb{E}(\mathbf{X}|U)) &= \mathbf{m}^{\otimes 2}q/(q-2) - \mathbf{m}^{\otimes 2}q^2/(q-1)^2 \\ &= \mathbf{m}^{\otimes 2}q/((q-1)^2(q-2)), \quad q > 2.\end{aligned}$$

This, (2.10) and (2.9) yield the variance-covariance matrix $\mathbf{X}$ as follows:

$$\mathrm{Var}(\mathbf{X}) = \frac{rq\mathbf{R}}{(r-2)(q-2)} + \frac{q\mathbf{m}^{\otimes 2}}{(q-1)^2(q-2)}, \quad q > 2, r > 2. \tag{2.11}$$

In particular, if $\mathbf{m} = 0$ one has

$$\mathrm{Var}(\mathbf{X}) = \frac{rq\mathbf{R}}{(r-2)(q-2)}, \quad q > 2, r > 2. \tag{2.12}$$

Hence the k-variate slash t has the same correlation matrix $\mathbf{R}$ as the k-variate t.

In the case of the standard t, there are convenient formulae for the moments. Given non-negative integers $p_1, \ldots, p_k$ such that $p = p_1 + \ldots + p_k < r/2$. If any of the $p_1, \ldots, p_k$ is odd, then

$$\mathbb{E}\big(T_1^{p_1} \cdots T_k^{p_k}\big) = 0.$$

If all of them are even and $r > p$, then

$$\mathbb{E}(T_1^{p_1}\cdots T_k^{p_k}) = \frac{r^{p/2}\prod_{j=1}^{k}\left[1\cdot 3\cdot 5\cdots(2p_j-1)\right]}{(r-2)(r-4)\cdots(r-p)} := c. \tag{2.13}$$

For details, see Kotz and Nadarajah (2004). Based on these formulae we have the following.

Theorem 1. *Let $\mathbf{X} = (X_1,\ldots,X_k)^\top \sim SLT_k(q,r)$. Assume $p_1,\ldots,p_k$ are nonnegative integers such that $p = p_1 + \ldots + p_k < r/2$.*

1) Suppose at leasts one of the $p_1,\ldots,p_k$ is odd. If $q > p$, then

$$\mathbb{E}(X_1^{p_1}\cdots X_k^{p_k}) = 0.$$

2) Suppose all $p_1,\ldots,p_k$ are even. If $q > p$ then

$$\mathbb{E}(X_1^{p_1}\cdots X_k^{p_k}) = cq/(q-p),$$

where c is given in (2.13). Otherwise if $q \le p$ then $\mathbb{E}(X_1^{p_1}\cdots X_k^{p_k})$ diverges.

PROOF. By the density formula (2.1) and using the substitution $\mathbf{t} = v\mathbf{x}$ we have

$$\begin{aligned}\mathbb{E}(X_1^{p_1}\cdots X_k^{p_k}) &= \int x_1^{p_1}\cdots x_k^{p_k}\left\{\int_0^1 qv^{k+q-1}t_k(vx_1,\ldots,vx_k;r)\,dv\right\}dx_1\cdots dx_k\\ &= \int_0^1 qv^{q-1-p}\,dv\left\{\mathbb{E}(T_1^{p_1}\cdots T_k^{p_k})\right\}.\end{aligned}$$

Note that the integral $\int_0^1 qv^{q-1-p}\,dv$ converges to $q/(q-p)$ if $q-p > 0$ and diverges otherwise. These and (2.13) yield the desired results. □

The marginal distributions Since the marginal distributions of a k-variate t are still t, the marginal distributions of a k-variate slash t are slash t.

Theorem 2. *The marginal distributions of a k-variate slash t distribution are still slash t.*

PROOF. It suffices to show without loss of generality that for every $0 \le s \le k$,

$$\int\cdots\int f_k(x_1,\ldots,x_s,x_{s+1},\ldots,x_k)\,dx_{s+1}\cdots dx_k = f_s(x_1,\ldots,x_s),\quad x_1,\ldots,x_s\in\mathbb{R},$$

where $f_k(\mathbf{x}) = f_k(\mathbf{x};q,r,\mathbf{m},\mathbf{R})$. Substitution of the density (2.1) in the left hand of the above equality gives

$$\int\cdots\int f_k(\mathbf{x})\,dx_{s+1}\cdots dx_k = \int_0^1 qv^{k+q-1}\int\cdots\int t_k(v\mathbf{x})dx_{s+1}\cdots dx_k dv,$$

where $t_k(\mathbf{t}) = t_k(\mathbf{t};r,\mathbf{m},\mathbf{R})$. By substitution $y_{s+1} = vx_{s+1},\ldots,y_k = vx_k$ one derives

$$\int\cdots\int t_k(v\mathbf{x})dx_{s+1}\cdots dx_k = v^{s-k}\int\cdots\int t_k(vx_1,\ldots,vx_s,y_{s+1},\ldots,y_k)\,dy_{s+1}\cdots dy_k.$$

Because the marginals of the k-variate t distribution are still t, we have

$$\int\cdots\int t_k(vx_1,\ldots,vx_s,y_{s+1},\ldots,y_k)\,dy_{s+1}\cdots dy_k = t_s(vx_1,\ldots,vx_k;r,\mathbf{m}_1,\mathbf{R}_{11}),$$

where $\mathbf{m} = (\mathbf{m}_1^\top, \mathbf{m}_2^\top)^\top$ with $\mathbf{m}_1 \in \mathbb{R}^s$ and $\mathbf{R}$ is partitioned into the 2×2 block matrix with $\mathbf{R}_{11}$ being the $s \times s$ matrix at the position of $(1,1)$-block. See pages 15-16 of Kotz and Nadarajah (2004). Combining the last two equalities yields the desired equality. □

Linear combinations Since the distribution of a linear function of a k-variate t variable is still t, it immediatly yields the following.

Theorem 3. *Let* $\mathbf{A}$ *be a nonsingular nonrandom matrix. If* $\mathbf{X} \sim SLT_k(q, r, \mathbf{m}, \mathbf{R})$, *then* $\mathbf{AX} \sim SLT_k(q, r, \mathbf{Am}, \mathbf{ARA}^\top)$.

The characteristic function Several authors derived the formulas for the characteristic functions $\varphi_{\mathbf{T}}$ of the k-variate t rv $\mathbf{T}$, see e.g. Joarder and Ali (1996), Dreier and Kotz (2002). Based on these formulas we can obtain the characteristic functions of the k-variate slash t using the following formula. For $\mathbf{X} \sim SLT_k(q, r, \mathbf{m}, \mathbf{R})$, it can be expressed as the ratio $\mathbf{X} = \mathbf{T}/U^{1/q}$ of two independent rv's, so that its characteristic function $\varphi_{\mathbf{X}}$ can be written as

$$\varphi_{\mathbf{X}}(\mathbf{t}) = \mathbb{E}(\exp(i\mathbf{t}^\top \mathbf{X})) = \int_0^1 \mathbb{E}(\exp(i\mathbf{t}^\top \mathbf{T} u^{-1/q}))\, du = \int_0^1 \varphi_{\mathbf{T}}(\mathbf{t} u^{-1/q})\, du \tag{2.14}$$

for $\mathbf{t}$ in some neighborhood $\mathbb{N}$ of the origin in which the above integral converges.

Example 1. For the standard univariate and bivariate slash t distributions $SLT_k(q, r), k = 1, 2$, their densities are given by

$$f_1(x; q, r) = \int_0^1 q v^q t_1(vx; r)\, dv, \quad x \in \mathbb{R},$$

and

$$f_2(\mathbf{x}; q, r) = \int_0^1 q v^{q+1} t_2(v\mathbf{x}; r)\, dv, \quad \mathbf{x} \in \mathbb{R}^2.$$

Displayed in Figure 2 are the density curves and contours. On the left panel are the density curves of the normal, t and the standard slash t with $q = 1$ and $r = 3$ degrees of freedom. The curves are calibrated so that they have the same height at the origin. Observe that the slash t has the fattest tail whereas the normal has the slimmest one. On the right panel are the contours of the bivariate slash t with $q = 3$ and $r = 5$. Clearly the contours are symmetric.

3 The multivariate skew-slash *t* distributions

In this section, we first recall the skew normal and skew t. In subsection 3.1, we define the standard skew-slash t distribution, study its relationships with other distributions and give the moments and characteristic function. In subsection 3.2, we use hidden truncation or selective sampling model and the order statistics to characterize the skew, slash and skew-slash normal and t distributions. In subsection 3.2, we define the general skew-slash t distribution, study its linear transformation and give an example in the end.

Azzalini and Dalla Valle (1996) introduced the skew normal distribution. A k-variate standard skew normal distribution has the density given by

$$2\phi_k(\mathbf{z})\Phi(\boldsymbol{\lambda}^\top \mathbf{z}), \quad \boldsymbol{\lambda}, \mathbf{z} \in \mathbb{R}^k,$$

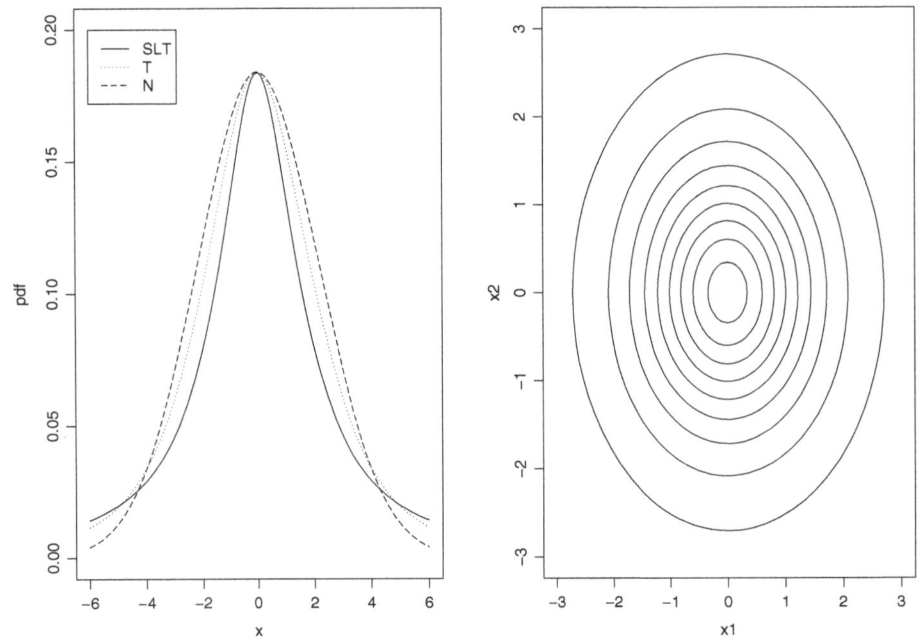

Figure 2 Density curves and contours of slash *t*. Left panel: Density curves of normal (N), student *t* (T), and the slash *t* ($SLT_1(1, 3)$). Right panel: Contours of the bivariate slash *t* distribution $SLT_2(3, 5)$.

where ϕ_k is the pdf of the k-variate standard normal $N_k(\mathbf{0}, \mathbf{I})$ and Φ is the cdf of the standard normal $N(0, 1)$. Denote it by $SKN_k(\boldsymbol{\lambda})$.

Kotz and Nadarajah (2004) wrote "...that the possibilities of constructing skewed multivariate t distributions are practically limitless". The two authors surveyed the definitions given by Gupta (2003), Sahu et al. (2003), Jones (2002) and Azzalini and Capitanio (2003).

Based on these definitions, we may define the skew-slash t distributions in different ways. In this article, however, we will take the following approach. First, we will define the standard skew-slash t based on the standard skew t, a common special case of Gupta (2003), Azzalini and Capitanio (2003), and others. We then introduce a general skew-slash t distribution by introducing location and scale parameters. Our definition may lose some nice interpretations. But we think that this definition is natural, concise and, in particular, convenient in applications.

According to Gupta (2003), a k-variate skew t distribution with parameters $\boldsymbol{\mu} \in \mathbb{R}^k$, Σ (correlation matrix $\mathbf{R}$), $\boldsymbol{\lambda} \in \mathbb{R}^k$ and $r > 0$ has the density

$$h_k(\mathbf{t}; \boldsymbol{\lambda}, r, \Sigma) = 2t_k(\mathbf{t}; r, 0, \mathbf{R})\Psi\left(\frac{\boldsymbol{\lambda}^\top \mathbf{t}}{\sqrt{1 + \mathbf{t}^\top \Sigma^{-1}\mathbf{t}/r}}; r + k\right), \quad \mathbf{t} \in \mathbb{R}^k, \tag{3.15}$$

where Ψ is the cdf of the univariate standard t distribution with $r + k$ degrees of freedom. Denote this distribution by $SKT_k(\boldsymbol{\lambda}, r, \Sigma)$. This form of the density given here is slightly different from that of Gupta (2003). Several constant parameters appeared in his density formual are not explicitly expressed in our form of the density. We have incorporated them in the parameters in the above density. Accordingly parameters of the same names may have different values.

When Σ is the identity matrix $\mathbf{I}_k$, $SKT_k(\boldsymbol{\lambda}, r, \mathbf{I}_k)$ is referred to as the *standard skew t* and denoted by $SKT_k(\boldsymbol{\lambda}, r)$ with density $h_k(\mathbf{t}; \boldsymbol{\lambda}, r)$ given by

$$h_k(\mathbf{t}; \boldsymbol{\lambda}, r) = 2t_k(\mathbf{t}; r)\Psi\left(\frac{\boldsymbol{\lambda}^\top \mathbf{t}}{\sqrt{1+\mathbf{t}^\top \mathbf{t}/r}}; r+k\right), \quad \mathbf{t} \in \mathbb{R}^k,$$

where $t_k(\mathbf{t}; r)$ is the density of the standard k-variate distribution $t_k(r)$ with degrees of freedom r. This is a common special case of the skew t shared by Gupta (2003), Azzalini and Capitanio (2003), and others.

3.1 The standard multivariate skew-slash *t* distribution

We begin with the definition, followed the moments and characteristic function.

Definition 2. A k-variate continuous random vector $\mathbf{W}_0$ is said to have a standard multivariate skew-slash t distribution with skewness parameter $\boldsymbol{\lambda}$, tail parameter q and degrees of freedom r, written $\mathbf{W}_0 \sim SSLT_k(\boldsymbol{\lambda}, q, r)$, if it can be written as the ratio of two independent rv's § $\sim SKT_k(\boldsymbol{\lambda}, r)$ and $U \sim U(0,1)$ as follows:

$$\mathbf{W}_0 = \mathbf{S}/U^{1/q}.$$

The standard skew-slash t generalizes the proposed standard k-variate slash t, the standard skew t of Gupta (2003) and Azzalini and Capitanio (2003), the standard slash normal of Kafadar (1988), and the standard skew-slash normal of Wang and Genton (2006). This is stated below.

Remark 2. *The limiting distribution of the standard skew-slash t distribution* $SSLT_k(\boldsymbol{\lambda}, q, r)$ *is, as* $q \to \infty$, *the standard skew t distribution* $SKT_k(\boldsymbol{\lambda}, r)$. *The limiting distribution of* $SSLT_k(\boldsymbol{\lambda}, q, r)$ *is, as* $r \to \infty$, *the standard skew-slash normal* $SSLN_k(\boldsymbol{\lambda}, q)$, *which includes as special cases the standard skew normal* $SKN_k(\boldsymbol{\lambda})$ $(q = \infty)$ *and the standard slash normal* $SLN_k(q)$ $(\boldsymbol{\lambda} = 0)$. *As* $\boldsymbol{\lambda} = 0$, $SSLT_k(\boldsymbol{\lambda}, q, r)$ *reduces to the slash t distribution* $SLT_k(q, r)$.

In a similar way to the derivation of the standard slash t density in Section 2, we can obtain the density of the standard skew-slash t distribution $SSLT_k(\boldsymbol{\lambda}, q, r)$ as follows: for $\mathbf{w} \in \mathbb{R}^k$,

$$g_k(\mathbf{w}; \boldsymbol{\lambda}, q, r) = 2q\int_0^1 u^{k+q-1} t_k(u\mathbf{w}; r)\Psi\left(\frac{u\boldsymbol{\lambda}^\top \mathbf{w}}{\sqrt{1+u^2\mathbf{w}^\top \mathbf{w}/r}}; k+r\right) du, \tag{3.16}$$

The moments Using the results of Azzalini and Dalla Valle (1996) or Kotz and Nadarajah (2004)(p.100-101) for the mean vector and covariance of the k-variate skew t distribution (i.e. by setting their $\boldsymbol{\alpha}$ equal to $\boldsymbol{\lambda}/\sqrt{1+p/\nu}$ with their $p = k$ and $\nu = r$ here), we obtain the mean vector and covariance matrix of $\mathbf{W}_0 \sim SSLT_k(\boldsymbol{\lambda}, q, r)$ as follows:

$$\mathbb{E}(\mathbf{W}_0) = \mathbb{E}(\mathbf{S})\mathbf{E}(U^{-1/q}) = \frac{\sqrt{2}qr}{\sqrt{\pi}(q-1)(r-2)} \frac{\boldsymbol{\lambda}}{\sqrt{r+k+r\boldsymbol{\lambda}^\top \boldsymbol{\lambda}}}, \quad q > 1, r > 2,$$

$$\text{Var}(\mathbf{W}_0) = \frac{qr}{(q-2)(r-2)(r-4)}\left(\mathbf{I}_k - \frac{2(r+4)}{\pi(r-2)} \frac{r\boldsymbol{\lambda}\boldsymbol{\lambda}^\top}{r+k+r\boldsymbol{\lambda}^\top \boldsymbol{\lambda}}\right), \quad q > 2, r > 4.$$

The characteristic function As in deriving the characteristic function for the multivariate slash t in (2.14), one can obtain the characteristic function $\varphi_{\mathbf{W}_0}$ of $\mathbf{W}_0$ as follows:

$$\varphi_{\mathbf{W}_0}(\mathbf{t}) = \mathbb{E}(\exp(i\mathbf{t}^\top \mathbf{W}_0)) = \int_0^1 \varphi_{\mathbf{S}}(\mathbf{t}u^{-1/q})\, du, \quad \mathbf{t} \in \mathbb{N}$$

for some neighborhood $\mathbb{N}$ of the origin in which the above integral converges, where $\varphi_{\mathbf{S}}$ is the characteristic function of the standard skew t distribution $\mathbf{S} \sim SKT_k(\boldsymbol{\lambda}, r)$.

3.2 Hidden truncation and order-statistics characterization

In this subsection, we characterize the skew t, skew-slash normal and skew-slash t distributions using the hidden truncation or selective sampling model and the order statistics of the components of a bivariate normal or t variable.

Hidden truncation or selective sampling We first give a fact about conditional pdf. Let $\mathbf{X}$ be a continuous random vector that has pdf f. Let X_0 be a random variable with cdf F_0. Let a be a measurable function of $\mathbf{X}$ such that $P(A) > 0$ with $A = \{a(\mathbf{X}) \geq X_0\}$. Suppose $\mathbf{X}$ and X_0 are independent. Then for every $\mathbf{x}$,

$$\begin{aligned} P(\mathbf{X} \leq \mathbf{x} | a(\mathbf{X}) \geq X_0) &= P(\mathbf{X} \leq \mathbf{x}, a(\mathbf{X}) \geq X_0)/P(A) \\ &= E(\mathbf{1}[\,\mathbf{X} \leq \mathbf{x}]\, F_0(a(\mathbf{X})))/P(A) = \int_{-\infty}^{\mathbf{x}} f(\mathbf{y}) F_0(a(\mathbf{y}))\, d\mathbf{y}/P(A). \end{aligned}$$

Hence the conditional pdf of $\mathbf{X}$ given A is

$$f(\mathbf{x}|A) = f(\mathbf{x}) F_0(a(\mathbf{x}))/P(A). \tag{3.17}$$

Using this we immediately derive the following results.

Proposition 1. *Suppose* $\mathbf{T} \sim t_k(r)$ *and* $T_0 \sim t_1(r+k)$ *are independent. Then the conditional pdf of* $\mathbf{T}$ *given* A *is*

$$f(\mathbf{t}|A) = t_k(\mathbf{t}; r)\Psi(a(\mathbf{t}); r+k)/P(A), \quad \mathbf{t} \in \mathbb{R}^k. \tag{3.18}$$

Consequently, if $U \sim U(0,1)$ *is independent of both* $\mathbf{T}$ *and* T_0 *then the conditional pdf of* $\mathbf{T}/U^{1/q}$ *for* $0 < q < 1$ *given* A *is*

$$g(\mathbf{t}|A) = 2q \int_0^1 u^{k+q-1} t(u\mathbf{t}; r)\Psi(a(u\mathbf{t}); r+k)\, du/2P(A), \quad \mathbf{t} \in \mathbb{R}^k. \tag{3.19}$$

In particular, both (3.18) and (3.19) hold for $a(\mathbf{t}) = (\tau_0 + \boldsymbol{\tau}^\top \mathbf{t})/\sqrt{1 + \mathbf{t}^\top \mathbf{t}/r}$ *where* $\tau_0, \boldsymbol{\tau}$ *are arbitrary constants. In this case,*

$$f(\mathbf{t}|A) = 2t_k(\mathbf{t}; r)\Psi\left(\frac{\tau_0 + \boldsymbol{\tau}^\top \mathbf{t}}{\sqrt{1 + \mathbf{t}^\top \mathbf{t}/r}}; r+k\right)/2P(A), \quad \mathbf{t} \in \mathbb{R}^k, \tag{3.20}$$

$$g(\mathbf{t}|A) = 2q \int_0^1 u^{k+q-1} t(u\mathbf{t}; r)\Psi\left(\frac{\tau_0 + u\boldsymbol{\tau}^\top \mathbf{t}}{\sqrt{1 + u^2 \mathbf{t}^\top \mathbf{t}/r}}; r+k\right) du/2P(A). \tag{3.21}$$

Remark 3. *The density function in (3.21) is the conditional pdf of the k-variate slash t rv* $\mathbf{X} = \mathbf{T}/U^{1/q}$ *given A. It is noteworthy that the hidden truncation model yields the pdf (3.20) and (3.21), the former is proportional to the pdf (3.15) of the skew t distribution and the latter is proportional to the pdf (3.16) of the proposed skew-slash t distribution.*

For more discussion see e.g. Chapter 6 of Genton (2004) and the references therein. The skew-slash normal of Wang and Genton (2006) is the special case of $r = \infty$.

Order statistics characterization Generalizing (c) of Theorem 1 in Arnold and Lin (2004), we give the following fact. Let (Y_1, Y_2) be a bivarate rv with pdf f and cdf F. Let $Y_{1,2} = \min(Y_1, Y_2)$ and $Y_{2,2} = \max(Y_1, Y_2)$ be the order statistics of the components of the bivarate random vector. Then

$$P\left(Y_{1,2} > y_1\right) = P(Y_2 \geq Y_1 > y_1) + P(Y_1 > Y_2 > y_1)$$
$$= \int_{y_1}^{\infty} dx_1 \int_{x_1}^{\infty} f(x_1, x_2)\, dx_2 + \int_{y_1}^{\infty} dx_2 \int_{x_2}^{\infty} f(x_1, x_2)\, dx_1.$$

Thus the pdf $f_{(1)}(y_1)$ of $Y_{1,2}$ is given by

$$f_{(1)}(y_1) = \int_{y_1}^{\infty} f(y_1, x_2)\, dx_2 + \int_{y_1}^{\infty} f(x_1, y_1)\, dx_1. \tag{3.22}$$

Let f_1, f_2 be the respective marginal pdf of Y_1, Y_2. Let

$$F_1(y_1|y_2) = P(Y_1 \leq y_1|Y_2 = y_2), \quad F_2(y_2|y_1) = P(Y_2 \leq y_2|Y_1 = y_1).$$

Then using $\int_{y_1}^{\infty} f(y_1, x_2)\, dx_2 = f_1(y_1) \int_{y_1}^{\infty} f(y_1, x_2)/f_1(y_1)\, dx_2 = f_1(y_1) \bar{F}_2(y_1)$ and (3.22) we derive

$$f_{(1)}(y_1) = f_1(y_1) \bar{F}_2(y_1|y_1) + f_2(y_1) \bar{F}_1(y_1|y_1). \tag{3.23}$$

Similarly we derive the pdf $f_{(2)}(y_2)$ of $Y_{2,2}$ below:

$$f_{(2)}(y_2) = f_2(y_2) F_1(y_2|y_2) + f_1(y_2) F_2(y_2|y_2). \tag{3.24}$$

In their Theorem 1, Arnold and Lin (2004) showed that the order statistics of the components of a random vector from a bivariate normal distribution obey the skew-normal law. Using (3.23) and (3.24) we can show that the order statistics of the components of a random vector from a bivariate t distribution obey the skew t law. Thus we extend their result from the normal to t distribution as stated below.

Proposition 2. *Let (T_1, T_2) have a bivariate t distribution $t_2(r, \mathbf{0}, \mathbf{R})$ with degrees of freedom r, mean zero vector and correlation matrix $\mathbf{R} = (1, \rho, \rho, 1)$ with $-1 < \rho < 1$. Define $T_{1,2} = \min(T_1, T_2)$ and $T_{2,2} = \max(T_1, T_2)$. Then the pdf $t_{(1)}, t_{(2)}$ of $T_{1,2}, T_{2,2}$ are given by*

$$t_{(1)}(t_1) = 2t_1(t_1; r)\Psi\left(\frac{-\lambda t_1}{\sqrt{1 + t_1^2/r}}; r+1\right), \quad t_1 \in \mathbb{R}, \tag{3.25}$$

$$t_{(2)}(t_2) = 2t_1(t_2; r)\Psi\left(\frac{\lambda t_2}{\sqrt{1 + t_2^2/r}}; r+1\right), \quad t_2 \in \mathbb{R}, \tag{3.26}$$

where $\lambda = \lambda(r, \rho) = \sqrt{(1 + 1/r)(1 - \rho)/(1 + \rho)}$.

Remark 4. *The density functions in (3.25) and (3.26) reduce to the result (c) of Theorem 1 of Arnold and Lin (2004) when the df $r = \infty$ as $\lambda(\infty, \rho) = \sqrt{(1 - \rho)/(1 + \rho)}$ is equal to their skewness parameter γ in the skew-normal distribution.*

PROOF OF PROPOSITION 2. Note first that T_1, T_2 have the same standard univariate t distribution $t_1(r)$. Using the formula for the conditional pdf of t distribution (see e.g. page 16 of Kotz and Nadarajah (2004)), the conditional cdf of T_2 given $T_1 = t_1$ can be written as

$$P(T_2 \leq t_2 | T_1 = t_1) = \Psi(b(t_1, t_2; r, \rho); r+1), \quad t_1, t_2 \in \mathbb{R},$$

where

$$b(t_1, t_2; r, \rho) = \frac{(t_2 - \rho t_1)\sqrt{1+1/r}}{\sqrt{(1+t_1^2/r)(1-\rho^2)}}.$$

Similarly,

$$P(T_1 \leq t_1 | T_2 = t_2) = \Psi(b(t_2, t_1; r, \rho); r+1), \quad t_1, t_2 \in \mathbb{R}.$$

We now apply (3.24), with both f_1 and f_2 equal to the pdf of $t_1(r)$ and both F_1 and F_2 equal to the cdf $\Psi(;r+1)$ of the t distribution $t(r+1)$, to obtain the pdf $t_{(2)}$ given in (3.26), noting in this case $b(t_2, t_2; r, \rho) = \lambda t_2/\sqrt{1+t_2^2/r}$. Aanloguously we can prove (3.25) in view of the equality $\bar{\Psi}(t;r) = \Psi(-r;r)$ by the symmetry of the univariate t distribution. This completes the proof. □

As a corollary of Proposition 2, we obtain a characterization of the skew-slash normal and t distributions through the order statistics of the components of a random vector from a bivariate t distribution as stated below.

Corollary 1. *Let (T_1, T_2) be given in Proposition 2. Assume $U \sim U(0,1)$ is independent of (T_1, T_2). Then the pdf $g_{(1)}, g_{(2)}$ of $T_{1,2}/U^{1/q}, T_{2,2}/U^{1/q}$ for $0 < q < 1$ are given by*

$$g_{(1)}(t_1) = 2q\int_0^1 u^q t_1(ut_1; r)\Psi\left(\frac{-u\lambda t_1}{\sqrt{1+u^2t_1^2/r}}; r+1\right) du, \quad t_1 \in \mathbb{R}, \tag{3.27}$$

$$g_{(2)}(t_2) = 2q\int_0^1 u^q t_1(ut_2; r)\Psi\left(\frac{u\lambda t_2}{\sqrt{1+u^2t_2^2/r}}; r+1\right) du, \quad t_2 \in \mathbb{R}. \tag{3.28}$$

Remark 5. *The density functions in (3.27) and (3.28) are (i) the pdf of the order statistics of the components of the random vector $(T_1, T_2)/U^{1/q}$ from the bivariate slash t distribution $t_2(r, q)$, and (ii) reduce to the case of the skew-slash normal of Wang and Genton (2006) when the df $r = \infty$.*

PROOF OF COROLLARY 1. The desired (3.27) follows from (3.25) and the equalities

$$P(T_{1,2}/U^{1/q} \leq t_1) = \int_0^1 P(T_{1,2} \leq u^{1/q}t_1)\, du = 2q\int_{-\infty}^{t_1}\int_0^1 v^q t_{(1)}(vs)\, dv\, ds,$$

where the independence of $T_{1,2}$ and U is used to claim the first equality while the second equality follows from a change of variables. Similarly (3.28) can be proved and this finishes the proof. □

The multivariate skew-slash t distributions We now introduce a general multivariate skew-slash t distribution by incorporating location and scale parameters.

Definition 3. A continuous k-variate rv $\mathbf{W}$ has a *multivariate skew-slash t distribution* with location $\boldsymbol{\mu}$, scale Σ, skewness parameter $\boldsymbol{\lambda} \in \mathbb{R}^k$, tail parameter q and degrees of freedom r, written $\mathbf{W} \sim SSLT_k(\boldsymbol{\lambda}, q, r, \boldsymbol{\mu}, \Sigma)$, if it can be represented as a linear transformation of the standard multivariate skew-slash t rv $\mathbf{W}_0 \sim SSLT_k(\boldsymbol{\lambda}, q, r)$ as follows:

$$\mathbf{W} = \boldsymbol{\mu} + \Sigma^{1/2}\mathbf{W}_0,$$

where $\Sigma^{1/2}$ is the the choleski decomposition of the positive definite covariance matrix Σ.

By a change of variables in (3.16), we derive the density of $\mathbf{W}$ given by

$$g_k(\mathbf{w}; \boldsymbol{\lambda}, q, r, \boldsymbol{\mu}, \Sigma) = 2q|\Sigma|^{-1/2} \int_0^1 u^{k+q-1} t_k(u\Sigma^{-1/2}(\mathbf{w} - \boldsymbol{\mu}); r) \times \Psi\left(\frac{u\boldsymbol{\lambda}^\top \Sigma^{-1/2}(\mathbf{w} - \boldsymbol{\mu})}{\sqrt{1 + u^2 Q(\mathbf{w}; \boldsymbol{\mu}, \Sigma)/r}}; k + r\right) du, \quad \mathbf{w} \in \mathbb{R}^k, \tag{3.29}$$

where $Q(\mathbf{w}; \boldsymbol{\mu}, \Sigma) = (\mathbf{w} - \boldsymbol{\mu})^\top \Sigma^{-1}(\mathbf{w} - \boldsymbol{\mu})$.

As in the case of the standard skew-slash t, one notices that the skew-slash t generalizes the slash t, the skew t of Azzalini and Capitanio, the slash normal of Kafadar, the skew normal of Azzalini and Dalla Valle and the skew-slash normal of Wang and Genton. This is stated below.

Remark 6. *The limiting distribution of the multivariate skew-slash t distribution $SSLT_k(\boldsymbol{\lambda}, q, r, \boldsymbol{\mu}, \Sigma)$, as q tends to infinity, is the skew t distribution $SKT_k(\boldsymbol{\lambda}, r, \boldsymbol{\mu}, \Sigma)$. The limiting distribution of $SSLT_k(\boldsymbol{\lambda}, q, r, \boldsymbol{\mu}, \Sigma)$ is, as r tends to infinity, the skew-slash normal $SSLN_k(\boldsymbol{\lambda}, q, \boldsymbol{\mu}, \Sigma)$, which include as special cases the k-variate skew normal $SKN_k(\boldsymbol{\lambda}, \boldsymbol{\mu}, \Sigma)$ $(q = \infty)$ and the k-variate slash normal $SLN_k(q; \boldsymbol{\mu}, \Sigma)$ $(\boldsymbol{\lambda} = 0)$. As $\boldsymbol{\lambda} = 0$, $SSLT_k(\boldsymbol{\lambda}, q, r, \boldsymbol{\mu}, \Sigma)$ simplifies to the k-variate slash t distribution SLT_k $(q, r, \boldsymbol{\mu}, \Sigma)$.*

Linear combinations Since the distribution of a linear function of a k-variate skew t variable is still skew t (see e.g. Section 5.9 of Kotz and Nadarajah (2004)), it immediatly yields the following result. Note that the relationship between our skewness parameter $\boldsymbol{\lambda}$ and their $\boldsymbol{\alpha}$ is $\boldsymbol{\lambda} = \sqrt{1 + k/r}\boldsymbol{\alpha}$. Let $\mathbf{D} = \mathrm{diag}(\sigma_{1,1}, \ldots, \sigma_{k,k})$ denote the diagonal matrix consisting of the diagonal entries of $\Sigma = (\sigma_{i,j})$ and $\mathbf{R} = \mathbf{D}^{-1/2}\Sigma\mathbf{D}^{-\top/2}$ be the correlation matrix.

Theorem 4. *Let $\mathbf{A}$ be a nonsingular matrix. If $\mathbf{W} \sim SSLT_k(\boldsymbol{\lambda}, q, r, \boldsymbol{\mu}, \Sigma)$, then $\mathbf{AW} \sim SSLT_k(\tilde{\boldsymbol{\lambda}}, q, r, \mathbf{A}\boldsymbol{\mu}, \tilde{\Sigma})$ where $\tilde{\Sigma} = \mathbf{A}\Sigma\mathbf{A}^\top$ and*

$$\tilde{\boldsymbol{\lambda}} = \frac{\tilde{\Sigma}^{-1/2}\mathbf{B}^\top\boldsymbol{\lambda}}{\sqrt{1 + (1 + k/r)^{-1}\boldsymbol{\lambda}^\top(\mathbf{R} - \mathbf{B}\tilde{\Sigma}^{-1}\mathbf{B}^\top)\boldsymbol{\lambda}}}, \qquad \mathbf{B} = \mathbf{D}^{-1/2}\Sigma\mathbf{A}.$$

To give graphical view of the skewness and tail behaviors of the skew-slash t distributions, we plot the density curves of the univariate standard skew-slash t and contours of the bivariate standard skew-slash t below.

Example 2. For the univariate and bivariate standard skew-slash t distributions $SSLT_1(\lambda, q, r)$ and $SSLT_2(\lambda_1, \lambda_2, q, r)$, the densities are given by

$$g_1(w; \lambda, q, r) = 2q \int_0^1 u^q t_1(uw; r) \Psi\left(\frac{u\lambda w}{\sqrt{1 + u^2 w^2/r}}; 1 + r \right) du, \quad w \in \mathbb{R}, \tag{3.30}$$

and, with $\mathbf{w} = (w_1, w_2)^\top \in \mathbb{R}^2$ and $\boldsymbol{\lambda} = (\lambda_1, \lambda_2)^\top$,

$$g_2(\mathbf{w}; \boldsymbol{\lambda}, q, r) = 2q \int_0^1 u^{q+1} t_2(u\mathbf{w}; r) \Psi\left(\frac{u(\lambda_1 w_1 + \lambda_2 w_2)}{\sqrt{1 + u^2(w_1^2 + w_2^2)/r}}; 2 + r \right) du. \tag{3.31}$$

Displayed in Figure 3 are the density curves of the standard skew-slash t distribution $SSLT_1(3, 1, 2)$, the skew t distribution $SKT_1(3, 2)$, and the slash t distribution $SLT_1(1, 2)$ distributions. The curves are shifted and rescaled for comparison. Observe that the skew-slash t has the fattest tail and is skewed to the right most.

Displayed in Figure 4 are the contour plots of the bivariate standard skew-slash t distribution $SSLT_2(\boldsymbol{\lambda}, 3, 5)$ for different values of the skewness parameter vector $\boldsymbol{\lambda}$. Clearly the contours are more skewed as the skewness parameter vector $\boldsymbol{\lambda}$ gets longer.

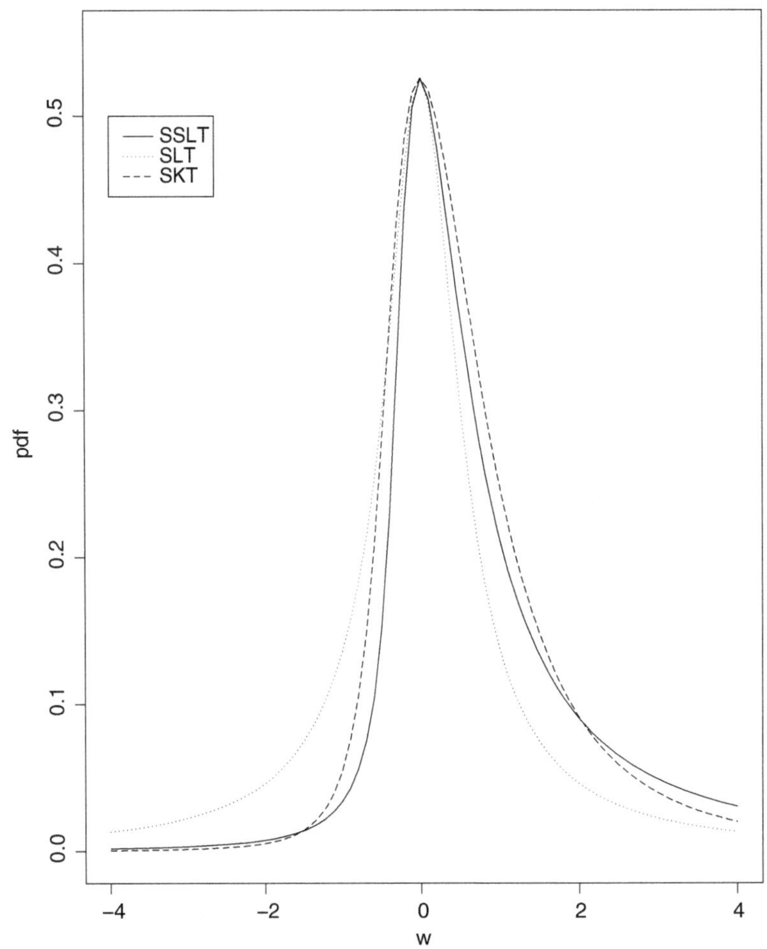

Figure 3 Density curves of skew-slash t $SSLT_1(3, 1, 2)$, slash t $SLT_1(1, 2)$, and skew t $SKT_1(3, 2)$.

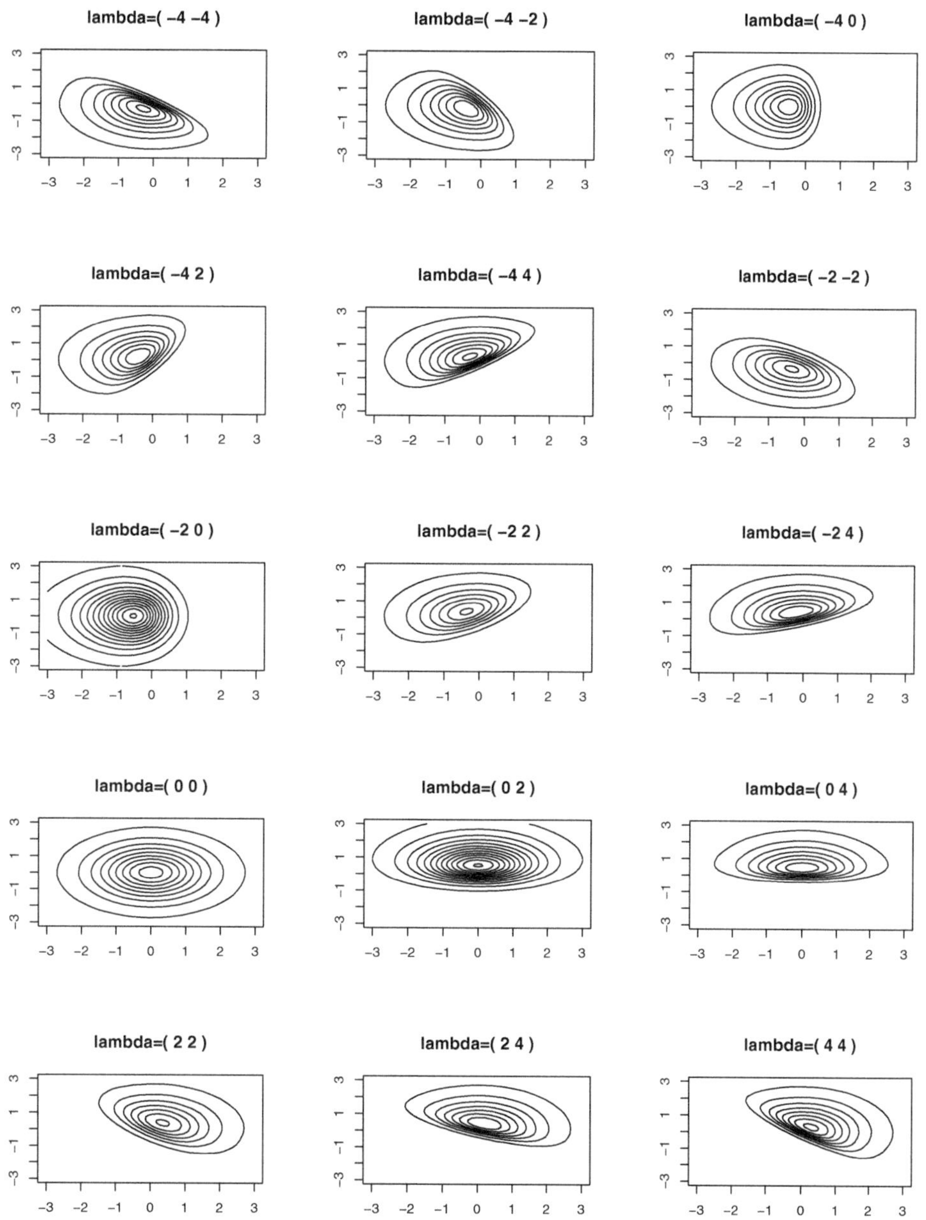

Figure 4 Contours of the standard 2-dimensional skew-slash *t* distribution $SSLT_2(\lambda, 3, 5)$.

4 Statistical inference

In this section, we discuss maximum likelihood estimation and the Bayesian method and provide the approximate sampling distribution of the estimates.

The likelihood approach Let $p(\mathbf{z}; \boldsymbol{\theta})$ denote either the slash t density in (2.1) or the skew-slash t density in (3.29), where $\boldsymbol{\theta}$ denotes the corresponding parameter vector, i.e. $\boldsymbol{\theta} = (q, r, \mathbf{m}, \mathbf{R})$ or $\boldsymbol{\theta} = (\boldsymbol{\lambda}, q, r, \boldsymbol{\mu}, \Sigma)$. As the degrees of freedom r is unknown, we estimate it by the MLE treating it as a positive real number. Let $\mathbf{Z}_1, \ldots, \mathbf{Z}_n$ be a random sample from the density p. Based on the sample, the parameter $\boldsymbol{\theta}$ can be estimated by the MLE

$\hat{\boldsymbol{\theta}}$. This is the parameter value $\hat{\boldsymbol{\theta}}$ which maximizes $L(\boldsymbol{\theta}) = \frac{1}{n}\sum_{i=1}^{n}\log p(\mathbf{Z}_i;\boldsymbol{\theta})$ of the log likelihood functions of the sample over the parameter space Θ, that is,

$$L(\hat{\boldsymbol{\theta}}) = \max_{\boldsymbol{\theta}\in\Theta}\frac{1}{n}\sum_{i=1}^{n}\log p(\mathbf{Z}_i;\boldsymbol{\theta}). \tag{4.32}$$

Let $\mathbf{S}_i(\boldsymbol{\theta}) = \frac{\partial}{\partial\boldsymbol{\theta}}\log p(\mathbf{Z}_i;\boldsymbol{\theta})$ be the score for the observation $\mathbf{Z}_i$. Under suitable conditions, the MLE $\hat{\boldsymbol{\theta}}$ is the solution to the score equation

$$\sum_{i=1}^{n}\mathbf{S}_i(\boldsymbol{\theta}) = 0. \tag{4.33}$$

Let $\mathbf{J}(\boldsymbol{\theta}) = E(\mathbf{S}_1(\boldsymbol{\theta})\mathbf{S}_1(\boldsymbol{\theta})^\top)$ be the information matrix. Clearly the information matrix can be estimated by

$$\hat{\mathbf{J}} = \frac{1}{n}\sum_{i=1}^{n}\mathbf{S}_i(\hat{\boldsymbol{\theta}})\mathbf{S}_i(\hat{\boldsymbol{\theta}})^\top. \tag{4.34}$$

Under suitable conditions, the sampling distribution of $\hat{\boldsymbol{\theta}}$ can be approximated by the normal distribution with mean vector $\boldsymbol{\theta}$ and variance-covariance matrix $\hat{\mathbf{J}}^{-1}$, that is, approximately

$$\hat{\boldsymbol{\theta}} \sim \mathcal{N}(\boldsymbol{\theta}, n^{-1}\hat{\mathbf{J}}^{-1}). \tag{4.35}$$

Based on this normal approximation, one can perform hypothesis testing, construct confidence intervals and, in particular, calculate the standard error (SE) of each component $\hat{\boldsymbol{\theta}}_j$ of the MLE $\hat{\boldsymbol{\theta}}$ as follows:

$$SE(\hat{\boldsymbol{\theta}}_j) = \sqrt{n^{-1}\hat{\mathbf{J}}^{jj}}, \quad j = 1,\ldots,k, \tag{4.36}$$

where $\hat{\mathbf{J}}^{jj}$ is the (j,j)- entry of the estimated inverse information matrix $\hat{\mathbf{J}}^{-1}$.

The numerical value of the MLE $\hat{\boldsymbol{\theta}}$ can be found by solving the score equation (4.33) using the newton's method. Alternatively, one can directly search the solution of the maximization problem (4.32), for example, using the subroutine *optim* in the R package.

As for initial values of the newton's algorithm, one can use the moment estimates of the parameters or other available consistent estimates. One technical issue here is that the estimate $\hat{\Sigma}$ of Σ must be positive definite. What we did in our applications was that we estimated the entries of Σ and then verified the positive definiteness of $\hat{\Sigma}$.

The Bayesian approach Given observed data $\mathbf{D}$, the likelihood function $L(\mathbf{D}|\boldsymbol{\theta})$ can be obtained from the proposed multivariate slash or skew-slash t distribution with parameter vector $\boldsymbol{\theta}$. The posterior density then satisfies $p(\boldsymbol{\theta}|\boldsymbol{D}) \propto L(\boldsymbol{D}|\boldsymbol{\theta})\pi(\boldsymbol{\theta})$, where $\pi(\boldsymbol{\theta})$ is the joint prior density of $\boldsymbol{\theta}$ based on the available prior information on it. We choose a prior density for each component of $\boldsymbol{\theta}$ and take the joint prior density of $\boldsymbol{\theta}$ to be equal to the product of the marginal prior densities. The resulting full Bayesian model has the hierarchical structure with the conditional density of $\boldsymbol{D}|\boldsymbol{\theta}$ and the prior distribution $\boldsymbol{\theta} \sim \pi(\boldsymbol{\theta})$ in the proposed model. One can obtain a random sample from the joint posterior density by the Markov Chain Monte Carlo (MCMC) method, and a parametric Baysian analysis of the model can be implemented using the Gibbs sampling method in R or JAGS.

5 Simulations and applications

In this section, we use simulations to compare the performance of the skew-slash t and the skew-slash normal distributions. Simulations are also used to demonstrate parameter estimation using both the maximum likelihood criterion and Bayesian paradigm. Then the skew-slash t, with the density given in (3.29), and the skew-slash normal, with the density given below, are applied to fit two real datasets.

$$\eta_k(\mathbf{w}; \boldsymbol{\lambda}, q, \boldsymbol{\mu}, \Sigma) = 2q \int_0^1 u^{k+q-1} \phi_k(u\mathbf{w}; u\boldsymbol{\mu}, \Sigma) \times \Phi\left(u\boldsymbol{\lambda}^\top \Sigma^{-1/2}(\mathbf{w} - \boldsymbol{\mu})\right) du, \quad \mathbf{w} \in \mathbb{R}^k, \tag{5.37}$$

where ϕ_k is the density of the k-variate normal distribution $N_k(\boldsymbol{\mu}, \Sigma)$ with mean vector $\boldsymbol{\mu}$ and covariance matrix Σ.

5.1 Simulation study

Comparison between the skew-slash t and skew-slash normal To compare the two distributions, we first generated data from the 2-variate skew-slash t then fitted it with both 2-variate skew-slash t and skew-slash normal, and vice versa (i.e. generated data from the 2-variate skew-slash normal then fitted it with the two distributions). Reported in Table 2 are the average AIC values and average MLE's of the parameters based on the sample size $n = 250$ and repetitions $M = 200$.

Notice that for data generated from both the skew-slash t and skew-slash normal, the average AIC values of the skew-slash t fitting were lower than those of the skew-slash normal, indicating a better overall model fitting of the former to the data than the latter.

Parameter estimation by the Bayesian method We now conducted simulations to study the behaviors of the MLE's and Bayesian estimates of the parameters in the data-generation models for the standard univariate and bivariate slash and skew-slash t distributions. Here the prior distributions are the standard normal $N(0, 1)$ for the skewness parameter λ (or the components of the parameter vector $\boldsymbol{\lambda}$), the exponential distribution with rate 0.1 truncated at 2 for the tail parameter q and truncated at 4 for the degrees of freedom r. It follows from Sahu et al. (2003) and Fernandez and Steel (1998) that the resulting distributions have finite variances. We generated 100 random samples of size 200 from the standard univariate slash and skew-slash t and the standard bivariate slash and skew-slash t. Reported in Table 3 are the average estimates of the parameters based on the maximum likelihood criterion and Bayesian paradigm. Observe that the two types of estimates are close for all the simulation setups.

5.2 Applications

Model fitting to the GAD Data Gestational age at delivery (GAD) is a variable widely studied in epidemiology, see, for example, Longnecker et al. (2001). We applied the skew-slash t and skew-slash normal distributions to fit the log transformed GAD of $n = 100$ observations. Figure 1 is the histogram superimposed with the fitted density curves, while Table 1 reports the MLE's, the standard errors (SE) of the parameter estimates and the AIC. We can see from Figure 1 that the skew-slash t distribution was able to better capture

Table 3 Maximum likelihood and Bayesian estimates comparison

Parameter	True Value	SLT MLE	SLT Bayes	SSLT MLE	SSLT Bayes
		Univariate			
λ	3			2.496	2.511
q	5	6.196	6.236	6.689	6.479
r	8	8.015	7.084	7.650	7.194
		Bivariate			
λ_1	-5			-5.543	-3.288
λ_2	3			3.333	1.967
q	5	6.410	4.113	5.899	4.118
r	8	7.480	8.468	7.726	8.609

The average maximum likelihood and Bayesian estimates based on 100 samples of size 200. SLT=Slash t and SSLT=Skew-Slash t.

the peak of the histogram, giving a better estimation of the density to the majority of data points. In the mean time, the AIC values in Table 1 indicated that the skew-slash t fitting was better than the skew-slash normal.

Model fitting to the AIS data Azzalini and Dalla Valle (1996) used their skew normal distribution to fit (LBM-lean body mass, BMI-body mass index) pairs of the athletes from Australian Institute of Sport (AIS), where the data of $n = 202$ observations were reported in Cook and Weisberg (1994). Wang and Genton (2006) used their skew-slash normal distribution to re-fit the data. Here we applied the proposed skew-slash t distribution to re-fit the (LBM, BMI) pairs in the AIS data. Before fitting we standardized the variables.

Figure 5 is the scatter plot superimposed with the fitted skew-slash t and skew-slash normal contours in the scale of LBM and BMI. The skew-slash normal contour is similar

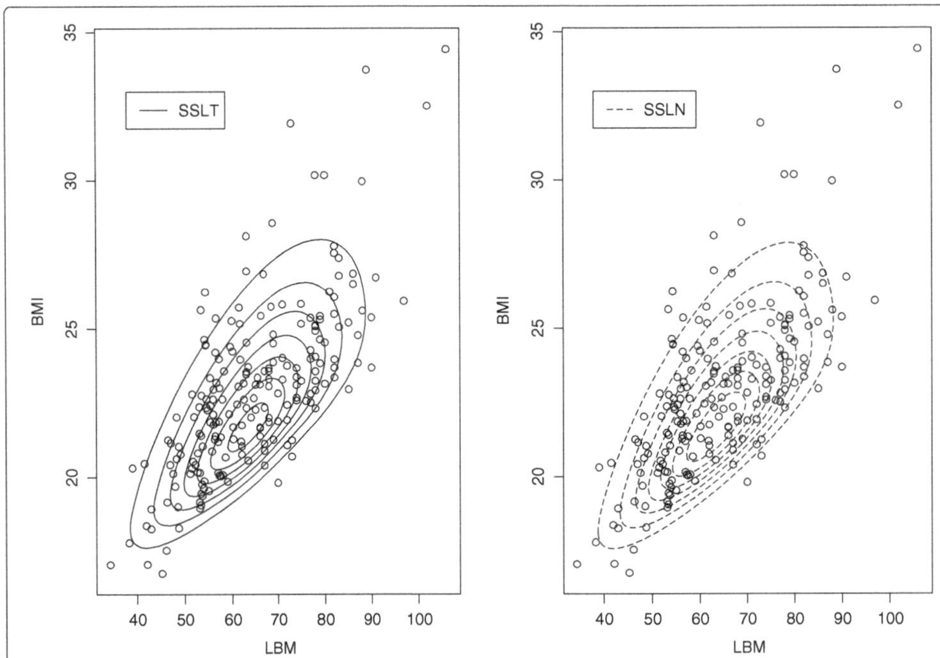

Figure 5 Scatter plot and fitted contours.

to what was reported in Wang and Genton (2006). The comparison between the two contours seemed to indicate that the fittings based on two models were close.

Reported in Table 4 are the MLE's, standard errors and AIC. A smaller AIC value of the skew-slash normal fitting than the skew-slash t fitting indicated that the former was a better fit to this data. When the skew-slash normal was used to fit the data we were able to obtain the MLE's. But the row and column corresponding to parameter q in the Hessian matrix were zero, suggesting a simpler skew normal fitting to this data. The skew-slash normal fitting by Wang and Genton (2006) led to the same conclusion though they did not report the standard errors of the parameter estimates.

In conclusion, the proposed slash and skew-slash t are competitive candidate models for fitting skewed and heavy tailed data. The parameters can be estimated under either the frequentist method or Bayesian paradigm. Although for a particular dataset the skew-slash t may not be the final model, it is a good choice to start with in model selection due to its flexibility and the fact that it takes the skew normal, skew t and hence the usual normal and t as its submodels.

6 Concluding remarks

In this article, we defined the multivariate slash and skew-slash distributions in a pursuit of providing additional distributions to simulate and fit skewed and heavy tailed data. We investigated the heavy tail behaviors and tractable properties of these distributions which are useful in simulations and applications to real data. We characterized the skew t, the skew-slash normal and the skew-slash t distributions using both the hidden truncation or selective sampling model and the order statistics of the components of a bivariate normal or t variable. We demonstrated that the proposed skew-slash t model takes as sub-models the slash t, the slash normal, the skew-slash normal, the skew normal, the skew t and hence the usual normal and t. This nested property can be used in hypothesis testing.

Our simulations and applications to real data indicated that the proposed skew-slash t fitting outperformed the skew-slash normal fitting. Even though the skew-slash normal contains a tail parameter q, the fitting with it to the GAD data was unsatisfactory as the SE of the MLE of the tail parameter q was large, see Table 1. This suggests that not all heavy-tail properties in data can be explained by the tail parameter q in the slash distributions. Thus it makes sense for us to further search for distributions which can be used to fit

Table 4 The skew-slash normal and *t* fitting to the AIS data

	SSLT		SSLN
	MLE	SE	MLE
AIC	981.018		978.426
μ_1	-0.210	0.151	-0.210
μ_2	-0.936	0.110	-0.954
σ_{11}^2	0.566	0.097	0.831
σ_{12}	0.465	0.086	0.708
σ_{22}^2	2.042	0.454	1.486
λ_1	-1.745	0.531	-2.225
λ_2	3.771	1.016	2.091
q	121.021	660.502	22.534
r	25.124	17.756	

SSLN=Skew-Slash Normal and SSLT=Skew-Slash t.

heavy tailed data. Our proposed slash and skew-slash t distributions can be considered as an example in this attempt. We complete our remarks by pointing out that the degrees-of-freedom parameter r and the tail parameter q would explain different types of fat tail behaviors existed in data.

Competing interests

The authors declare that they have no competing interests.

Authors' contributions

FT contributed to the entire work, HP to the theoretical part, and YT to the inference, simulations and applications. All authors read and approved the final manuscript.

Acknowledgements

The authors gratefully thank two anonymous reviewers for their suggestions that substantially improved the article.

Author details

[1]Department of Mathematical Sciences, Indiana University Purdue University Indianapolis, N. Blackford Street, IN 46202, Indianapolis, USA. [2]Department of Statistics, Florida State University, N. Woodward Ave., FL 32306, Tallahassee, USA.

References

Arnold, BC, Lin, GD: Characterization of the Skew-normal and Generalized Chi Distributions. Sankhyā. **66**, 593–606 (2004)

Azzalini, A, Capitanio, A: Distributions generated by perturbation of symmetry with emphasis on a multivariate skew t distribution. J. R. Stat. Soc. Ser. B. **65**, 367–389 (2003)

Azzalini, A, Dalla Valle, A: The multivariate skew-normal distribution. Biometrika. **83**, 715–726 (1996)

Cook, RD, Weisberg, S: An Introduction to Regression Graphics. Wiley, New York (1994)

Dreier, I, Kotz, S: A note on the characteristic function of the t-distribution. Statist. & Probabil. Lett. **57**, 221–224 (2002)

Fernandez, C, Steel, MFJ: On Bayesian Modeling of Fat Tails and Skewness. J. Am. Statist. Assoc. **93**, 359–371 (1998)

Genton, MG (ed.): Skew-elliptical distributions and their applications: a journey beyond normality. Chapman and Hall/CRC, Boca Raton (2004)

Gupta, AK: Multivariate skew t-distribution, Statistics: A Journal of Theoretical and Applied Statistics. **37**, 359–363 (2003). doi:10.1080/715019247

Joarder, AH, Ali, MM: On the characteristic function of the multivariate student t distribution. Pak. J. Statist. **12**, 55–62 (1996)

Johnson, N, Kotz, S: Distributions in Statistics: Continuous Multivariate Distributions. John Wiley & Sons, New York (1972)

Jones, MC: Marginal replacement in multivariate densities, with application to skewing spherically symmetric distributions. J. Multivar. Anal. **81**, 85–99 (2002)

Kafadar, K: Slash Distribution. In: Johnson, NL, Kotz, S, Read, C (eds.) Encyclopedia of Statistical Sciences. vol. 8, pp. 510–511. Wiley, New York, (1988)

Kotz, S, Nadarajah, S: Multivariate t distributions and their applications. Cambridge University Press, Cambridge (2004)

Longnecker, MP, Klebanoff, MA, Zhou, H, Brock, JW: Association between maternal serum concentration of the ddt metabolite dde and preterm and small-for-gestational-age babies at birth. Lancet. **358**, 110–114 (2001)

Sahu, SK, Dey, DK, Branco, MD: A new class of multivariate skew distributions with applications to Bayesian regression models. Can. J. Stat. **31**, 129–150 (2003)

Wang, J, Genton, M: The multivariate skew-slash distributions. J. Statist. Plann. Inferr. **136**, 209–220 (2006)

11

A new multivariate two-sample test using regular minimum-weight spanning subgraphs

David M Ruth

Correspondence: druth@usna.edu
Department of Mathematics, United States Naval Academy, 572-C Holloway Rd, Annapolis, MD 21402, USA

Abstract

A new nonparametric test is proposed for the multivariate two-sample problem. Similar to Rosenbaum's cross-match test, each observation is considered to be a vertex of a complete undirected weighted graph; interpoint distances are edge weights. A minimum-weight, *r*-regular subgraph is constructed, and the mean cross-count test statistic is equal to the number of edges in the subgraph containing one observation from the first group and one from the second, divided by *r*. Unequal distributions will tend to result in fewer edges that connect vertices between different groups. The mean cross-count test is sensitive to a wide range of distribution differences and has impressive power characteristics. We derive the first and second moments of the mean cross-count test, and note that simulation studies suggest this test statistic is asymptotically normal regardless of underlying data distributions. A small simulation study compares the power of the mean cross-count test to Hotelling's T^2 test and to the cross-match test. This new test is a more powerful generalization of Rosenbaum's test (the cross-match test is the case $r = 1$) and constitutes a noteworthy addition to the class of multivariate, nonparametric two-sample tests.

Keywords: Distribution-free test; Graph-theoretic procedure; Change point

1 Background

1.1 Objective

Consider $N = m + n$ independent multivariate observations $\mathbf{Y}_1, \ldots, \mathbf{Y}_m$ and $\mathbf{Y}_{m+1}, \ldots, \mathbf{Y}_N$, where each $\mathbf{Y}_i$ is drawn from distribution F for $1 \le i \le m$ and from distribution G for $m + 1 \le i \le N$. The dimension of the observations does not depend on N. The covariates may be quantitative or categorical; there need only exist some function, d, that measures distance between observations. The null hypothesis is that $F = G$. The objective is a two-sample test that has little or no dependence on the underlying distribution of the data. Furthermore, this test should have sufficient power to be useful for applications.

1.2 Motivation

We follow in the vein of graph-theoretic tests for homogeneity: Consider each observation to be a vertex of a complete, undirected, weighted graph, $\mathcal{G}$, and assign interpoint distances as edge weights. The distribution of these distances is sensitive to departures from homogeneity; Maa et al. (1996) prove that two distributions are identical if and only if the distributions of inter-point distances within and between observations sampled from the two populations are the same. Friedman and Rafsky (1979, 1981) fit a minimum spanning tree to $\mathcal{G}$ and count the number of edges in the tree that connect vertices from different groups to test whether the sampling distributions are the same. Schilling (1986), Henze (1988), and Hall and Tajvidi (2002) examine properties of nearest-neighbor subgraphs of $\mathcal{G}$ to test for homogeneity.

Rosenbaum (2005) provides a novel approach to this problem: Suppose N is even. Find a minimum-weight non-bipartite matching on $\mathcal{G}$, which is the lowest-weight spanning subgraph for which the degree of each vertex with respect to the subgraph is one and which consists of $N/2$ non-adjacent edges. Rosenbaum's cross-match statistic, A_1, counts the number of edges in the matching that include one vertex from each of the two groups. Under the null hypothesis of no group difference each vertex is equally likely to be paired with any other vertex. Rosenbaum (2005) shows that the exact null distribution of A_1 is found by combinatorial argument to be

$$P(A_1 = a_1) = \frac{2^{a_1}(N/2)!}{\binom{N}{m}\left(\frac{m-a_1}{2}\right)!a_1!\left(\frac{n-a_1}{2}\right)!} \tag{1}$$

for $a_1 \in \{0, 2, \ldots, \min(m, n)\}$ and m and n even, or $a_1 \in \{1, 3, \ldots, \min(m, n)\}$ and m and n odd; $P(A_1 = a_1) = 0$ otherwise. In the denominator of (1), $\frac{1}{2}(m-a_1)$ is the number of edges in the matching where both vertices are in the group of size m and $\frac{1}{2}(n-a_1)$ is the number of edges in the matching where both vertices are in the group of size n.

When the two groups are drawn from different distributions the number of within-group pairs tends to be higher than for the null case, so the null hypothesis of homogeneity is rejected if A_1 is sufficiently small. For odd N, this procedure may be simply modified by introducing a pseudo-observation, $\mathbf{Y}_0$ such that $d(\mathbf{Y}_0, \mathbf{Y}_i) = 0$ for all $i \in \{1, \ldots, N\}$, and randomly assigning it to one of the two groups. Then find a minimum-weight non-bipartite matching on this resulting graph with $N + 1$ vertices and compute A_1 with respect to observations $\mathbf{Y}_0, \ldots, \mathbf{Y}_N$.

That the exact null distribution of A_1 is known, regardless of the underlying data distribution, is a particularly attractive property for a multivariate two-sample test. Furthermore, the asymptotic normality of A_1 facilitates testing for large-sample problems. However, the cross-match test has relatively low power. Since only a single non-bipartite matching is considered in this test, information contained in the proximity of many pairs of points is ignored. Friedman and Rafsky (1979, 1981) observe that the power of their single-tree test is enhanced by evaluating successive disjoint low-weight spanning trees. Similarly, Ruth and Koyak (2011) show that ensembles of disjoint low-weight non-bipartite matchings carry significant information regarding whether a distributional change occurs over a sequence of independent observations. A drawback associated with examining collections of such subgraphs is that null distributions are extremely difficult to determine. Mindful of this caveat, we offer an extension of the cross-match test which exploits the information contained in the distances between many pairs of points.

2 Methods

2.1 Illustrating example

Consider the bivariate sample of size $N = 20$ listed in Table 1 and displayed in Figure 1. The sample consists of independent observations in groups 1 (○) and 2 (△); observations

Table 1 Bivariate data for illustrating example

Observation number	Group	Covariate 1	Covariate 2
1	1	-0.323	-1.389
2	1	1.020	-2.078
3	1	-0.269	-1.020
4	1	0.296	-0.144
5	1	0.602	1.021
6	1	0.814	-0.508
7	1	-0.475	-0.690
8	1	-0.079	1.360
9	1	-0.228	0.926
10	1	-0.481	1.958
11	2	1.269	1.275
12	2	0.954	2.133
13	2	-0.103	2.763
14	2	-0.581	-0.428
15	2	2.367	0.222
16	2	0.980	1.870
17	2	0.494	1.981
18	2	0.293	0.236
19	2	1.535	0.981
20	2	1.993	-0.120

within groups are identically distributed. For the purposes of this example, these data were simulated from distributions whose locations differ by one unit in each dimension. Figure 1 also shows the minimum-weight non-bipartite matching associated with this sample with respect to Euclidean distance. The present goal is to identify the distribution difference between these groups, making no assumptions about the underlying distributions.

The cross-match test is applicable here; for this example the value of the cross-match statistic is $A_1 = 4$ with a corresponding p-value = 0.433. So, the cross-match test is insufficiently powerful to identify a distribution difference in this case. In the next section, we introduce an extension of the cross-match test that enhances test power significantly.

2.2 The mean cross-count (MCC) test

As before, we assume an even number N of observations forming a complete, undirected, weighted graph, $\mathcal{G}$. Rather than find a minimum-weight non-bipartite matching, we find a *minimum-weight r-regular spanning subgraph of* $\mathcal{G}$, where $1 \le r \le N-2$, denoted $\mathcal{G}_r^*$. That is, $\mathcal{G}_r^*$ is a subgraph of $\mathcal{G}$ with the following properties:

a) Every vertex in $\mathcal{G}$ is also in $\mathcal{G}_r^*$.
b) Every vertex in $\mathcal{G}_r^*$ has degree r.
c) The total weight of all edges in $\mathcal{G}_r^*$ is the lowest among all subgraphs of $\mathcal{G}$ which satisfy properties (a) and (b).

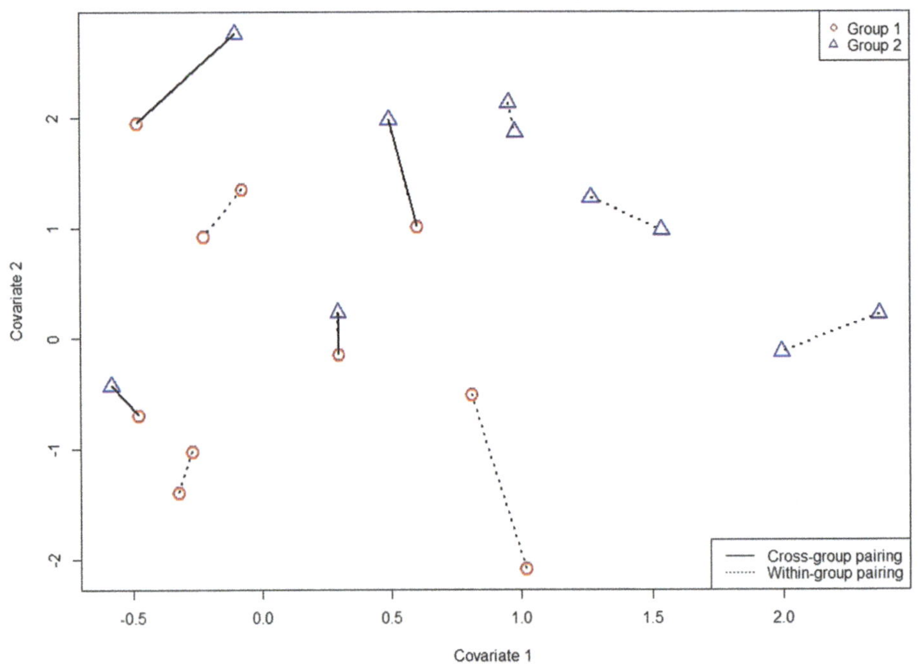

Figure 1 Bivariate data for illustrating example with optimal non-bipartite matching on groups 1 (○) and 2 (△) with $m = n = 10$. Cross-group pairs are connected by solid lines; within-group pairs by dotted lines.

In graph theory, an r-regular spanning subgraph of $\mathcal{G}$ is sometimes called an *r-factor of $\mathcal{G}$*. Note that $\mathcal{G}_1^*$ is the special case of a minimum-weight non-bipartite matching used by Rosenbaum (2005), and $\mathcal{G}_{N-1}^*$ is identical to $\mathcal{G}$. In practice, we are mainly interested in $2 \leq r \leq N/2$, although the theoretical details are not so constrained. Minimum-weight r-factors may be computed as follows: For any subgraph of $\mathcal{G}$, let x_{ij} be an indicator variable equal to 1 if the edge connecting vertices i and j is included in the subgraph and let d_{ij} be the distance between vertex i and vertex j. Then the edges of $\mathcal{G}_r^*$ solve following the combinatorial optimization problem:

$$\begin{aligned} &\min_{\mathbf{x}} \sum_{j=2}^{N} \sum_{i=1}^{j-1} d_{ij} x_{ij} \\ &\text{subject to} \sum_{i=1}^{k-1} x_{ik} + \sum_{j=k+1}^{N} x_{kj} = r \quad \forall k \in \{1, \ldots, N\} \\ &x_{ij} \in \{0,1\} \ \forall j \in \{i+1, \ldots, N\},\ \forall i \in \{1, \ldots, N-1\}. \end{aligned} \tag{2}$$

Anderson (1972) assures the existence of a solution for $r \leq N/2$. Solutions for $r > N/2$ are guaranteed by the fact that the complement of an r-regular subgraph of $\mathcal{G}$ is an $(N - 1 - r)$ -regular graph. For this paper, solutions are found in R using the package "lpSolve" for $N \leq 400$. For $N > 400$, solutions are found in R using the package "gurobi" due to the computational complexity of larger problems.

Similar to the cross-match test, we count the number of edges A_r in $\mathcal{G}_r^*$ that include a vertex from each group. We call $T_r = A_r/r$ the *mean cross-count* (*MCC*) *statistic*. The idea here is that the number of within-group edges in $\mathcal{G}_r^*$ will be higher for cases of a distribution difference than for the null case. So, small values of T_r are evidence against the null hypothesis. Note that $T_1 = A_1$ is the cross-match statistic as before. One could use the total cross-count, A_r, as an equivalent test-statistic; however, we choose to scale this value to give some notion of "average cross-count per vertex degree" (hence the name "mean cross-count"). For odd N, randomly introduce a pseudo-observation in the same manner as the $r = 1$ case discussed in Section 1.2.

2.3 Illustrating example, continued

Figure 2 shows a minimum-weight 3-factor, $\mathcal{G}_3^*$, for the data in Table 1 with respect to Euclidean distance. Cross-group edges are shown with solid lines. For this case, $A_3 = 12 \Rightarrow T_3 = 4$, so the test statistic value here is the same as Rosenbaum's cross-match test statistic. A discussion of the distribution of T_r is in Section 3.1; for this example, we estimate the p-value for T_r by permutation test on the observation vertex labels. Using 10,000 permutations yields an estimated p-value = 0.146. While not enough evidence to conclude a group difference, this reduction in p-value relative to the $r = 1$ case (p-value = 0.433) suggests that considering minimum-weight r-factors for $r > 1$ may improve test power. In Section 3.2 we demonstrate significant power advantages that are realized for the MCC statistic.

3 Results and discussion

3.1 MCC moments and normal approximation

For the following discussion we assume N is even, adopting the convention that if the number of observations is odd then we will consider N to be the number of observations including a pseudo-observation as previously discussed. To find the mean and variance of T_r under the null hypothesis, we proceed as follows: Let $\mathcal{G}$ be the complete

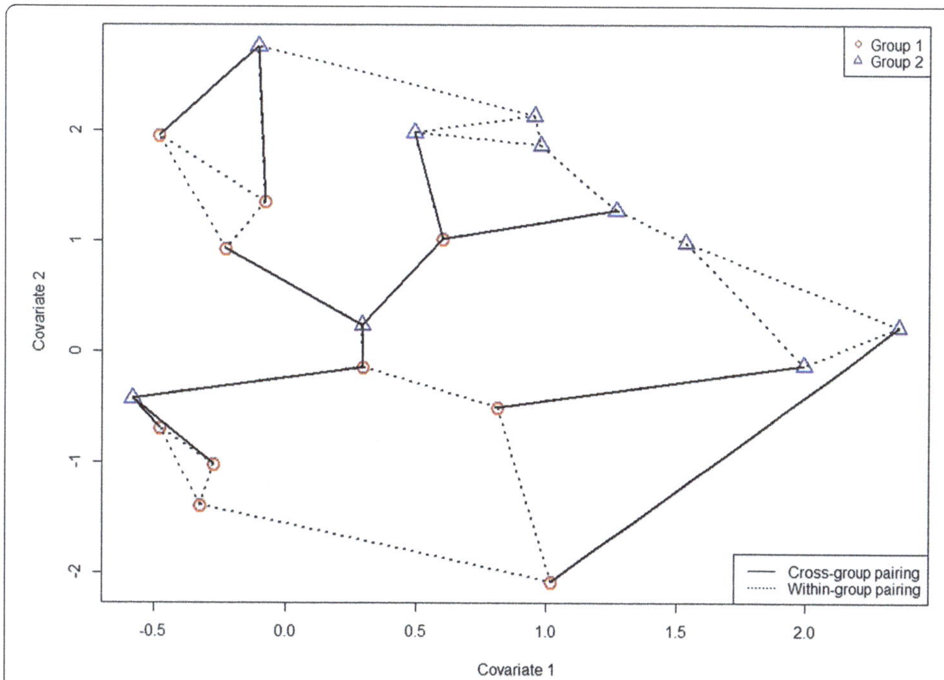

Figure 2 Bivariate data for illustrating example with optimal 3-factor on groups 1 (○) and 2 (△) with $m = n = 10$. Cross-group pairs are connected by solid lines; within-group pairs by dotted lines.

undirected graph $(\mathbb{Z}_N, E_N)$ where the vertex set $\mathbb{Z}_N$ consists of the indices $1, 2, \ldots, N$ and the edge set consists of all $N(N - 1)/2$ pairs of vertices; by convention, write the pairs with smaller vertex first, so $E_N = \{(i, j) : 1 \leq i < j \leq N\}$. Partition $\mathbb{Z}_N$ into two sets S and T, with $|S| = m$ and $|T| = n$, so $m + n = N$. Denote $E_N^{(S,T)}$ as the set of all edges with one vertex in S and the other in T. Let X_{ij} be the random variable that indicates whether edge (i, j) is included in a minimum-weight r-regular subgraph, $\mathcal{G}_r^*$, with $1 \leq r \leq N - 2$. By the r-regularity of $\mathcal{G}_r^*$, for each $i \in \mathbb{Z}_N$ we have $r = \sum_{j=1}^{i-1} X_{ji} + \sum_{j=i+1}^{N} X_{ij}$, and so

$$\begin{aligned} r = E[r] &= E\left[\sum_{j=1}^{i-1} X_{ji} + \sum_{j=i+1}^{N} X_{ij}\right] = \sum_{j=1}^{i-1} E[X_{ji}] + \sum_{j=i+1}^{N} E[X_{ij}] \\ &= \sum_{j=1}^{i-1} P(X_{ji} = 1) + \sum_{j=i+1}^{N} P(X_{ij} = 1). \end{aligned} \tag{3}$$

But under the null hypothesis, each edge is equally likely to be included in $\mathcal{G}_r^*$, so $r = (N - 1)P(X_{12} = 1)$. Therefore, for all $(i, j) \in E_N$

$$E[X_{ij}] = P(X_{ij} = 1) = \frac{r}{N-1} \tag{4}$$

and

$$\mathrm{Var}[X_{ij}] = P(X_{ij} = 1)P(X_{ij} = 0) = \frac{r(N-1-r)}{(N-1)^2}. \tag{5}$$

The total cross-count, A_r, may be written

$$A_r = \sum_{(i,j) \in E_N^{(S,T)}} X_{ij} \tag{6}$$

resulting in

$$\begin{aligned} E[T_r] &= \frac{1}{r} E[A_r] = \frac{1}{r} E\left[\sum_{(i,j) \in E_N^{(S,T)}} X_{ij}\right] = \frac{mn}{r} E[X_{ij}] \\ &= \frac{mn}{N-1}. \end{aligned} \tag{7}$$

Finding the variance of T_r is slightly more involved. First take

$$\mathrm{Var}[A_r] = \mathrm{Var}\left[\sum_{(i,j) \in E_N^{(S,T)}} X_{ij}\right] = \sum_{(i,j) \in E_N^{(S,T)}} \mathrm{Var}[X_{ij}] + \sum_{\substack{(i,j),(k,l) \in E_N^{(S,T)} \\ (i,j) \neq (k,l)}} \mathrm{Cov}[X_{ij}, X_{kl}]. \tag{8}$$

The sum of variances is computed directly as

$$\sum_{(i,j) \in E_N^{(S,T)}} \mathrm{Var}[X_{ij}] = mn\mathrm{Var}[X_{ij}] = \frac{mnr(N-1-r)}{(N-1)^2}. \tag{9}$$

The sum of covariances may be partitioned into terms that include pairs of adjacent edges and terms that include disjoint (i.e., non-adjacent) edges:

$$\sum_{\substack{i,k\in S\\ j,l\in T\\ (i,j)\neq(k,l)}} \mathrm{Cov}\left[X_{ij},X_{kl}\right] = \sum_{\substack{i\in S\\ j,l\in T\\ j\neq l}} \mathrm{Cov}\left[X_{ij},X_{il}\right] + \sum_{\substack{i,k\in S\\ i\neq k\\ j\in T}} \mathrm{Cov}\left[X_{ij},X_{kj}\right] + \sum_{\substack{i,k\in S\\ i\neq k\\ j,l\in T\\ j\neq l}} \mathrm{Cov}\left[X_{ij},X_{kl}\right] \quad (10)$$

For any two adjacent edges (k, l) and (i, j),

$$\begin{aligned} P(X_{ij}X_{kl}=1) &= P(X_{kl}=1|X_{ij}=1)P(X_{ij}=1) = \frac{(r-1)r}{(N-2)(N-1)} \\ &= E\left[X_{ij}X_{kl}\right], \end{aligned} \quad (11)$$

so

$$\begin{aligned} &\sum_{\substack{i\in S\\ j,l\in T\\ j\neq l}} \mathrm{Cov}\left[X_{ij},X_{il}\right] + \sum_{\substack{i,k\in S\\ i\neq k\\ j\in T}} \mathrm{Cov}\left[X_{ij},X_{kj}\right] \\ &= (mn(n-1)+m(m-1)n)\left(\frac{(r-1)r}{(N-2)(N-1)} - \left(\frac{r}{N-1}\right)^2\right) \\ &= -\frac{mnr(N-1-r)}{(N-1)^2}. \end{aligned} \quad (12)$$

For any two disjoint edges (k, l) and (i, j),

$$\begin{aligned} P(X_{ij}X_{kl}=1) &= P(X_{kl}=1|X_{ij}=1)P(X_{ij}=1) = \frac{(r(N-4)+2)}{(N-3)(N-2)}\frac{r}{(N-1)} \\ &= E\left[X_{ij}X_{kl}\right], \end{aligned} \quad (13)$$

So

$$\begin{aligned} \sum_{\substack{i,k\in S\\ i\neq k\\ j,l\in T\\ j\neq l}} \mathrm{Cov}\left[X_{ij},X_{kl}\right] &= m(m-1)n(n-1)\left(\frac{(r(N-4)+2)}{(N-3)(N-2)}\frac{r}{(N-1)} - \left(\frac{r}{N-1}\right)^2\right) \\ &= \frac{2m(m-1)n(n-1)(N-1-r)r}{(N-3)(N-2)(N-1)^2}. \end{aligned} \quad (14)$$

Combining terms yields

$$\begin{aligned} \mathrm{Var}[A_r] &= \sum_{\substack{i\in S\\ j\in T}} \mathrm{Var}\left[X_{ij}\right] + \sum_{\substack{i\in S\\ j,l\in T\\ j\neq l}} \mathrm{Cov}\left[X_{ij},X_{il}\right] + \sum_{\substack{i,k\in S\\ i\neq k\\ j\in T}} \mathrm{Cov}\left[X_{ij},X_{kj}\right] + \sum_{\substack{i,k\in S\\ i\neq k\\ j,l\in T\\ j\neq l}} \mathrm{Cov}\left[X_{ij},X_{kl}\right] \\ &= \frac{mnr(N-1-r)}{(N-1)^2} - \frac{mnr(N-1-r)}{(N-1)^2} \\ &+ \frac{2m(m-1)n(n-1)(N-1-r)r}{(N-3)(N-2)(N-1)^2} \\ &= \frac{2m(m-1)n(n-1)(N-1-r)r}{(N-3)(N-2)(N-1)^2}. \end{aligned} \quad (15)$$

Therefore,

$$\mathrm{Var}[T_r] = \frac{1}{r^2}\mathrm{Var}[A_r] = \frac{2m(m-1)n(n-1)(N-1-r)}{r(N-3)(N-2)(N-1)^2} \ . \tag{16}$$

We note in particular that the first and second moment results in (7) and (16) match the results in Rosenbaum (2005) for the special case $r = 1$.

Simulation suggests that the null distribution of T_r is negatively skewed, but that for sufficiently large N and possibly certain conditions on r this distribution is asymptotically normal, independent of distribution function F. Rosenbaum (2005) proves that T_r is asymptotically normal for $r = 1$ for any distribution function; proof of this conjecture for $r > 1$ remains an open problem. This conjecture is supported by the normal QQ-plots shown in Figures 3 and 4 for 1,000 simulated values of $(T_r - E[T_r])/\sqrt{\mathrm{Var}(T_r)}$ at $r = 5$, 50 and $N = 150$, 600 with $m/n = 1/2$, under sampling from uniform distributions on $[-1, 1]^5$ and $[-1, 1]^{20}$. For the smaller sample size ($N = 150$), negative skewness is stronger for lower dimension and for higher r, with $r = 50$ and Dim = 5 being the most strongly skewed case shown. For the larger sample size ($N = 600$), skewness effects appear to vanish for all but the $r = 50$ and Dim = 5 case, and even in this case skewness is vastly diminished compared to the smaller sample size. Other distribution families and other values of m/n produce similar results.

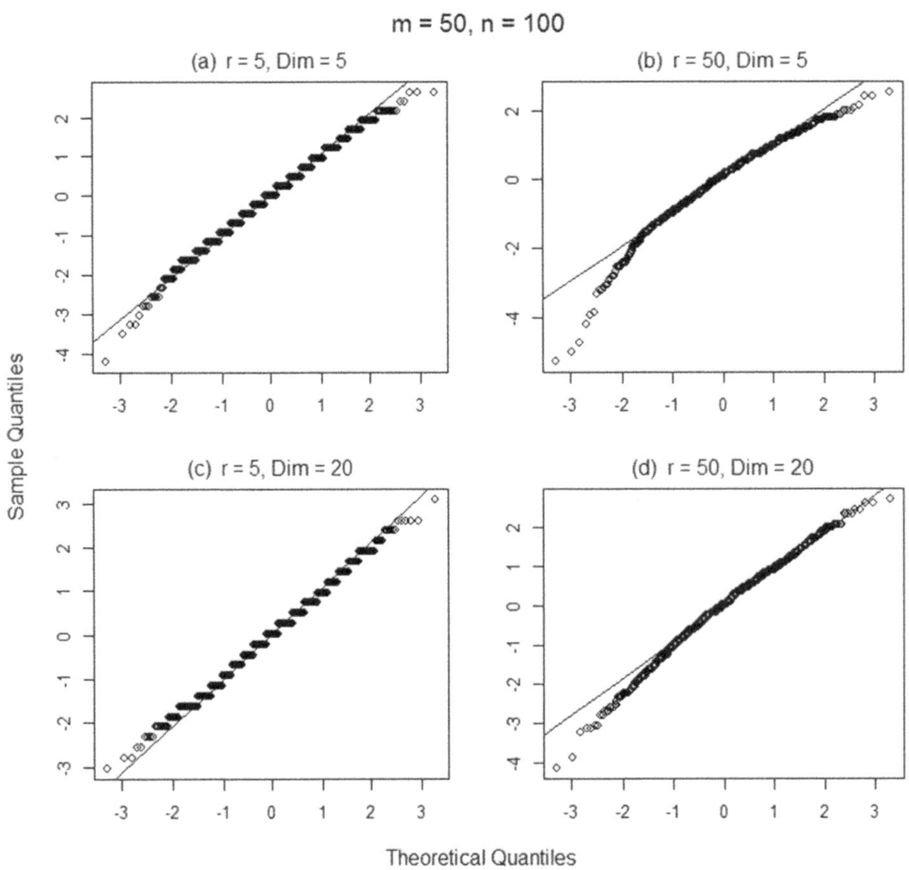

Figure 3 Normal QQ-plots of 1,000 simulated values of T_r for $m = 50$, $n = 100$ and $r = 5$, 50. Panels **(a)** and **(b)** are from independent samples of Unif $[-1, 1]^5$ variates. Panels **(c)** and **(d)** are from independent samples of Unif $[-1, 1]^{20}$ variates.

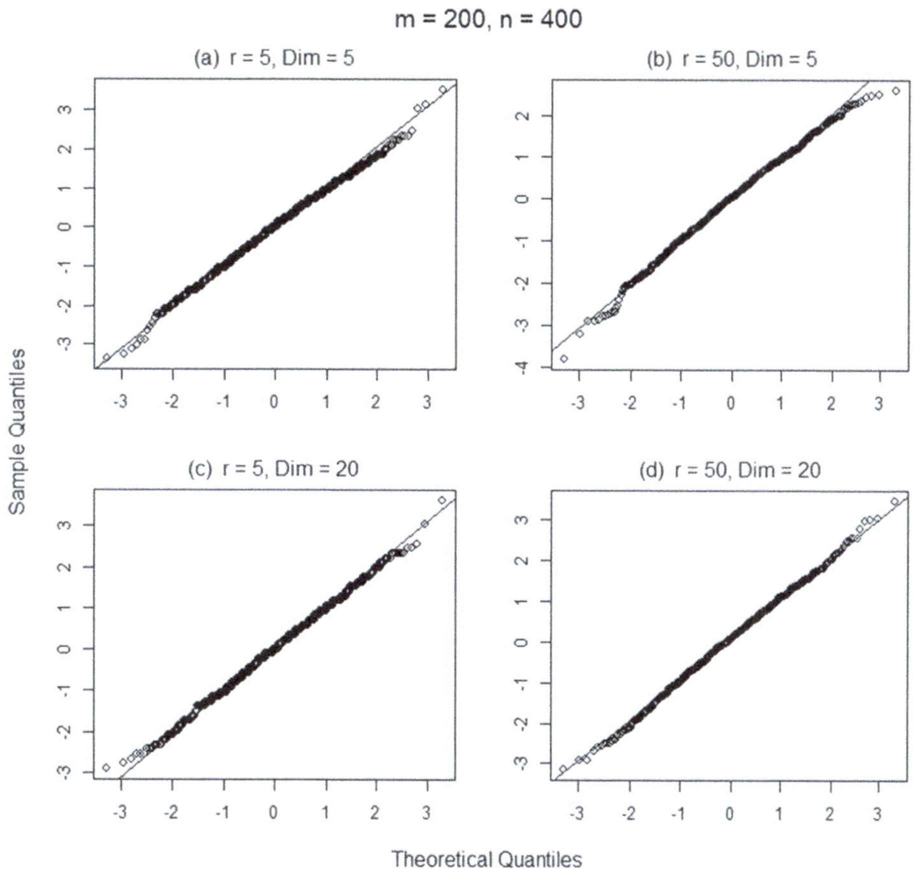

Figure 4 Normal QQ-plots of 1,000 simulated values of T_r for $m = 200$, $n = 400$ and $r = 5$, 50. Panels **(a)** and **(b)** are from independent samples of Unif $[-1, 1]^5$ variates. Panels **(c)** and **(d)** are from independent samples of Unif $[-1, 1]^{20}$ variates.

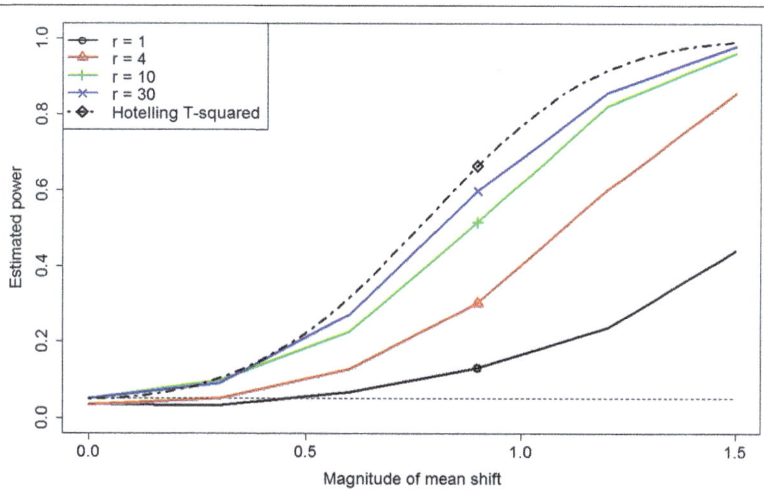

Figure 5 Power estimates for the MCC test at $r = 1$, 4, 10, and 30 and exact power for Hotelling's T^2 statistic for 5-variate normal mean alternatives with $m = 20$ and $n = 40$. The horizontal dashed line is test level $\alpha = .05$.

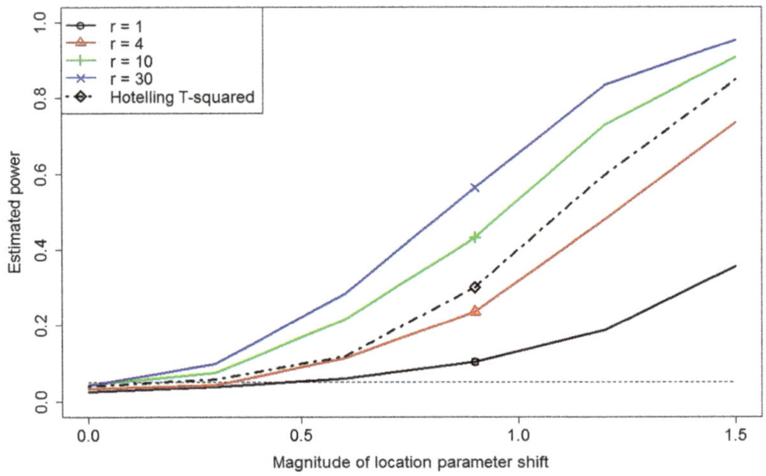

Figure 6 Power estimates for the MCC test at r = 1, 4, 10, and 30 and estimated power for Hotelling's T^2 statistic for 5-variate lognormal location parameter alternatives with m = 20 and n = 40. The horizontal dashed line is test level $\alpha = .05$.

Future work remains to bound rejection region probabilities in terms of sample size, dimension, and choice of r. In the absence of such theoretical bounds, for practical purposes a permutation test on observation indices serves as a suitable method to estimate p-values for the MCC test in cases where a normal approximation cannot be justified.

3.2 Small simulation study

We compare power characteristics of tests for two different location-shift scenarios. For each case, 1000 simulations are conducted for each shift in location parameter, group sizes are $m = 20$ and $n = 40$, and tests are conducted at significance level $\alpha = .05$. Distances are Euclidean. Estimated power is shown for MCC tests with $r = 1, 4, 10$, and 30 (where $r = 1$ is the cross-match test), and the performance of these tests is compared directly to that of Hotelling's T^2 test. Critical values for the MCC test were estimated through simulation for this study. All simulations were performed in R.

For the first example, the smaller group is drawn from a multivariate normal distribution with mean vector $\mathbf{0}$, identity covariance matrix, and dimension 5. The larger group is drawn from the same family, but the location vector of the second group is $\boldsymbol{\Delta}$, where $\boldsymbol{\Delta}$ ranges in magnitude from 0 to 1.5 by increments of 0.3. Hotelling's T^2 test is known to be the uniformly most powerful invariant test for location shift under these conditions (Bilodeau and Brenner 1999) and the exact power of the test is known for all location alternatives.

Figure 5 displays the estimated power results. We notice immediately that a modest increase of $r = 1$ to $r = 4$ substantially improves on the power of the cross-count test. As r continues to increase, MCC performance is even more impressive; the $r = 30 = N/2$ case performs nearly as well as Hotelling's T^2 test. Power estimates for cases $r > N/2$ are not shown. Not surprisingly, test power generally decreases as r increases beyond $N/2$ toward $N - 1$; in the extreme case T_{N-1} takes the fixed value $\frac{mn}{N-1}$ and hence the MCC test with $r = N - 1$ has power equal to zero against all alternatives.

For the second example, the first group is drawn from the multivariate log-normal distribution, where each of the 5 dimensions consists of independent, univariate log-normal draws with location parameter 0 and scale parameter 1. As before, the second group is drawn from the same family, but the location parameter vector for the second group is $\mathbf{\Delta}$, where the magnitude of $\mathbf{\Delta}$ ranges from 0 to 1.5 by increments of 0.3 and each dimension of $\mathbf{\Delta}$ takes equal value. The lognormal distribution is considered here to examine the effects of a skewed distribution on the tests in question. Since the underlying distributions are no longer multivariate normal, the power of Hotelling's T^2 test is estimated by simulation for this example.

Figure 6 displays the estimated power results. As before, we see that the power of the MCC test with $r = 4$ is much better than for $r = 1$. It is particularly noteworthy that for sufficiently large r the MCC test outperforms Hotelling's T^2 test.

4 Conclusions

The mean cross-count test is a powerful, non-parametric multivariate two-sample test that is applicable to any case where a notion of distance between observations exists. While this paper considers only location shifts, other simulations show that the MCC test has power in a variety of alternative cases as well. A shortcoming of the MCC test is that the null distribution for T_r is not simple (and perhaps not possible) to compute for all $r > 1$ and is not exactly distribution-free in these cases; in contrast, the test upon which it is based, the cross-match test with $r = 1$, has a known distribution that is independent of the distribution being tested.

It is known that T_1 is asymptotically normal, and while the mean and variance of T_r are derived herein and simulation suggests that the normal approximation for T_r is appropriate for sufficiently large N with $r > 1$, this property remains to be proven. This proof is part of ongoing work, as is sharpening the normal approximation via Edgeworth expansion based on higher moments of T_r. Likewise, finding useful criteria for choosing r is another area for future work. This choice is subject to competing factors: On the one hand, the power of T_r appears to improve as r increases to $N/2$ when group sizes are equal (i.e., $m = n = N/2$); therefore, $r = N/2$ seems a good choice for equal group sizes. On the other hand, the normal approximation appears to worsen as r increases; thus it may be desirable to restrict the size of r for this sake. Furthermore, an additional effect exists when group sizes are different. For example, assume $m < n$. If $r \geq m$, then at least one edge in $\mathcal{G}_r^*$ contains a vertex from each group and contributes to the cross-count, increasing the value of T_r. This is true even if the two groups are very different. Since a higher cross-count weakens the evidence against a group difference, this consideration suggests choosing $r < \min(m, n)$. A similar effect exists for multimodal distributions, suggesting that the size of r might be restricted as the number of modes grows. In any case, the best choice of r in practice clearly depends upon application specifics.

Competing interest
The author declares that he has no competing interests.

References

Anderson, I: Perfect matchings of a graph sufficient conditions for matchings. Proc Edinburgh Math Soc **18**, 129–136 (1972). Ser. 2

Bilodeau, M, Brenner, D: Theory of Multivariate Statistics. Springer, New York (1999)

Friedman, J, Rafsky, L: Multivariate generalizations of the Wald-Wolfowitz and Smirnov two-sample tests. Ann Stat **7**, 697–717 (1979)

Friedman, J, Rafsky, L: Graphics for the multivariate two-sample problem. JASA **76**, 277–287 (1981)

Hall, P, Tajvidi, N: Permutation tests for equality of distributions in high-dimensional settings. Biometrika **89**, 359–374 (2002)

Henze, N: A multivariate two-sample test based on the number of nearest neighbor type coincidences. Ann Stat **16**, 772–783 (1988)

Maa, J, Pearl, D, Bartoszynski, R: Reducing multidimensional two-sample data to one-dimensional interpoint comparisons. Ann Stat **24**, 1069–1074 (1996)

Rosenbaum, P: An exact distribution-free test comparing two multivariate distributions based on adjacency. JRSS B **67**, 515–530 (2005)

Ruth, D, Koyak, K: Nonparametric tests for homogeneity based on non-bipartite matching. JASA **106**, 1615–1625 (2011)

Schilling, M: Multivariate two-sample tests based on nearest neighbors. JASA **81**, 799–806 (1986)

12

Approximating the distributions of runs and patterns

Brad C Johnson[*†] and James C Fu[†]

*Correspondence:
brad.johnson@umanitoba.ca
[†]Equal Contributors
Department of Statistics, University of Manitoba, Winnipeg, Canada

Abstract

The distribution theory of runs and patterns has been successfully used in a variety of applications including, for example, nonparametric hypothesis testing, reliability theory, quality control, DNA sequence analysis, general applied probability and computer science. The exact distributions of the number of runs and patterns are often very hard to obtain or computationally problematic, especially when the pattern is complex and *n* is very large. Normal, Poisson and compound Poisson approximations are frequently used to approximate these distributions. In this manuscript, we (i) study the asymptotic relative error of the normal, Poisson, compound Poisson and finite Markov chain imbedding and large deviation approximations; and (ii) provide some numerical studies to comparing these approximations with the exact probabilities for moderately sized *n*. Both theoretical and numerical results show that, in the relative sense, the finite Markov chain imbedding approximation performs the best in the left tail and the large deviation approximation performs best in the right tail.

AMS Subject Classification: Primary 60E05; Secondary 60J10

Keywords: Finite Markov chain imbedding; Rate functions; Multi-state trials; Runs and patterns

Introduction and notation

Let $\{X_i\}_{i=1}^n$ be a sequence of m-state trials ($m \geq 2$) taking values in the set $\mathcal{S} = \{s_1, \ldots, s_m\}$ of m symbols. For simplicity, $\{X_i\}_{i=1}^n$ will be denoted $\{X_i\}$ and n will be allowed to be ∞. A *simple pattern* $\Lambda = s_{i_1}s_{i_2}\cdots s_{i_\ell}$, of length ℓ, is the juxtaposition of ℓ (not necessarily distinct) symbols from $\mathcal{S}$. Given a simple pattern Λ, we let $X_n(\Lambda)$ denote the number of either non-overlapping or overlapping occurrences of Λ in the sequence $\{X_i\}_{i=1}^n$, where the method of counting will be made clear by the context. The waiting time $W(\Lambda, x)$ until the x'th occurrence of the simple pattern Λ in $\{X_i\}_{i=1}^n$ is thus defined by

$$W(\Lambda, x) = \inf\{n \in \mathbb{N} : X_n(\Lambda) = x\},$$

and, by convention, the waiting time for the first occurrence is denoted $W(\Lambda) = W(\Lambda, 1)$. Finally, we define the inter arrival times

$$W_i(\Lambda) = W(\Lambda, i) - W(\Lambda, i-1), \qquad \text{for } i = 1, 2, \ldots,$$

where $W(\Lambda, 0) := 0$.

We say that two patterns Λ_1 and Λ_2 are distinct if neither Λ_1 appears in Λ_2 nor Λ_2 appears in Λ_1. If $\Lambda_1, \ldots, \Lambda_r$ are pairwise distinct simple patterns, we define the compound pattern $\Lambda = \bigcup_{i=1}^{r} \Lambda_i$, where an occurrence of any Λ_i is considered an occurrence of Λ. For a compound pattern $\Lambda = \Lambda_1 \cup \cdots \cup \Lambda_r$, we similarly define

$$X_n(\Lambda) = \sum_{j=1}^{r} X_n(\Lambda_j).$$

The waiting times $W(\Lambda, x)$, $W(\Lambda)$ and $W_i(\Lambda)$ are then defined as above, and often referred to as *sooner* waiting times.

From these definitions it is easy to see that, for any simple or compound pattern Λ, x and n, the events $\{X_n(\Lambda) < x\}$ and $\{W(\Lambda, x) > n\}$ are equivalent and hence

$$\mathbb{P}\{X_n(\Lambda) < x\} = \mathbb{P}\{W(\Lambda, x) > n\}, \tag{1}$$

which provides a convenient way of studying the exact and approximate distribution of $X_n(\Lambda)$ through the waiting time distributions of $W(\Lambda, x)$.

Throughout this paper, unless specified otherwise, we assume that the trials $\{X_i\}$ are either independent and identically distributed (i.i.d.) or first order Markov dependent; the pattern Λ is either simple or compound; and the counting of occurrences of Λ is in a non-overlapping fashion.

The distribution of the number of runs and patterns in a sequence of multi-state trials or random permutations of a set of integers have been successfully used in various fields in applied probability, statistics and discrete mathematics. Examples include reliability theory, quality control, DNA sequence analysis, psychology, ecology, astronomy, nonparametric tests, successions, and the Eulerian and Simon-Newcomb numbers (the latter 3 being defined for permutations). Two recent books, Balakrishnan and Koutras (2002) and Fu and Lou (2003), provide some scope of the distribution theory of runs and patterns and Martin et al. (2010) and Nuel et al. (2010) provides some extensions to sets of sequences.

Given a pattern Λ, the exact distribution of $X_n(\Lambda)$ traditionally has been determined using combinatoric analysis on a case by case basis. The formulae for these distributions are often very complex and computationally problematic. Even for many simple patterns, their distributions in terms of combinatoric analysis remains unknown, especially when the $\{X_i\}$ are Markov dependent multi-state trials.

The waiting time $W(\Lambda)$ for the first occurrence of certain types of runs and patterns have been studied by many authors. See, for example, Blom and Thorburn (1982), Gerber and Li (1981), Schwager (1983), and Solov'ev (1966). More recently, Fu and Koutras (1994) developed a method for determining the exact distributions of $X_n(\Lambda)$ and $W(\Lambda)$ for any simple or compound Λ in either i.i.d. or Markov dependent trials (see also Fu and Lou 2003). The method was referred to as the Finite Markov Chain Imbedding (FMCI) technique, which can be easily described as follows: given a simple or compound pattern Λ, there exists a finite Markov chain $\{Y_i\}$ defined on a finite state space, say $\Omega = \{1, \ldots, d, \alpha\}$, with an absorbing state α and transition probability matrix of the form

$$\mathbf{P} = \begin{array}{c} \Omega - \alpha \\ \alpha \end{array} \left[\begin{array}{c|c} \mathbf{N} & \mathbf{c} \\ \hline \mathbf{0} & 1 \end{array} \right], \tag{2}$$

where c is a column vector. The distribution of the waiting time for Λ is given by

$$\mathbb{P}\{W(\Lambda) = n\} = \xi_0 \mathbf{N}^{n-1}(\mathbf{I} - \mathbf{N})\mathbf{1}' \tag{3}$$

where ξ_0 is the initial distribution, $\mathbf{N}$ is the *essential transition probability matrix* (i.e. the sub-stochastic matrix consisting of only the transient states of $\{Y_i\}$) as defined in (2), $\mathbf{I}$ is a $d \times d$ identity matrix and $\mathbf{1} = (1, 1, \ldots, 1)$ is a $1 \times d$ row-vector. Furthermore, the random variable $X_n(\Lambda)$, the number of occurrences of Λ in $\{X_i\}$, is also finite Markov chain imbeddable and its distribution is given by

$$\mathbb{P}\{X_n(\Lambda) < x\} = \mathbb{P}\{W(\Lambda, x) > n\} = \xi_0 \mathbf{N}_x^n \mathbf{1}', \tag{4}$$

where the essential transition probability matrix $\mathbf{N}_x$ has the form

$$\mathbf{N}_x = \begin{bmatrix} \mathbf{N} & \mathbf{C} & & & \\ & \mathbf{N} & \mathbf{C} & \mathbf{0} & \\ & & \ddots & \ddots & \\ & \mathbf{0} & & \mathbf{N} & \mathbf{C} \\ & & & & \mathbf{N} \end{bmatrix}, \tag{5}$$

the matrix $\mathbf{N}$ is given by (2), and the matrix $\mathbf{C}$ defines the "continuation" transition probabilities from one occurrence to the next and depends on $\mathbf{c}$ in (2).

If the pattern Λ is long and complex and n is very large, then the computation of $\mathbb{P}\{X_n(\Lambda) = x\}$ can become problematic and, to overcome this problem, various asymptotic approximations have been developed for these probabilities.

In real applications, if the exact distribution is not available or is hard to compute, it is important to know which approximations perform well and are easy to compute. Furthermore, it is important to know how these approximations perform with respect to each other and the exact distribution from both a theoretical and numerical standpoint. The aims of this manuscript are two-fold: (i) we first study the asymptotic relative error of the normal, Poisson (or compound Poisson), and FMCI approximations with respect to the exact distribution; and (ii) we then provide a numerical study of these three approximations with the exact probabilities in cases where x is fixed and $n \to \infty$ and when n is fixed and x varies. As an important byproduct, the FMCI technique allows the normal and Poisson approximations to be applied in more cases, for example, the distribution of compound patterns and patterns in Markov dependent trials.

The approximations

Normal approximation

The normal approximation is one of the most popular for approximating the distribution of the number of runs or patterns $X_n(\Lambda)$ in Statistics. In general, when Λ is simple or compound, the trials are i.i.d., and the counting is non-overlapping, by appealing to (1) and renewal arguments, it has been shown that $X_n(\Lambda)$ is asymptotically normally distributed (cf. Fu and Lou 2007; Karlin and Taylor 1975). The form of the approximation is

$$\lim_{n\to\infty} \mathbb{P}\left\{ \frac{X_n(\Lambda) - n/\mu_W}{\sqrt{n\sigma_W^2 \mu_W^{-3}}} \le u \right\} = \Phi(u), \tag{6}$$

where $\Phi(\cdot)$ denotes the standard normal distribution function and μ_W and σ_W^2 are the mean and variance of $W(\Lambda)$ respectively, which are given by

$$\mu_W = \boldsymbol{\xi}_0(\mathbf{I}-\mathbf{N})^{-1}\mathbf{1}', \quad \text{and} \tag{7}$$
$$\sigma_W^2 = \boldsymbol{\xi}_0(\mathbf{I}+\mathbf{N})(\mathbf{I}-\mathbf{N})^{-2}\mathbf{1}' - \mu_W^2. \tag{8}$$

Given a pattern Λ, it is well known that the mean μ_W and the variance σ_W^2 are difficult to obtain via combinatoric arguments, especially when Λ is a compound pattern or the trials are Markov dependent. For example, as pointed out in Karlin (2005) and Kleffe and Borodovski (1992), approximate values of μ_W and σ_W^2 must sometimes be used. Since $W(\Lambda)$ is finite Markov chain imbeddeble, (7) and (8), provide the exact values.

The limit in (6) is appropriate when the sequence of inter arrival times $\{W_i(\Lambda)\}$ are i.i.d., which is the case for simple and compound patterns when the $\{X_i\}$ are i.i.d. and counting is non-overlapping. When occurrences of Λ correspond to a delayed renewal process, which can occur for Markov dependent trials and/or overlapping counting, we could use the mean and variance of $W_2(\Lambda)$ for the normalizing constants, which are easily obtained by modifying $\boldsymbol{\xi}_0$ in (7) and (8). Even more general cases can be handled by making use of a functional central limit theorem for Markov chains (see, for example, (Meyn and Tweedie 1993, §17.4) and (Asmussen 2003, Theorem 7.2, pg. 30) for the details).

Poisson and compound poisson approximations

It is well known that, in a sequence of Bernoulli(p) trials, if $np \to \lambda$ as $n \to \infty$, then the probability of k successes in n trials can be approximated by a Poisson probability with parameter λ, denoted $\mathscr{P}(\lambda)$. This idea has been extended to certain patterns Λ and, under certain conditions, the distribution of $X_n(\Lambda)$ can be approximated by a Poisson distribution with parameter μ_n in the sense that

$$d_{\mathrm{TV}}(\mathscr{L}(X_n(\Lambda)), \mathscr{P}(\mu_n)) < \varepsilon_n, \tag{9}$$

where $\mathscr{L}(\cdot)$ denotes the distribution (law) of a random variable and $d_{\mathrm{TV}}(\cdot,\cdot)$ denotes the total variation distance.

The primary tool used to obtain μ_n and the bound ε_n is the Stein-Chen method (Chen 1975), and this method has been refined by various authors Arratia et al. (1990), Barbour and Eagleson (1983), Barbour and Eagleson (1984), Barbour and Eagleson (1987), Barbour and Hall (1984), Godbole (1990a), Godbole (1990b), Godbole (1991), Godbole and Schaffner (1993), and Holst et al. (1988). This method has also been extended to compound Poisson approximations for the distributions of runs and patterns and Barbour and Chryssaphinou (2001) provides an excellent theoretical review of these approximations.

In practice, $\mu_n = \mathbb{E}X_n(\Lambda)$ or the expectation of a closely related run statistic is used (cf. Balakrishnan and Koutras 2002, §5.2.3) so that, in the former case,

$$\mathbb{P}\{X_n(\Lambda) = x\} \approx \frac{(\mathbb{E}X_n(\Lambda))^x}{x!} \exp\{-\mathbb{E}X_n(\Lambda)\}. \tag{10}$$

Finding $\mathbb{E}X_n(\Lambda)$ and the bound ε_n is usually done on a case by case basis. For the mathematical details, the books (Barbour et al. 1992a) and (Balakrishnan and Koutras 2002) are recommended.

Let $\mathscr{P}_c(\lambda, \nu)$ denote the compound Poisson distribution, that is, the distribution of the random variable $\sum_{j=1}^{M} Y_j$ where the random variable M has a Poisson distribution with parameter λ and the Y_j are i.i.d. having distribution ν. A compound Poisson distribution for approximating nonnegative random variables was suggested in Barbour et al. (1992b) (see also Barbour et al. (1995,1996)). The approximation is formulated similarly to the Poisson approximation:

$$d_{\mathrm{TV}}(\mathscr{L}(X_n(\Lambda)), \mathscr{P}_c(\lambda, \nu)) < \varepsilon_n. \tag{11}$$

The distribution of $N_{n,k}$, the number of non-overlapping occurrences of k consecutive successes in n i.i.d. Bernoulli trials, is one of the most important in this area and one of the most studied in the literature. Reversing the roles of S (success) and F (failure), the reliability of consecutive-k-out-of-n system, denoted $C(k, n{:}F)$, is given by $\mathbb{P}\{N_{n,k} = 0\}$. Even in this simple case (i.e. $\Lambda = SS\cdots S$), there are several ways to apply the Poisson approximation techniques. For example, (Godbole 1991, Theorem 2) shows that approximating $N_{n,k}$ with a $\mathscr{P}(\mathbb{E}N_{n,k})$ distribution works well if certain conditions hold. Godbole and Schaffner (Godbole and Schaffner 1993, pg. 340) suggests an improved Poisson approximation for word patterns.

The primary difficulty in applying the Poisson approximation is the determination of the optimal parameter μ_n, which is higly dependent on the structure of the pattern Λ. In particular, if Λ is long and has several uneven overlapping sub-patterns, then finding μ_n by their method can be very tedious. In the sequel, we show that even the (asymptotic) best choice for μ_n for Poisson approximations does not perform well in the relative sense.

FMCI approximations

Approximations based on the FMCI approach depend on the spectral decomposition of the essential transition probability matrix $\mathbf{N}$.

Let $\mathbf{N}$ be a $w \times w$ essential transition probability matrix associated with a finite Markov chain $\{Y_n : n \geq 0\}$ corresponding to the distribution of the waiting time $W(\Lambda)$. Let $1 > \lambda_1 \geq |\lambda_2| \geq \cdots \geq |\lambda_w|$ denote the ordered eigenvalues of $\mathbf{N}$, repeated according to their algebraic multiplicities, with associated (right) eigenvectors $\eta'_1, \eta'_2, \cdots, \eta'_w$. When the geometric multiplicity of λ_i is less than its algebraic multiplicity, we will use vectors of 0's for the unspecified eigenvectors. The fact that λ_1 can be taken as a positive real number and that η_1 can be taken to be non-negative are consequences of the Perron-Frobenious Theorem for non-negative matrices (*cf.* Seneta 1981).

Definition 1. We will say that $\{Y_n : n \geq 0\}$, or equivalently, $\mathbf{N}$, satisfies the *FMCI Approximation Conditions* if

(i) there exists constants $a_1, \ldots, a_w$ such that

$$\mathbf{1}' = \sum_{i=1}^{w} a_i \eta'_i, \tag{12}$$

(ii) λ_1 has algebraic multiplicity g and $\lambda_1 > |\lambda_j|$ for all $j > g$.

Verifying these conditions is usually straightforward. They certainly hold if $\mathbf{N}$ is irreducible and aperiodic, but also hold in many other cases as well. For example, (12) requires only that $\mathbf{1}'$ is in the linear space spanned by $\{\eta'_1, \eta'_2, \cdots, \eta'_w\}$, which can hold even when $\mathbf{N}$ is defective (not diagonizable). Condition (ii) requires that the communication classes corresponding λ_1 are aperiodic. That is, if Ψ is a communication class and $\mathbf{N}[\Psi]$ corresponds to the substocastic matrix $\mathbf{N}$ restricted to the states in Ψ, with largest eigenvalue $\lambda_1[\Psi]$, then all Ψ such that $\lambda_1[\Psi] = \lambda_1$ should be aperiodic. We also mention that the algebraic multiplicity of λ_1 is the number of communication classes Ψ such that $\lambda_1[\Psi] = \lambda_1$.

Fu and Johnson (2009) give the following theorem.

Theorem 1. *Let $\{X_i\}$ be a sequence of i.i.d. trials taking values in $\mathcal{S}$, let Λ be a simple pattern of length ℓ with $d \times d$ essential transition probability matrix $\mathbf{N}$ and let $X_n(\Lambda)$ be the number of non-overlapping occurrences of Λ in $\{X_i\}$. If $\mathbf{N}$ satisfies the FMCI approximation conditions then, for any fixed $x \geq 0$,*

$$\mathbb{P}\{X_n(\Lambda) = x\} \sim a^{x+1} \binom{n - x(\ell - 1)}{x} (1 - \lambda_1)^x \lambda_1^{n-x}, \tag{13}$$

where $a = \sum_{j=1}^{g} a_j(\boldsymbol{\xi}_0 \boldsymbol{\eta}_j')$. If $g = 1$, as is usually the case, then $a = a_1(\boldsymbol{\xi}_0 \boldsymbol{\eta}_1')$.

Given a pattern Λ, the approximation in (13) requires finding the Markov chain imbedding associated with the waiting time $W(\Lambda)$, the essential transition probability matrix $\mathbf{N}$ as well as its eigenvalues and associated eigenvectors. Usually, these steps are rather simple and can be easily automated together with (13). Even for very large n and large ℓ, say $n = 1{,}000{,}000$ and $\ell = 50$, the CPU time is negligible. Fu and Johnson (2009) also provide details on extending these results to compound patterns, overlapping counting and Markov dependent trials.

For the purpose of comparing these approximations, we prefer to write (13) as

$$\mathbb{P}\{X_n(\Lambda) = x\} \sim a^{x+1} \left(\frac{1 - \lambda_1}{\lambda_1}\right)^x \binom{n - x(\ell - 1)}{x} \exp\{n \ln \lambda_1\} \tag{14}$$

Note that the approximation havs three parts: a constant part; a polynomial in n of degree x; and a third (dominant) part which converges to 0 exponentially fast as $n \to \infty$.

More precisely, the FMCI approximation in (13) may be written as

$$\begin{aligned} \mathbb{P}\{X_n(\Lambda) = x\} &= a^{x+1} \left(\frac{1 - \lambda_1}{\lambda_1}\right)^x \binom{n - x(\ell - 1)}{x} \\ &\quad \times \exp\{n \ln \lambda_1\} \left[1 + o\left(\left|\frac{\lambda_{g+1}}{\lambda_1}\right|^{n/(x+1)-\ell}\right)\right]. \end{aligned} \tag{15}$$

Since $|\lambda_{g+1}| < \lambda_1$, the term $|\lambda_{g+1}/\lambda_1|^{n/(x+1)-\ell}$ tends to 0 exponentially as $n \to \infty$ and hence is negligible if $n/(x + 1) - \ell$ is moderate or large (say ≥ 50).

Large deviation approximation

Fu et al. (2012) provide the following large deviation approximation for right-tail probabilities for the number of non-overlapping occurrences for simple patterns Λ. The reasons for providing only the right-tail large deviation approximation are (i) all of the above mentioned approximations fail to approximate the extreme right-tail probabilities and (ii) the FMCI approximation provides an accurate approximation for left-tail probabilities.

Theorem 2. *Let $\varepsilon = x\mu_W^2/(1 + x\mu_W)$ and let*

$$\varphi_W(t) = 1 + (e^t - 1)\boldsymbol{\xi}(\mathbf{I} - e^t\mathbf{N})^{-1}\mathbf{1}', \tag{16}$$

be the moment generating function of $W(\Lambda)$. Then

$$\mathbb{P}\{X_n(\Lambda) \geq \mathbb{E}X_n(\Lambda) + nx\} = e^{-n\beta(\varepsilon,\Lambda)} \frac{1}{\sqrt{n}} \left\{b_0 + b_1 n^{-1} + \cdots + b_m n^{-m} + \mathcal{O}(n^{-m-1})\right\}, \tag{17}$$

where

$$\beta(x,\Lambda) = \left(\frac{1}{\mu_W} + x\right) h(\varepsilon,\tau) = \left(\frac{1}{\mu_W} + x\right)\left[-\frac{\tau\mu_W}{1 + x\mu_W} - \ln\varphi_{W(\Lambda)}(-\tau)\right], \tag{18}$$

$h(\varepsilon, t) = \varepsilon t - \ln\varphi_{\mu_W - W(\Lambda)}(t)$, τ is the solution to $h'(\varepsilon,\tau) = 0$, and

$$\begin{aligned} b_0 &= \frac{1}{\sigma\tau\sqrt{2\pi(\mu^{-1}+x)}} \\ b_1 &= \frac{1}{\sigma\tau\sqrt{2\pi(\mu^{-1}+x)^3}}\left\{-\frac{1}{\sigma^2\tau^2} + \frac{h^{(3)}(\varepsilon,\tau)}{2\tau\sigma^4} - \frac{h^{(4)}(\varepsilon,\tau)}{8\sigma^4} - \frac{5(h^{(3)}(\varepsilon,\tau))^2}{24\sigma^6}\right\} \\ \sigma &= \sqrt{-h''(\varepsilon,\tau)}. \end{aligned} \tag{19}$$

Comparisons and relative error

For a given n, x and pattern Λ, we define the relative error of an approximation with respect to the exact probability $\mathbb{P}\{X_n(\Lambda) = x\}$ as

$$R(x : E, A) = \operatorname{sgn}(A - E)\left[\max\left(\frac{E}{A}, \frac{A}{E}\right) - 1\right],$$

where A stands for the approximate probability and E stands for the exact probability $\mathbb{P}\{X_n(\Lambda) = x\}$. This quantity, $R(x : E, A)$, goes from $-\infty$ to ∞ and treats the importance of overestimation the same as underestimation. It is clear that $R(x : E, A) > 0$ implies that the approximation is overestimating the exact probability and that $R(x : E, A) < 0$ implies that the approximation is underestimating the exact probability. Since, for fixed x, the probability $\mathbb{P}\{X_n(\Lambda) = x\}$ converges to 0 exponentially fast as $n \to \infty$, it follows that $R(x : E, A) \to \pm\infty$ implies that the approximation tends to 0 with the wrong rate. If $R(x : E, A)$ is near 0 then the approximation is close to the exact probability $\mathbb{P}\{X_n(\Lambda) = x\}$.

Note that $R(x : E, A)$ is a function of x, n and the method of approximation used. The following theorem provides the asymptotic relative error for the Normal approximation (N), the Poisson approximation ($P(\mu_n)$) and the finite Markov chain imbedding approximation (F).

Theorem 3. *Let $\{X_i\}$ be a sequence of i.i.d. multi-state trials taking values in $\mathcal{S}$ and let Λ be a simple pattern defined on $\mathcal{S}$. Then, for every fixed x, we have,*

(i) $$\lim_{n\to\infty} R(x : E, F) = 0; \tag{20}$$

(ii) $$\lim_{n\to\infty} R(x : E, P(\mu_n)) = \begin{cases} \infty, & \textit{if } \limsup_n \mu_n/n < -\ln\lambda_1; \\ c(x), & \textit{if } \lim_n \mu_n/n = -\ln\lambda_1; \\ -\infty, & \textit{if } \liminf_n \mu_n/n > -\ln\lambda_1; \end{cases} \tag{21}$$

$$(iii) \qquad \lim_{n\to\infty} R(x:E,N) = \begin{cases} \infty, & \text{if } \mu_W/2\sigma_W^2 \le -\ln\lambda_1; \\ -\infty, & \text{if } \mu_W/2\sigma_W^2 > -\ln\lambda_1; \end{cases} \tag{22}$$

where the exact probability is computed using (4) and

$$c(x) = a^{x+1}\left(\frac{\lambda_1 - 1}{\lambda_1 \ln\lambda_1}\right)^x - 1.$$

Proof. Given a pattern Λ and x, for the finite Markov chain imbedding approximation we have

$$\lim_{n\to\infty} \frac{\mathbb{P}\{X_n(\Lambda) = x\}}{a^{x+1}\left(\frac{1-\lambda_1}{\lambda_1}\right)^x \binom{n - x(\ell-1)}{x} \exp\{n\ln\lambda_1\}} = 1$$

and hence (i) follows immediately from the definition of $R(x:E,A)$ and Theorem 1.

For the Poisson approximation we have, since $E/F \sim 1$ by (i),

$$\frac{E}{P(\mu_n)} = \frac{E}{F} \times \frac{F}{P(\mu_n)} \sim \frac{F}{P(\mu_n)}$$

and hence

$$\frac{E}{P(\mu_n)} = \frac{\mathbb{P}\{X_n(\Lambda) = x\}}{\frac{\mu_n^x}{x!}\exp\{-\mu_n\}}$$
$$\sim \frac{a^{x+1}\left(\frac{1-\lambda_1}{\lambda_1}\right)^x \binom{n-x(\ell-1)}{x} \exp\{n\ln\lambda_1\}}{\frac{\mu_n^x}{x!}\exp\{-\mu_n\}}. \tag{23}$$

If $\liminf_n \mu_n/n > -\ln\lambda_1$ then $\exp\{n\ln\lambda_1 + \mu_n\}$ tends to 0 exponentially fast which overrides the polynomial term and hence $R(x:E,P(\mu_n)) \to -\infty$ as $n\to\infty$ for all fixed x. Similarly, if $\limsup_n \mu_n/n < -\ln\lambda_1$, then $R(x:E,P(\mu_n)) \to \infty$ as $n\to\infty$ for all fixed x. Furthermore, if $\lim_n \mu_n/n = -\ln\lambda_1$, then the ratio yields

$$\lim_{n\to\infty} R(x:E,P(-n\ln\lambda_1)) = a^{x+1}\left(\frac{\lambda_1-1}{\lambda_1\ln\lambda_1}\right)^x - 1$$

and this completes the proof of (ii). Note also that, if $\limsup_n \mu_n/n > -\ln\lambda_1$ and $\liminf_n \mu_n/n < -\ln\lambda_1$, then $\lim_n R(x:E,P(\mu_n))$ will not exist.

For the normal approximation we have that $X_n(\Lambda)$ is approximately normal with mean n/μ_W and variance $n\sigma_W^2/\mu_W^3$ and hence

$$\mathbb{P}\{X_n(\Lambda) = x\} \approx N = \int_{x-1/2}^{x+1/2} \frac{1}{\sqrt{2\pi n\sigma_W^2\mu_W^{-3}}} \exp\left\{-\frac{(t-n/\mu_W)^2}{2n\sigma_W^2\mu_W^{-3}}\right\} dt$$

Hence, provided $n > \mu_W(x+1/2)$, we have

$$N \le \frac{1}{\sqrt{2\pi n\sigma_W^2\mu_W^{-3}}} \exp\left\{-\frac{(x+1/2-n/\mu_W)^2}{2n\sigma_W^2\mu_W^{-3}}\right\}.$$

Therefore, as in the proof of (ii), we are interested in the asymptotics of F/N, which yields

$$\frac{F}{N} \sim \sqrt{\frac{2\pi n\sigma_W^2}{\mu_W^3}}\, a^{x+1}\left(\frac{1-\lambda_1}{\lambda_1}\right)^x \binom{n-x(\ell-1)}{x} \times \exp\left\{n\ln\lambda_1 + \frac{(x+1/2-n/\mu_W)^2}{2n\sigma_W^2\mu_W^{-3}}\right\}. \tag{24}$$

We may rewrite the argument of the exponential function as

$$n\left[\ln\lambda_1 + \frac{\mu_W}{2\sigma_W^2}\left(\frac{\mu_W(x+1/2)}{n}-1\right)^2\right],$$

making it clear that (24) converges to ∞ if $\mu_W/2\sigma_W^2 \geq -\ln\lambda_1$ and 0 otherwise. Therefore, $R(x:E,N) \to \infty$ if $\mu_W/2\sigma_W^2 \geq -\ln\lambda_1$ and $R(x:E,N) \to -\infty$ if $\mu_W/2\sigma_W^2 < -\ln\lambda_1$ and the proof of (iii) is complete. □

Theorem 3 (ii) implies that asymptotically (for fixed x and $n \to \infty$), the Poisson approximation performs poorly (in the relative sense) regardless of the value μ_n used. When Λ is simple and does not have overlapping sub-patterns, taking $\mu_n = \mathbb{E}X_n(\Lambda)$ is normally recommended for the Poisson approximation (*cf.* Arratia et al. 1990). In this case, non-overlapping and overlapping counting is equivalent. The following corollary shows that, for fixed x, the Poisson approximation will (asymptotically) always overestimate the exact probability in the following sense.

Corollary 1. *Let Λ be a simple pattern defined on an i.i.d. sequence of multi-state trials. For $\mu_n = \mathbb{E}X_n(\Lambda)$, we have*

$$\lim_{n\to\infty} R(x:E,P(\mu_n)) = \infty$$

for all fixed x.

Proof. Recall that, in this case, $X_n(\Lambda)$ is a renewal process with i.i.d. inter-renewal times with mean $\mu_W = \mathbb{E}W(\Lambda)$ and hence, by the elementary renewal theorem, we have $\mathbb{E}X_n(\Lambda)/n \to 1/\mu_W$ so that $\mathbb{E}X_n(\Lambda) \sim n/\mu_W$. Therefore, by Theorem 3 (ii), it is sufficient to show that $n/\mu_W < -n\ln\lambda_1$ for all sufficiently large n, or

$$e^{-1/\mu_W} > \lambda_1.$$

Now, since $0 < \lambda_1 \in \mathbb{R}$ is a dominant eigenvalue of $\mathbf{N}$, it follows that: $0 < (1-\lambda_1)^{-1} \in \mathbb{R}$ is a dominant eigenvalue of the matrix $(\mathbf{I}-\mathbf{N})^{-1} = \mathbf{A} = (a_{ij})$; $a_{ij} \geq 0$ with at least one $a_{ij} > 0$; and $\mathbf{A}\mathbf{1}' = (\mathbf{I}-\mathbf{N})^{-1}\mathbf{1}' \leq \mu_W\mathbf{1}'$. Hence, by a simple corollary to the Perron-Frobenius Theorem for nonnegative matrices (cf. Karlin and Taylor 1975, Corollary 2.2, pg. 551), we have

$$\frac{1}{1-\lambda_1} = \limsup_{n\to\infty}\left(\max_{i,j}|a_{ij}^{(n)}|\right)^{1/n} \leq \mu_W,$$

where $a_{ij}^{(n)} = (\mathbf{A}^n)_{ij}$. Therefore, provided $\mu_W < \infty$,

$$e^{-1/\mu_W} > 1 - \frac{1}{\mu_W} \geq \lambda_1,$$

which completes the proof. □

Corollary 1 implies that, if $\mu_n \sim \mathbb{E}X_n(\Lambda)$, then the Poisson approximation will always overestimate the exact probability as $n \to \infty$. Together with Theorem 3 (ii), this implies that using $\mu_n \sim -n \ln \lambda_1$ results in the best Poisson approximation as $n \to \infty$.

We also comment that, for the normal approximation, both $\mu_W/2\sigma_W^2 < -\ln \lambda_1$ and $\mu_W/2\sigma_W^2 \geq -\ln \lambda_1$ are possible. As a simple example, suppose we have a sequence of i.i.d. Bernoulli(p) trials and $\Lambda = SSS$. If $p = 1/2$, we obtain

$$\mu_W = 14, \quad \sigma_W^2 = 142 \quad \text{and} \quad \lambda_1 = 0.9196434,$$

and

$$\frac{\mu_W}{2\sigma_W^2} = 0.04929577 < -\ln \lambda_1 = 0.08376932.$$

However, with $p = 0.9$, we obtain

$$\mu_W = 3.717421, \quad \sigma_W^2 = 2.145694 \quad \text{and} \quad \lambda_1 = 0.5419067;$$

and

$$\frac{\mu_W}{2\sigma_W^2} = 0.8662513 > -\ln \lambda_1 = 0.6126614.$$

Thus, $R(x : E, N) \to \pm\infty$ are both possible depending on x, the pattern, and the probability structure of the $\{X_i\}$.

Numerical comparisons

In the previous section we showed that, for fixed x and $n \to \infty$, the approximation based on the finite Markov chain imbedding technique outperforms the Poisson and normal approximations. In practice, however, one is interested in the performance of these approximations not only when x is fixed and $n \to \infty$, but also when n is fixed (at some moderate value) and x varies. The reason we consider only large or moderate n in our numerical study is that, for small n, the FMCI technique easily gives the exact results. In this section we present some numerical experiments to illustrate the advantages (and disadvantages) of the methods discussed.

The approximations we compare are: the finite Markov chain approximation in (13) (FMCI); the Poisson approximation with $\mu_n = n/\mu_W$ ($\sim \mathbb{E}X_n(\Lambda)$) where μ_W is calculated using (7) (Poisson); The normal approximation given in (6) (Normal); and the large deviation approximation given in Theorem 2 (LD), which is only for right-tail probabilities.

Reliability of $C(k, n{:}F)$ systems

A consecutive-k-out-of-n:F system is a system of n independent and linearly connected components, each with common (continuous) lifetime distribution F, in which the system fails if k consecutive components fail. At a given time $t > 0$, the probability a component is working is $p = 1 - F(t)$ and the probability a single component has failed is $q = 1 - p$ and hence the probability the system has failed is equivalent to the probability that k (or more) consecutive components have failed, which is equivalent to the probability of k consecutive failures in a sequence of n Bernoulli trials with success probability p. Barbour et al. (1995) present a table of various bounds for system reliability based on a Poisson approximation and a compound approximation and compare these to bounds found in Fu (1985). Table 1 shows the exact probabilities and relative errors for the FMCI and Poisson

Table 1 Approximation errors for $C(k, n:F)$ systems

n	k	q	Exact	FMCI	Poisson	CP
5	2	0.01	0.99960	0.00000	-0.00010	0.00000
5	2	0.10	0.96309	0.00000	-0.00788	0.00119
5	2	0.25	0.79980	-0.00002	-0.02697	0.04654
10	2	0.01	0.99911	0.00000	-0.00010	0.00000
10	2	0.10	0.91975	0.00000	-0.00728	0.00312
10	2	0.25	0.61180	0.00000	-0.00869	0.12266
10	4	0.01	1.00000	0.00000	0.00000	0.00000
10	4	0.10	0.99936	0.00000	-0.00026	0.00000
10	4	0.25	0.97855	0.00000	-0.00776	0.00038
50	2	0.01	0.99516	0.00000	-0.00010	0.00000
50	2	0.10	0.63633	0.00000	-0.00251	0.01871
50	2	0.25	0.07173	0.00000	0.14441	0.96838
50	4	0.01	1.00000	0.00000	0.00000	0.00000
50	4	0.10	0.99577	0.00000	-0.00026	0.00000
50	4	0.25	0.86897	0.00000	-0.00663	0.00312
100	2	0.01	0.99024	0.00000	-0.00010	0.00000
100	2	0.10	0.40151	0.00000	0.00343	0.03854
100	2	0.25	0.00492	0.00000	0.36933	2.97133
100	4	0.01	1.00000	0.00000	0.00000	0.00000
100	4	0.10	0.99129	0.00000	-0.00026	0.00001
100	4	0.25	0.74908	0.00000	-0.00523	0.00656
500	4	0.20	0.52721	0.00000	-0.00086	0.00611
1,000	4	0.20	0.27696	0.00000	0.00183	0.01232
10,000	5	0.20	0.07710	0.00000	0.00183	0.00560

approximations as well as the compound Poisson approximation in Barbour et al. (1995) (CP).

The FMCI approximation performs very well for the parameters tested here. As expected, the Poisson and compound Poisson approximations perform well when nq^k is relatively small. When the reliability of the system is relatively low, the Poisson and compound Poisson approximations begin to degrade.

Approximating the distribution of $N_{n,k}$

Recall that $N_{n,k}$ is the number of non-overlapping occurrences of k consecutive successes in $\{X_i\}$ (i.e. $N_{n,k} = X_n(\Lambda)$ with $\Lambda = SS\cdots S$ of length k). By reversing the roles of success and failure, the reliability of $C(k, n:F)$ systems can be related to the distribution of $N_{n,k}$. In this section we present some examples of approximating $\mathbb{P}\{N_{n,k} = x\}$ with the approximations FMCI, Normal, Poisson and LD.

Figure 1 shows the relative error $R(x : E, A)$ in these approximations for (a) $N_{2000,4}$; (b) $N_{5000,4}$; and (c) $N_{250000,6}$ when the probability of success is $p = 0.3$. On all of the figures, the top axis is on a standard z-scale making use of the asymptotic mean and variance of $X_n(\Lambda)$ — namely,

$$z = \frac{x - n/\mu_W}{\sqrt{n\sigma_W^2\mu_W^{-3}}}.$$

We notice that the Finite Markov chain imbedding approximation (FMCI) performs very well in the left tail of the distribution in all cases. Its performance degrades as x gets

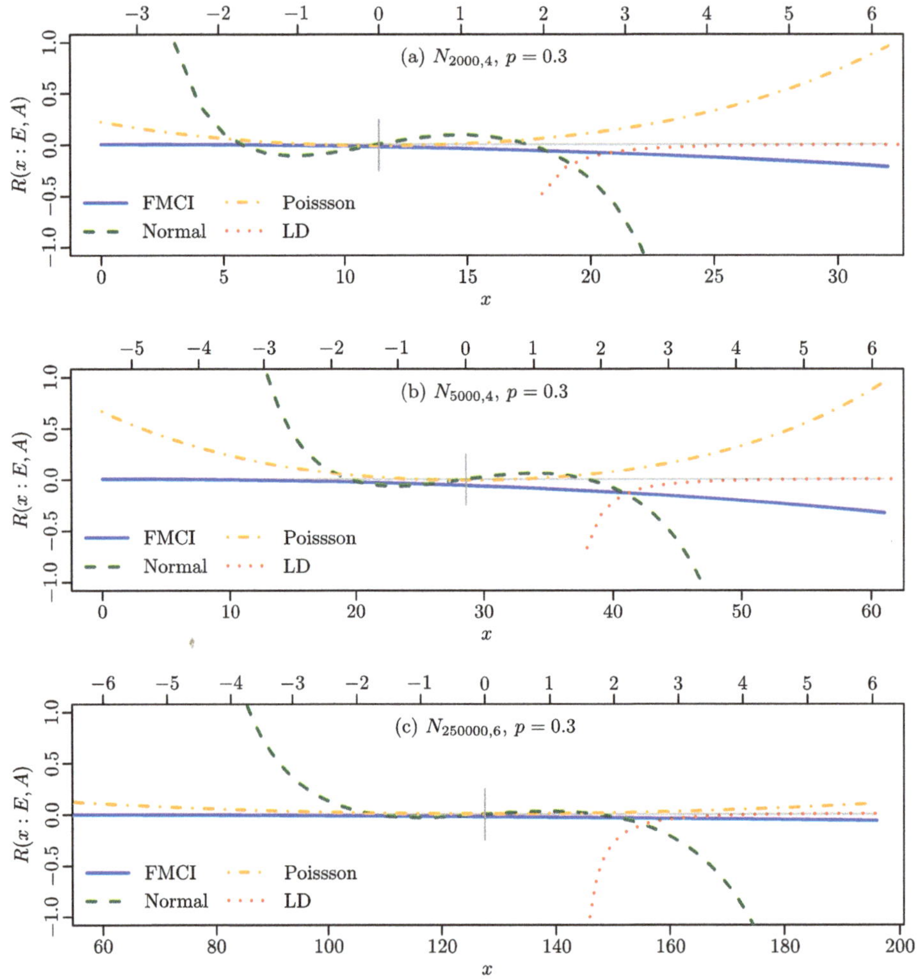

Figure 1 Relative errors of the FMCI, Normal, Poisson and LD approximations $N_{2000,4}$, $N_{5000,4}$ and $N_{250000,6}$ with $p = 0.3$.

large but its performance is more consistent than both the Poisson and Normal approximations in this case. The large deviation approximation performs well in the right tail in all cases. In (c), the FMCI approximation performs very well throughout most of the support. The Poisson approximations also perform well over most of the x considered. The normal approximation performs well in the neighbourhood of $\mathbb{E}X_n(\Lambda)$ but not in the tails.

As the probability of success p increases, the FMCI approximation still performs very well in the left tail, but it's performance tends to degrade more quickly as x increases. The Poisson approximations also quickly degrades as p increases since $\mathbb{E}N_{n,k}$ increases. For larger p, the Normal approximation tends to work better near the mean. In the far left tail, the FMCI approximation is preferred and in the far right tail, the LD approximation is preferred.

Biological sequences

Sequences of DNA nucleotides are of great interest (as are sequences of amino acids and other biological sequences). Figure 2 shows the relative errors for approximating $\mathbb{P}\{X_n(\Lambda) = x\}$ with Λ = ACG (n = 1,000 and 10,000) and Λ = CATTAG (n = 500,000).

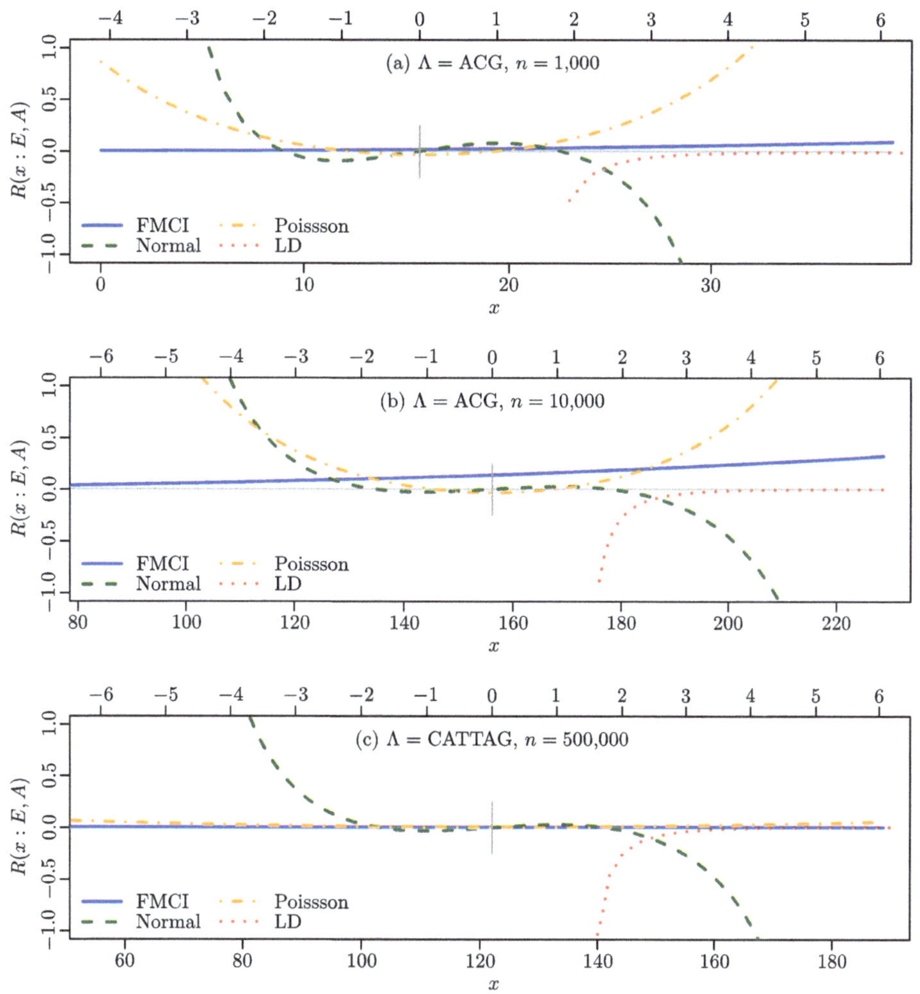

Figure 2 Relative errors of the FMCI, normal, Poisson and large deviation approximations for the patterns $\Lambda = ACG$ with $n = 1{,}000$ and 10,000 and $\Lambda =$ CATTAG with $n = 500{,}000$.

We see that the FMCI approximation again performs very well in the left tail, although, in (b), the performance degrades somewhat as x gets large. The large deviation approximation performs very well in the right tail, especially when x is greater than 3 standard deviations above the mean. While it is difficult to give a rule of thumb, the FMCI approximation seems to perform very well when $x \leq \mathcal{O}(n^{1/2})$. The normal approximation works best within a few standard deviations of the mean and performs best in this region when $\mathbb{E}X_n(\Lambda)$ is relatively large.

Discussion and conclusions

The finite Markov chain imbedding approximations (FMCI and LD) provide an alternative to the usual normal and Poisson approximations for the distributions of runs and patterns. While the FMCI approximation is simple, accurate and fast, it has one disadvantage over the normal and Poisson approximations — it requires the use of the FMCI technique, which is non-traditional and less known in the Statistics community, except in the area of system reliability (*cf.* Cui et al. 2010). On the other hand, the FMCI technique does not require the rather strong conditions necessary for the Poisson techniques, such as $np^k \to \lambda$. This condition is seldom satisfied in practical applications. For example, in DNA sequence analysis, the probabilities p_A, p_C, p_G and p_T do not tend to 0 as n increases. They may not all be in the neighbourhood of 1/4 but they are bounded away from 0.

For all of the numeric results in the previous section, the exact probabilities $\mathbb{P}\{X_n(\Lambda) = x\}$ are obtained via the FMCI technique and their CPU times were only a few seconds or less than a minute even in the case of Λ = CATTAG and n = 500,000. Based on our experience, if the length of the pattern is less than 20 and n is less than 1,000,000, the exact probability should be computed.

Competing interests
The authors declare that they have no competing interests.

Authors' contributions
BJ and JF contributed equally to the mathematical details. BJ performed the numerical comparisons and prepared the manuscript. Both authors read and approved the final manuscript.

Acknowledgements
This work was supported, in part, by the Natural Sciences and Engineering Research Council of Canada.

References
Arratia, R, Goldstein, L, Gordon, L: Poisson approximation and the Chen-Stein method. Stat. Sci. **5**(4), 403–434 (1990)
Asmussen, S: Applied Probability and Queues. 2nd edn. Applications of Mathematics, vol. 51, p. 438. Springer, New York (2003)
Balakrishnan, N, Koutras, MV: Runs and Scans with Applications. Wiley Series in Probability and Statistics. p. 452. Wiley-Interscience [John Wiley & Sons], New York (2002)
Barbour, AD, Eagleson, GK: Poisson approximation for some statistics based on exchangeable trials. Adv. Appl. Probab. **15**(3), 585–600 (1983)
Barbour, AD, Eagleson, GK: Poisson convergence for dissociated statistics. J. Roy. Statist. Soc. Ser. B. **46**(3), 397–402 (1984)
Barbour, AD, Eagleson, GK: An improved Poisson limit theorem for sums of dissociated random variables. J. Appl. Probab. **24**(3), 586–599 (1987)
Barbour, AD, Hall, P: On the rate of Poisson convergence. Math. Proc. Cambridge Philos. Soc. **95**(3), 473–480 (1984)
Barbour, AD, Chryssaphinou, O: Compound Poisson approximation: a user's guide. Ann. Appl. Probab. **11**(3), 964–1002 (2001)
Barbour, AD, Holst, L, Janson, S: Poisson Approximation. Oxford Studies in Probability (1992a). Oxford Science Publications
Barbour, AD, Chen, LHY, Loh, W-L: Compound Poisson approximation for nonnegative random variables via Stein's method. Ann. Probab. **20**(4), 1843–1866 (1992b)
Barbour, AD, Chryssaphinou, O, Roos, M: Compound Poisson approximation in reliability theory. IEEE T. Reliab. **44**(3), 398–402 (1995)
Barbour, AD, Chryssaphinou, O, Roos, M: Compound Poisson approximation in systems reliability. Naval Res. Logist. **43**(2), 251–264 (1996)
Blom, G, Thorburn, D: How many random digits are required until given sequences are obtained? J. Appl. Probab. **19**(3), 518–531 (1982)
Chen, LHY: Poisson approximation for dependent trials. Ann. Probab. **3**(3), 534–545 (1975)
Cui, L, Xu, Y, Zhao, X: Developments and applications of the finite Markov chain imbedding approach in reliability. IEEE T. Reliab. **59**(4), 685–690 (2010)
Fu, JC: Reliability of a large consecutive-k-out-of-n:F system. IEEE T. Reliab. **R-34**, 120–127 (1985)
Fu, JC, Johnson, BC: Approximate probabilities for runs and patterns in i.i.d. and Markov dependent multi-state trials. Adv. Appl. Probab. **41**(1), 292–308 (2009)
Fu, JC, Koutras, MV: Distribution theory of runs: a Markov chain approach. J. Amer. Statist. Assoc. **89**(427), 1050–1058 (1994)
Fu, JC, Lou, WYW: Distribution Theory of Runs and Patterns and Its Applications. p. 162. World Scientific Publishing Co. Inc, River Edge (2003)
Fu, JC, Lou, WYW: On the normal approximation for the distribution of the number of simple or compound patterns in a random sequence of multi-state trials. Methodol. Comput. Appl. Probab. **9**(2), 195–205 (2007)
Fu, JC, Johnson, BC, Chang, Y-M: Approximating the extreme right-hand tail probability for the distribution of the number of patterns in a sequence of multi-state trials. J. Stat. Plan. Infer. **142**(2), 473–480 (2012)
Gerber, HU, Li, S-YR: The occurrence of sequence patterns in repeated experiments and hitting times in a Markov chain. Stochastic Process. Appl. **11**(1), 101–108 (1981)
Godbole, AP: Degenerate and Poisson convergence criteria for success runs. Statist. Probab. Lett. **10**(3), 247–255 (1990a)
Godbole, AP: Specific formulae for some success run distributions. Statist. Probab. Lett. **10**(2), 119–124 (1990b)
Godbole, AP: Poisson approximations for runs and patterns of rare events. Adv. Appl. Probab. **23**(4), 851–865 (1991)
Godbole, AP, Schaffner, AA: Improved Poisson approximations for word patterns. Adv. Appl. Probab. **25**(2), 334–347 (1993)
Holst, L, Kennedy, JE, Quine, MP: Rates of Poisson convergence for some coverage and urn problems using coupling. J. Appl. Probab. **25**(4), 717–724 (1988)
Karlin, S: Statistical signals in bioinformatics. Proc. Natl. Acad. Sci. U. S. A. **102**(38), 13355–13362 (2005)

Karlin, S, Taylor, HM: A First Course in Stochastic Processes. 2nd edn., p. 557. Academic Press [A subsidiary of Harcourt Brace Jovanovich, Publishers], New York-London (1975)
Kleffe, J, Borodovski, M: First and second moment of counts of words in random text generated by Markov chains. Comp Applic Biosci. **8**, 443–441 (1992)
Martin, J, Regad, L, Camproux, A-C, Nuel, G: Finite Markov chain embedding for the exact distribution of patterns in a set of random sequences. In: Skiadas, CH (ed.) Advances in Data Analysis. Statistics for Industry and Technology, pp. 171–180. Birkhäuser, Boston (2010)
Meyn, SP, Tweedie, RL: Markov Chains and Stochastic Stability. Communications and Control Engineering Series (1993)
Nuel, G, Regad, L, Martin, J, Camproux, A-C: Exact distribution of a pattern in a set of random sequences generated by a Markov source: applications to biological data. Algorithm Mol. Biol. **5**(1), 1–18 (2010)
Schwager, SJ: Run probabilities in sequences of Markov-dependent trials. J. Amer. Statist. Assoc. **78**(381), 168–180 (1983)
Seneta, E: Non-negative Matrices and Markov Chains. 2nd edn. Springer, New York (1981)
Solov'ev, AD: A combinatorial identity and its application to the problem on the first occurrence of a rare event. Teor. Verojatnost. i Primenen. **11**, 313–320 (1966)

Permissions

All chapters in this book were first published in JSDA, by Springer; hereby published with permission under the Creative Commons Attribution License or equivalent. Every chapter published in this book has been scrutinized by our experts. Their significance has been extensively debated. The topics covered herein carry significant findings which will fuel the growth of the discipline. They may even be implemented as practical applications or may be referred to as a beginning point for another development.

The contributors of this book come from diverse backgrounds, making this book a truly international effort. This book will bring forth new frontiers with its revolutionizing research information and detailed analysis of the nascent developments around the world.

We would like to thank all the contributing authors for lending their expertise to make the book truly unique. They have played a crucial role in the development of this book. Without their invaluable contributions this book wouldn't have been possible. They have made vital efforts to compile up to date information on the varied aspects of this subject to make this book a valuable addition to the collection of many professionals and students.

This book was conceptualized with the vision of imparting up-to-date information and advanced data in this field. To ensure the same, a matchless editorial board was set up. Every individual on the board went through rigorous rounds of assessment to prove their worth. After which they invested a large part of their time researching and compiling the most relevant data for our readers.

The editorial board has been involved in producing this book since its inception. They have spent rigorous hours researching and exploring the diverse topics which have resulted in the successful publishing of this book. They have passed on their knowledge of decades through this book. To expedite this challenging task, the publisher supported the team at every step. A small team of assistant editors was also appointed to further simplify the editing procedure and attain best results for the readers.

Apart from the editorial board, the designing team has also invested a significant amount of their time in understanding the subject and creating the most relevant covers. They scrutinized every image to scout for the most suitable representation of the subject and create an appropriate cover for the book.

The publishing team has been an ardent support to the editorial, designing and production team. Their endless efforts to recruit the best for this project, has resulted in the accomplishment of this book. They are a veteran in the field of academics and their pool of knowledge is as vast as their experience in printing. Their expertise and guidance has proved useful at every step. Their uncompromising quality standards have made this book an exceptional effort. Their encouragement from time to time has been an inspiration for everyone.

The publisher and the editorial board hope that this book will prove to be a valuable piece of knowledge for researchers, students, practitioners and scholars across the globe.

List of Contributors

Peter Congdon
School of Geography, Queen Mary University of London, London E1 4NS, UK

Gwo Dong Lin
Institute of Statistical Science, Academia Sinica, Taiwan, Republic of China

Xiaoling Dou
Waseda University, Tokyo, Japan

Satoshi Kuriki
The Institute of Statistical Mathematics, Tokyo, Japan

Jin-Sheng Huang
Department of Mathematics and Statistics, University of Guelph, Guelph, Canada

Joseph W McKean
Department of Statistics, Western Michigan University, Kalamazoo, USA

John D Kloke
Department of Statistics, Western Michigan University, Kalamazoo, USA

Li Qi
Department of Statistics, Western Michigan University, Kalamazoo, USA

Yanqing Sun
Department of Mathematics and Statistics, The University of North Carolina at Charlotte, 28223 Charlotte, NC, USA

Shujiao Huang

Broderick O Oluyede
Department of Mathematical Sciences, Georgia Southern University, Statesboro, GA 30458, USA

Lipika Ghosh
Department of Mathematics and Statistics, University of North Carolina at Charlotte, Charlotte, NC 28223, USA

Jiancheng Jiang
Department of Mathematics and Statistics, University of North Carolina at Charlotte, Charlotte, NC 28223, USA

Yanqing Sun
Department of Mathematics and Statistics, University of North Carolina at Charlotte, Charlotte, NC 28223, USA

Haibo Zhou
Department of Biostatistics, University of North Carolina at Chapel Hill, Chapel Hill, NC 27599, USA.

José María Sarabia
Department of Economics, University of Cantabria, Avda. de los Castros s/n, 39005 Santander, Spain

Faustino Prieto
Department of Economics, University of Cantabria, Avda. de los Castros s/n, 39005 Santander, Spain

Vanesa Jordá
Department of Economics, University of Cantabria, Avda. de los Castros s/n, 39005 Santander, Spain

Tao Lu
Department of Epidemiology and Biostatistics, State University of New York, Albany, New York, USA

Min Wang
Department of Mathematical Sciences, Michigan Technological University, Houghton, Michigan, USA

Muhammad H Tahir
Department Statistics, Baghdad Campus, The Islamia University of Bahawalpur, 63100 Bahawalpur, Pakistan

Gauss M Cordeiro
Departamento de Estatística, Universidade Federal de Pernambuco, PE 50740-540 Recife, Brazil

Morad Alizadeh
Department of Statistics, Persian Gulf University of Bushehr, Bushehr, Iran

Muhammad Mansoor
Department Statistics, The Islamia University of Bahawalpur, 63100 Bahawalpur, Pakistan

Muhammad Zubair
Department Statistics, The Islamia University of Bahawalpur, 63100 Bahawalpur, Pakistan

Gholamhossein G Hamedani
Department of Mathematics, Statistics and Computer Science Marquette University, WI 53201-1881 Milwaukee, USA

Fei Tan
Department of Mathematical Sciences, Indiana University Purdue University Indianapolis, N. Blackford Street, IN 46202, Indianapolis, USA

Yuanyuan Tang
Department of Statistics, Florida State University, N. Woodward Ave., FL 32306, Tallahassee, USA

Hanxiang Peng
Department of Mathematical Sciences, Indiana University Purdue University Indianapolis, N. Blackford Street, IN 46202, Indianapolis, USA

David M Ruth
Department of Mathematics, United States Naval Academy, 572-C Holloway Rd, Annapolis, MD 21402, USA

Brad C Johnson
Department of Statistics, University of Manitoba, Winnipeg, Canada

James C Fu
Department of Statistics, University of Manitoba, Winnipeg, Canada